Principles of Cyber-Physical Systems

Principles of Cyber-Physical Systems

Rajeev Alur

The MIT Press

Cambridge, Massachusetts

London, England

This book was set using LATEX by the author.

Library of Congress Cataloging-in-Publication Data

Alur, Rajeev, 1966–

Principles of cyber-physical systems / Rajeev Alur.

p. cm

Includes bibliographical references and index.

ISBN: 978-0-262-02911-7 (hardcover: alk. paper)
ISBN: 978-0-262-54892-2 (paperback)

1. Automatic control. 2. System design. 3. Embedded internet devices.
4. Internet of things. 5. Formal methods (Computer science). I. Title.

TJ213.A365 2015

006.2'2—dc23

2014039755

Dedicated to the memory of my parents

Contents

Preface

A *cyber-physical system* consists of computing devices communicating with one another and interacting with the physical world via sensors and actuators. Increasingly, such systems are everywhere, from smart buildings to medical devices to automobiles. The challenge of developing design and analysis tools to ensure reliability of such systems has attracted researchers from academia as well as industry over the past decade resulting in a vibrant and multi-disciplinary field of study.

The goal of this textbook is to provide an introduction to the principles of design, specification, modeling, and analysis of cyber-physical systems. These principles are drawn from a diverse set of sub-disciplines including model-based design, concurrency theory, distributed algorithms, formal methods for specification and verification, control theory, real-time systems, and hybrid systems. I have attempted to provide a coherent introduction to selected ideas from these different topics that are relevant to the design and analysis of cyber-physical systems. Throughout the textbook, mathematical concepts of modeling, specification, and analysis are illustrated by representative case studies from distributed algorithms, network protocols, control design, and robotics.

The textbook is self-contained, and is suitable for a semester-long course aimed at upper level undergraduate or first-year graduate students in computer science, computer engineering, or electrical engineering. Chapter 1 discusses alternatives for selection of topics for the organization of such a course.

My interest in cyber-physical systems is rooted in the fruitful research collaboration with Tom Henzinger on hybrid systems dating back to 1990s. Furthermore, the organization of this textbook is based on the unpublished manuscript titled *Computer-Aided Verification* coauthored by Tom and me. Some of the examples and figures in chapters 2 and 3 are copied from this manuscript with Tom's permission. Thus Tom's contribution to this textbook is invaluable and I am deeply grateful to him.

My understanding of cyber-physical systems and the contents of this book are greatly influenced by my interactions with faculty and students in PRECISE, a research center focused on cyber-physical systems in Penn Engineering. I am grateful to my colleagues Vijay Kumar, Insup Lee, Rahul Mangharam, George Pappas, Linh Phan, Oleg Sokolsky, and Ufuk Topcu for continued collaborations and support. I am also thankful to DARPA and NSF for providing sustained funding to my research projects in cyber-physical systems.

For the past five years, I have used drafts of this textbook in the course titled *Principles of Embedded Computation* aimed primarily at the Embedded Systems Masters program at Penn. Teaching this course on a regular basis has been a key motivating factor for finishing this book, and the feedback from students has significantly improved its contents. Thanks to all my students and also to the wonderful teaching assistants: Sanjian Chen, Zhihao Jiang, Salar Moarref, Truong Nghiem, Nimit Singhania, and Rahul Vasist.

I have also been fortunate to receive feedback on drafts of this manuscript from researchers at other universities. In particular, chapters 6 and 9 are much improved based on the suggestions from Sriram Sankaranarayanan and Paulo Tabuada. Special thanks to Christos Stergiou for carefully proofreading a recent version and his help with Matlab simulations of the examples in chapter 9.

This is also an opportunity to thank my publisher, MIT Press, for supporting this project. In particular, Virginia Crossman, Marie Lufkin Lee, and Marc Lowenthal have offered help and encouragement throughout the process of publishing this book.

Writing a textbook takes many years, and would not have been possible without the support of my family. I am particularly grateful to my wife, Mona, for her friendship, love, and patience.

Rajeev Alur
University of Pennsylvania
Philadelphia, USA
January 2015

1

Introduction

1.1 What Is a Cyber-Physical System?

The original computer was a stand-alone device focused on number crunching and information processing. While we continue to use computers for these tasks today, the more ubiquitous use of computers is within *embedded systems*. An embedded system consists of hardware and software integrated within a mechanical or an electrical system designed for a specific purpose. From watches to cameras to refrigerators, almost every engineered product today is an embedded system with an integrated microcontroller and software. The concept of a cyber-physical system is a generalization of embedded systems. A *cyber-physical system* consists of a collection of computing devices communicating with one another and interacting with the physical world via sensors and actuators in a feedback loop. Increasingly, such systems are everywhere, from smart buildings to medical devices to automobiles.

As an illustrative example of a cyber-physical system, consider a team of autonomous mobile robots tasked with the identification and retrieval of a target inside a house with an unknown floor plan. To achieve this task, each robot must be equipped with multiple sensors that collect the relevant information about the physical world. Examples of on-board sensors include a GPS receiver to track a robot's location, a camera to take snapshots of its surrounds, and an infrared thermal sensor to detect the presence of humans. A key computational problem then is to construct a global map of the house based on all the data collected, and this requires the robots to exchange information using wireless links in a coordinated fashion. The current knowledge of the positions of the robots, obstacles, and target then can be used to determine a motion plan for each of the robots. Such a motion plan includes high-level commands for each of the robots of the form "move in the north-west direction at a constant speed of 5 mph." This directive then needs to be translated to low-level control inputs for the motors controlling the robot's motion. The design objectives include safe operation (for instance, a robot should not bump into obstacles or other robots), mission completion (for example, the target should be retrieved), and physical stability (for example, each robot should be stable as a dynamical system). Construction of the multi-robot system to meet these objectives requires design of strategies for *control, computing*, and *communication* in a synergistic manner.

Although certain forms of cyber-physical systems have been in industrial use since the 1980s, only recently has the technology for processors, wireless communication, and sensors matured to allow the production of components with impressive capabilities at a low cost. Realizing the full potential of these emerging computing platforms requires advances in tools and methodology for constructing reliable cyber-physical systems. This challenge of developing a systematic approach to integrated design of control, computation, and communication proved to be the catalyst for the formation of a distinct academic discipline of *cyber-physical systems* during the 2000s. The science of design for cyber-physical systems has been identified as a key research priority by government agencies as well as industries in automotive, avionics, manufacturing, and medical devices.

1.2   Key Features of Cyber-Physical Systems

Theory, methodology, and tools for assisting developers to build hardware and software systems in a systematic manner have been the central themes in computer science since its inception. The classical theory of computation, with its focus on computational complexity, and the methodology based on structured programming have both been instrumental in our ability to build today's complex software infrastructure. These principles for design of traditional software systems, however, are of no direct help in building cyber-physical systems due to significant differences in design concerns. The distinguishing characteristics of cyber-physical systems are discussed below.

Reactive Computation

In the classical model of computation, a computing device produces an output when supplied with an input. An example of such a computation is a program that, given an input list of numbers, outputs a sorted version of the input list. The notion of correctness for such a program can be captured mathematically as a *function* from input values to output values. Theory of computability and complexity gives us an understanding of which functions are computable and which functions are computable efficiently. The traditional programming abstractions of functions and procedures allow us to write programs for computing complex functions by composing simpler functions.

A *reactive* system, in contrast, interacts with its environment in an ongoing manner via inputs and outputs. As a typical example of reactive computation, consider a program for a cruise controller in a car. Such a program receives high-level input commands for turning the cruise controller on and off and for changing the desired cruising speed. The control program needs to respond to such inputs by changing its output, which corresponds to the force that is applied to the engine throttle. The behavior of such a system then is naturally described by a *sequence* of observed inputs and outputs, and the notion of correctness specifies which input/output sequences correspond to acceptable behaviors. Cyber-physical systems are reactive systems, and thus the focus of this book is on reactive computation.

Concurrency

In a traditional *sequential* model of computation, the computation consists of a sequence of instructions executed one at a time. In *concurrent* computation, multiples threads of computation, usually called components or processes, are executing concurrently, exchanging information with one another to achieve the desired goal of the computation. Concurrency is fundamental to cyber-physical systems. In our example of a team of autonomous mobile robots, the robots themselves are separate entities and are thus executing concurrently. Each robot has multiple sensors and processors, and computing tasks such as constructing a map of the environment based on vision data and motion planning based on the map of the environment can be executing on separate processors in parallel. The motion planning task can be subdivided into logically concurrent subtasks such as local planning to avoid obstacles and global planning for optimal progress toward the target.

Understanding models and design principles for distributed and concurrent computation thus is critical for cyber-physical systems. For sequential computation, the model of *Turing machines* is accepted as the canonical model of computation. No such agreement exists for concurrent computation with a rich variety of proposals of formal models. Broadly speaking, these models fall into two categories: in *synchronous* models, components execute in lock-step, and the computation progresses in a logical sequence of synchronized rounds; and in *asynchronous* models, components execute at independent speeds, exchanging information by sending and receiving messages. Both types of models are useful for the design of cyber-physical systems. In our example, the system of robots can be viewed as an asynchronous system consisting of individual robots exchanging messages, whereas for simplicity of design, the computation on a single robot can be divided into concurrent activities executing in a logically synchronous manner.

Feedback Control of the Physical World

A *control system* interacts with the physical world in a feedback loop by measuring the environment via sensors and influencing it via actuators. For example, a cruise controller is constantly monitoring the speed of the car and adjusts the throttle force so that the speed stays close to the desired cruising speed. Controllers are components of a cyber-physical system, and this integration of computing devices with the physical world sets cyber-physical systems apart from the traditional computers.

Design of controllers for the physical world requires modeling the dynamics of the physical quantities: to adjust the throttle force, a cruise controller needs a model of how the speed of the car changes with time as a function of the throttle force. The theory of dynamical control systems is a well-developed discipline with a rich set of mathematical tools for design and analysis, and a basic understanding of these principles is valuable to designers of cyber-physical systems. The traditional control theory focuses on *continuous-time* systems. In a cyber-physical system, the controller consists of discrete software comprising concurrent components operating in multiple possible modes of operation, interacting with the continuously evolving physical environment. Such systems with a mix of discrete and continuous dynamics are sometimes called *hybrid systems*, and the emerging principles of design and analysis of controllers for such systems will be studied in this book.

Real-Time Computation

Programming languages, and the supporting infrastructure of operating systems and processor architectures, typically do not support an explicit notion of real time. This offers a convenient abstraction for traditional computing applications such as document processing, but real-time performance is critical for cyber-physical systems. For example, for a cruise controller to satisfactorily control the speed of the car, its design should take into account the time it takes its subcomponents to execute the necessary computations and communicate the results.

Modeling timing delays, understanding their impact on the correctness requirements and system performance, timing-dependent coordination protocols, and resource-allocation strategies to ensure predictability have been the subject of study in the sub-discipline of *real-time systems*. A principled approach to design and implementation of cyber-physical systems thus builds on these techniques.

Safety-Critical Applications

While designing and implementing a cruise controller, we expect a high level of assurance in the correct operation of the system because errors can lead to unacceptable consequences such as loss of life. Applications where the *safety* of the system has a higher priority over other design objectives such as performance and development cost are called *safety-critical*. Computing devices now control aircrafts, automobiles, and medical devices and thus are all examples of cyber-physical systems for safety-critical applications. In this context, establishing that the system works correctly at design time is of paramount importance and sometimes mandatory due to government regulations for certification of systems.

The traditional route to system development is design and implementation, followed by extensive testing and validation to detect bugs. The more principled approach to system development involves writing mathematically precise requirements of the desired system, designing models of system components along with the environment in which the system is supposed to operate, and using analysis tools to check that the system model meets the requirements. Compared to the traditional approach, this methodology can detect design errors in early stages and ensure higher reliability. Such an approach based on formal models and verification is appealing in safety-critical applications, is being increasingly adopted by industry, and will be a central theme in this book.

1.3   Overview of Topics

The goal of this textbook is to provide an introduction to the principles of design, specification, modeling, and analysis of cyber-physical systems. Due to the distinguishing characteristics of cyber-physical systems, these principles are drawn from a diverse set of sub-disciplines including model-based design, concurrency theory, distributed algorithms, formal methods for specification and verification, control theory, real-time systems, and hybrid systems. Research conferences and textbooks are devoted to each of these sub-disciplines, and this book is aimed at explaining the core ideas relevant to system design and analysis in each of these in a coherent manner. The topics are discussed by interweaving the three themes of *formal models*, *model-based design*, and *specification and analysis* as discussed below.

Formal Models

The goal of modeling in system design is to provide mathematical abstractions to manage the complexity of design. In the context of reactive systems, the basic unit of modeling is a component that interacts with its environment via inputs and outputs. Different forms of interaction lead to different classes of models. We begin in chapter 2 by focusing on *synchronous* modeling, where all components execute in lock-step in a sequence of rounds. Then in chapter 4, we switch to *asynchronous* models, where different activities execute at independent speeds. In chapter 6, we study *continuous-time* models of dynamical systems that are suitable for capturing the evolution of the physical world. Chapter 7 introduces *timed* models, where the interaction among components is facilitated by the knowledge of concrete bounds on timing delays. Finally, chapter 9 considers *hybrid* systems by integrating models of discrete interaction and dynamical systems.

To describe models, we use a combination of block diagrams, code fragments, state machines, and differential equations. We define our models *formally*, that is, in a mathematically precise manner. The formal semantics allows us to answer questions such as, "what are the possible behaviors of a component" and "what does it mean to compose two components" rigorously. Examples of modeling concepts covered in this book include nondeterministic behavior, input-output interfaces of components, time-triggered and event-triggered communication, await dependencies for synchronous composition, communication using shared memory, atomicity of synchronization primitives, fairness for asynchronous systems, equilibria for dynamical systems, and Zeno behaviors for timed and hybrid systems.

Specification and Analysis

To check that the design (or system implementation) works correctly as intended, the designer first needs to express the requirements capturing correctness in a mathematically precise manner. Analysis tools then allow the designer to check that the system satisfies its requirements. This textbook covers a range of specification formalisms and associated techniques for formal verification.

Chapter 3 introduces *safety* requirements. A safety requirement asserts that "nothing bad ever happens" and can be formalized using invariants and monitors. We first consider the general technique of *inductive invariants* for proving that a system satisfies its safety specification and then *state-space exploration* algorithms for automatically establishing safety properties. Both *enumerative* and *symbolic* search algorithms are developed, including the symbolic exploration using the data structure of Ordered Binary Decision Diagrams (BDDs) commonly used in hardware verification. The presence of continuous-time dynamics of the physical quantities poses new challenges for safety verification of cyber-physical systems. For verifying systems with hybrid dynamics, we study the proof method based on *barrier certificates* and symbolic search algorithms for two special classes, namely, *timed automata* and *linear hybrid automata*.

Chapter 5 introduces *liveness* requirements: such a specification asserts that "something good eventually happens." We introduce the temporal logic Linear Temporal Logic (LTL) to formally express such correctness requirements and show how the notion of monitors can be generalized to Büchi automata so as to capture LTL requirements. The problem of automatically verifying LTL requirements of system models is known as *model checking*. Both enumerative and symbolic state-space exploration techniques are generalized to solve the model-checking problem, and the method of *ranking functions* is developed as a general proof principle for proving liveness requirements.

For dynamical systems, a basic design requirement concerns *stability*, which informally means that small perturbations to system inputs should not cause a disproportionate change in its observed behavior. This classical topic from control theory is studied in chapter 6, with a particular focus on *linear systems* for which tools from linear algebra are shown to be useful for mathematically establishing stability.

While implementing embedded systems, a key analysis problem is to establish that the time it takes to execute different tasks in the system model on a given computational platform is consistent with the model-level assumptions regarding the timing delays. *Real-time scheduling* theory is aimed at formalizing and solving this problem and is the subject of chapter 8. We focus primarily on understanding two fundamental scheduling algorithms: Earliest Deadline First (EDF) and Rate Monotonic.

Model-Based Design and Case Studies

Principles of modeling, specification, and analysis are illustrated by constructing solutions to representative design problems from distributed algorithms, network protocols, control design, and robotics. We illustrate how modeling differs from programming, for instance, by allowing the specification of *nondeterministic* behavior and including explicit models of how the *environment* behaves. While designing model-based solutions, we emphasize two principles:

1. *Structured* design: simple components can be composed to perform more complex tasks, and, conversely, a design problem can be decomposed into simpler subtasks.

2. *Requirements-based* design: correctness requirements are specified precisely up front and are used to guide the exploration among design alternatives and for debugging of the design in early stages.

We study the classical distributed coordination problems of *mutual exclusion*, *consensus*, and *leader election*. These problems are revisited throughout the textbook, highlighting how the power of synchronization primitives influences the design. Another set of problems focuses on message communication, including how to reliably transmit messages in the presence of a lossy network and how to synchronize a sender and a receiver in the presence of timing uncertainties introduced by imperfect clocks. The design of a cruise controller illustrates how to integrate synchronous design using block diagrams with the design of a low-level PID controller. We conclude with case studies representative of cyber-physical systems: design of a pacemaker monitoring and responding to timing patterns of heart pulsations, obstacle avoidance for a team of robots using coordination, and design of stabilizing controllers communicating over a multi-hop network.

1.4 Guide to Course Organization

This textbook is suitable for a semester-long course aimed at upper level undergraduate or first-year graduate students in computer science, computer engineering, or electrical engineering. This section gives some suggestions regarding the organization of such a course.

Prerequisites

The textbook emphasizes principles of modeling, design, specification, and analysis. These principles are drawn from a range of mathematical topics such as calculus, discrete mathematics, linear algebra, and logic. Most of the concepts

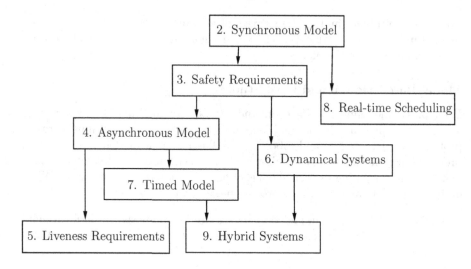

Figure 1.1: Dependencies among Chapters

are developed from the first principles, and thus courses in these topics are not
prerequisites. However, some basic level of mathematical maturity is essential
for understanding the course material. The course is suitable for students who
have completed the required theory courses either in the computer science curriculum (such as *Discrete Mathematics* and *Theory of Computation*) or in the
electrical engineering curriculum (such as *Signals and Systems* and *Dynamical
Systems*).

Throughout the textbook, we discuss design problems from a range of applications such as control systems, distributed coordination, network protocols,
and robotics. Analogous to the necessary mathematical principles, in each of
our case studies, the basic constraints of the application domain are explained
without assuming any explicit background knowledge. However, prior experience with design and implementation of software and systems is necessary
to appreciate these examples. This experience can be gained by completing
project-oriented undergraduate courses: for example, the courses on programming and operating systems in the computer science curriculum and those on
mechanical or control systems in the engineering curriculum.

Selection of Topics

Not all topics from the textbook can be covered during a semester-long course.
Figure 1.1 shows dependencies among chapters that can be used to guide the
selections of topics. In each chapter, the essential concepts can be covered even
if one skips the sections marked with an asterisk. We describe three possible
courses below.

A fast-paced course that aims to cover all the themes, namely, modeling, design, specification, and verification, is feasible and has been taught at University of Pennsylvania for many years. For such a course, we recommend that the following sections be skipped: 3.3, 5.3, 6.4, 7.2, and 9.3.

A course primarily focused on modeling and design can omit techniques for analysis and verification. In particular, sections 3.3, 3.4, 5.2, 5.3, 6.4, 7.3, and 9.3, can be skipped. However, it is recommended that even such a course should emphasize the role of formally specified requirements in principled design and, thus, should include the specification formalisms.

A third possibility to narrow the scope of the course is by omitting chapters 6, 8, and 9. Such a course focuses on modeling, design, specification, and verification of reactive systems but does not include modeling of the interaction with the physical world.

Homework and Projects

Each section has a number of exercises at its end, and students are expected to answer these questions with mathematical rigor. The more challenging exercises are marked with an asterisk.

Besides solving theoretical exercises, coursework should include design projects using software for modeling and analysis. The textbook discusses modeling concepts in a generic manner and is not tied to the concrete syntax of any specific tool. The following are some examples of design projects:

1. Synchronous modeling and symbolic safety verification (chapters 2 and 3): A project focused on synchronous hardware designs, such as arbiters and on-chip communication protocols, offers an opportunity to understand how to structure a design as a hierarchical composition of subcomponents. Industry-standard hardware description languages such as VHDL and Verilog (see `vhdl.org`) can be used for such a modeling project. Alternatively, the academic tool NuSmv (see `nusmv.fbk.eu`) can be used for modeling and verification of requirements using BDD-based symbolic state-space exploration.

2. Asynchronous modeling and model checking (chapters 4 and 5): A distributed protocol, such as a cache coherence protocol used for coordinating accesses to global shared memory on modern multiprocessor systems, can be modeled as communicating asynchronous processes in the modeling tool Spin (see `spinroot.com`). The tool allows the user to specify both safety and liveness requirements in temporal logic and to debug the correctness of the protocol using model checking.

3. Control design for dynamical systems (chapter 6): A traditional project in a course on control systems involves building a model of a physical system, designing a controller for it, and then establishing the stability

of the composed system by using tools of linear algebra. Such a project is quite suitable for this course also. The most commonly used software for such a project is MATLAB (produced by Mathworks, `mathworks.com`), and a typical problem is to design a controller to maintain a pendulum attached to a moving cart in a vertically inverted position.

4. Modeling and verification of timed systems (chapter 7): The modeling tool UPPAAL (see `uppaal.org`) supports modeling using interacting timed automata and verification of safety properties using symbolic state-space exploration. A suitable case study in this domain consists of requirements-based design and analysis of a control algorithm for a medical device such as an autonomous infusion pump or a detailed design of a pacemaker.

5. Modeling and simulation of hybrid systems (chapter 9): The modeling tools such as STATEFLOW and SIMULINK (see `mathworks.com`), MODELICA (see `modelica.org`), and PTOLEMY (see `ptolemy.org`) allow structured modeling of hybrid systems, and multi-robot coordination offers a fertile problem domain to set up projects in design and analysis of cyber-physical systems using these modeling tools. The goal of analysis in such a project is to understand the trade-offs among different design parameters using numerical simulation.

Supplementary Reading

The case for a new *science* of design for embedded and cyber-physical software systems, with an emphasis on high assurance, has been made by many researchers over the last decade (see [Lee00, SLMR05, KSLB03, HS06, SV07]). Now there is a vibrant academic research community in this sub-discipline, and the annual conferences *Embedded Systems Week* (see `esweek.org`) and *Cyber-Physical Systems Week* (see `cpsweek.org`) reflect the current trends in research in cyber-physical systems.

The textbook *Introduction to Embedded Systems* [LS11] is the closest to ours in terms of focus and selection of topics and, thus, is a valuable supplementary textbook. For comparisons, [LS11] covers a broader range of topics, for instance, it discusses processor architectures for embedded applications, whereas this textbook includes a more in-depth development of analysis and verification techniques as well as case studies.

Specialized books on each of the topics covered in this textbook are also useful for a deeper study of that topic. For principles of model-based design, [Hal93] is focused on synchronous models, [Mar03] is devoted to design of embedded systems, and [Pto14] highlights design by integrating heterogeneous modeling styles. Among the rich literature on distributed systems, [Lyn96] and [CM88] introduce a wide range of distributed algorithms emphasizing formal modeling, correctness requirements, and verification. For an introduction to formal logic and its applications to software verification, we recommend [HR04] and

[BM07]. Textbooks devoted to automated verification and model checking include [CGP00] and [BK08], and [Lam02] explains how logic can be used for specification and development of reactive systems. Dynamical systems, with a focus on design of controllers for linear systems, is a classical topic with many textbooks such as [AM06] and [FPE02]. For an introduction to real-time systems, with an emphasis on schedulability, see [But97] and [Liu00]. Finally, the research monographs [Tab09], [Pla10], and [LA14] focus on formal modeling, control, and verification of hybrid systems.

2

Synchronous Model

A *functional* component produces outputs when supplied with inputs, and its behavior can be mathematically described using a mapping between input and output values. A *reactive* component, in contrast, maintains an internal state and interacts with other components via inputs and outputs in an ongoing manner. We first focus on a *discrete* and *synchronous* model of reactive computation in which all components execute in a sequence of rounds. In each round, a reactive component reads its inputs; based on its current state and inputs, it computes outputs and updates the internal state.

2.1 Reactive Components

As a first example, consider the `Delay` component shown in figure 2.1. The component has a Boolean input variable *in*, a Boolean output variable *out*, and an internal state modeled by a Boolean variable x. To describe the behavior of the component, we first need to describe the initial values for the state variables. For `Delay`, assume that the initial value of x is 0. In each round of execution, the component sets the output variable *out* to the value of the state variable x at the beginning of the round and then updates the state to the value of the input variable in the current round. Thus, in the first round, the output will be 0, and in each subsequent round, the output will be equal to the input in the previous round.

2.1.1 Variables, Valuations, and Expressions

To explain the various aspects of the definition of a component precisely, we need a bit of mathematical notation concerning variables, expressions over variables, and assignments of values to variables. We use *typed variables* to describe components. The commonly used types are:

- **nat** denoting the set of natural numbers.

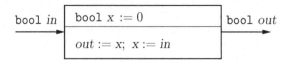

Figure 2.1: Reactive Component Delay

- int denoting the set of integers.

- real denoting the set of real numbers.

- bool denoting the set of Boolean values $\{0, 1\}$.

- An enumerated type contains a finite number of symbolic constants; an example of such a type is the set $\{on, off\}$ with two values.

Given a set V of typed variables, a *valuation* over V is a type-consistent assignment to all the variables in V. That is, a valuation over V is a function q with domain V such that for each variable $v \in V$, $q(v)$ is a value belonging to the type of v. We use Q_V to denote the set of all valuations over V. For example, if V contains two variables, the variable x of type bool and variable y of type nat, then a valuation q assigns a Boolean value to x and a natural number to y, and the set Q_V contains all such possible valuations.

A *typed expression* e over a set V of typed variables is constructed using variables in V, constants, and primitive operations over types corresponding to these variables. Over numerical types, such as nat, int, and real, we will use arithmetic operations such as addition and multiplication and comparison operations such as $=$ and $\leq$. To construct Boolean expressions, we use the following logical operators:

- *Negation* ($\neg$): the expression $\neg e$ evaluates to 1 precisely when e evaluates to 0;

- *Conjunction* ($\wedge$): the expression $e_1 \wedge e_2$ evaluates to 1 precisely when both e_1 and e_2 evaluate to 1; and

- *Disjunction* ($\vee$): the expression $e_1 \vee e_2$ evaluates to 1 precisely when at least one of e_1 or e_2 evaluates to 1.

- *Implication* ($\rightarrow$): the expression $e_1 \rightarrow e_2$ evaluates to 1 precisely when either e_1 evaluates to 0 or e_2 evaluates to 1.

2.1.2   Inputs, Outputs, and States

The component Delay of figure 2.1 has one input variable, one output variable, and one state variable. In general, a component C has a set I of typed input variables, a set O of typed output variables, and a set S of typed state variables.

All three sets should be *finite*. To avoid conflicts in variable names, these sets should also be *disjoint* from one another.

For the `Delay` component, $I = \{in\}$, $O = \{out\}$, and $S = \{x\}$.

In our illustrations, we draw components as rectangular boxes. For each input variable, there is an incoming arrow incident upon this box, and for each output variable, there is an outgoing arrow. These arrows are labeled with the names and types of the corresponding variables. The state variables are listed inside the component box.

An *input* to a reactive component C is a valuation over the set I of its input variables, and the set of all possible inputs is Q_I. An *output* of a component C is a valuation over the set O of its output variables, and the set of all possible outputs is Q_O. A *state* of a component C is a valuation over the set S of its state variables, and the set of its states is Q_S.

For the `Delay` component, an input is a Boolean value for the variable *in*, an output is a Boolean value for the variable *out*, and a state is a Boolean value for the variable x. Thus, each of the sets Q_I, Q_O, and Q_S contains two elements.

2.1.3   Initialization

To describe the dynamics of the component, we must specify the initial states and how the component reacts to a given input in each state. A variety of programming styles are used to describe this, ranging from imperative style (for instance, SYSTEMC and ESTEREL), declarative equational style (for instance, LUSTRE), and hierarchical state machines (for instance, STATEFLOW). To be analyzable by tools, the precise *syntax*—what are the legal code fragments for describing the initialization and update, and the precise *semantics*—what are the corresponding mathematical sets of initial states and reactions, needs to be formalized. This can be challenging for real-world languages and even for "toy" languages, defining semantics formally requires a potentially overwhelming level of mathematical notation. We will use a combination of common imperative constructs and state machines, introducing the features as needed, without rigorously formalizing the mathematical semantics.

The *initialization* of a component, denoted *Init*, specifies the initial values for all the state variables in S. Whenever a state variable is declared, the corresponding initial values are described using an assignment. For example, in the `Delay` component, the state variable x is initialized using the assignment $x := 0$. Sometimes we want to specify multiple possible initial values to allow modeling of situations where initial conditions are only partially known. For this purpose, we will use a new construct called `choose`, which returns an arbitrarily chosen value from its argument set. For the `Delay` component, consider an alternative declaration for the variable x given by

$$\texttt{bool } x := \texttt{choose } \{0, 1\}.$$

In this modified version, `choose` may return either 0 or 1; as a result, the initial value of the variable x may be either 0 or 1. Another example of initialization using the `choose` construct is the declaration

$$\texttt{real } x := \texttt{choose } \{z \mid 0 \leq z \leq 2\}.$$

This means that the variable x is real-valued, and its initial value can be any real number between 0 and 2.

A state q of the component is called an *initial state* if, for every state variable x, the value $q(x)$ is consistent with the initialization of the variable x. The set of all initial states is denoted $[\![Init]\!]$. Thus, the initialization *Init* is a syntactic description of how the component initializes state variables, and the corresponding set $[\![Init]\!]$ is its mathematical semantics.

For the component `Delay`, the set $[\![Init]\!]$ contains the single initial state that assigns the value 0 to x.

In our illustrations of components, we split the box representing the component by a horizontal line, and the top part lists all the state variables along with their types, followed by their initialization.

2.1.4 Update

The computation of a component in response to an input in each round is given by its *reaction description*, denoted *React*. If the component in state s, when supplied with input i, can produce output o and update its state to t, we write $s \xrightarrow{i/o} t$. Such a response is called a *reaction*.

A natural way to describe the reactions is using code that assigns values to the output variables and updates the values of the state variables. This code can use the values of the input variables and the state variables at the beginning of the round. The set of all possible reactions of the component is the semantics of the reaction description and is denoted $[\![React]\!]$.

For the `Delay` component, the reaction description *React* is a sequence of two assignment statements:

$$out := x; \ x := in.$$

That is, in a state s, given an input i, the component copies the state to the output and updates the state to the current input. In this case, the component has four possible reactions:

$$0 \xrightarrow{0/0} 0; \quad 0 \xrightarrow{1/0} 1; \quad 1 \xrightarrow{0/1} 0; \quad 1 \xrightarrow{1/1} 1.$$

In our illustrations, the reaction description is given in the lower half of the component box.

The reaction description typically is a sequence of statements, where each statement is either an assignment statement or a conditional statement. An assignment statement is of the form $x := e$, where e is an expression of the same type as the type of the variable x. We allow the right-hand side of an assignment to use the **choose** construct to specify a set of possible values, thereby permitting multiple responses to the same input in a given state. A conditional statement is of the form

$$\textbf{if } b \textbf{ then } stmt_1 \textbf{ else } stmt_2$$

where b is a Boolean expression, and $stmt_1$ and $stmt_2$ are code fragments given as sequences of assignment and conditional statements. To execute the conditional statement, the expression b is evaluated first. If it evaluates to 1, the code for $stmt_1$ is executed; otherwise the code for $stmt_2$ is executed. We will use curly braces { and } to group statements together when needed. It is possible that the **else**-branch of a conditional statement is missing.

Given a state s and an input i, to find the possible reactions of the component, we execute the reaction description code. If we can execute the code without errors, and if the values assigned to all the output variables give the output valuation o, while the updated values of all the state variables give the state valuation t, then this contributes the reaction $s \xrightarrow{i/o} t$ to the set $[\![React]\!]$. Such a successful execution may not be possible for different reasons. Two common cases are discussed below.

First, when the code tries to execute an assignment statement $x := e$, we expect x to be either an output variable or a state variable; an attempt to assign a value to an input or an undeclared variable is an error. To be able to evaluate the expression e, it should refer only to state variables, input variables, and those output variables assigned values by previously executed assignment statements. In the description of the **Delay** component, if we replace the reaction description by

$$x := out; \ out := in$$

then the first statement cannot be executed as no value of the output variable *out* is known, and for this description, the corresponding set of reactions is the empty set.

Second, if the code does not assign values to all the output variables, then this execution cannot define a valid reaction. For example, in the **Delay** component, if we replace the reaction description by the conditional assignment

$$\textbf{if } (x \neq in) \textbf{ then } out := x$$

then the statement updates the output only when the input differs from the current state. The modified component then has only two reactions: $0 \xrightarrow{1/0} 1$ and $1 \xrightarrow{0/1} 0$. This can be interpreted as a specification of a component that rejects the input 0 in state 0 and rejects the input 1 in state 1.

The reaction description can also use *local* variables, that is, auxiliary variables used to store results of intermediate computation. Suppose given integer values of input variable in_1 and in_2, we want to compute the output variable *out* with value equal to the difference of the squares of the two input values. The code below achieves this using only one multiplication:

```
local int x, y;
    x := in₁ + in₂;
    y := in₁ − in₂;
    out := x * y.
```

In this description, the values of the variables x and y are not available to the other components and are not stored across rounds.

We summarize the discussion so far by presenting a formal definition of a reactive component below:

SYNCHRONOUS REACTIVE COMPONENT

A *synchronous reactive component* C is described by

- a finite set I of typed input variables defining the set Q_I of inputs,

- a finite set O of typed output variables defining the set Q_O of outputs,

- a finite set S of typed state variables defining the set Q_S of states,

- an initialization *Init* defining the set $[\![Init]\!] \subseteq Q_S$ of initial states, and

- a reaction description *React* defining the set $[\![React]\!]$ of reactions of the form $s \xrightarrow{i/o} t$, where s, t are states, i is an input, and o is an output.

2.1.5 Executions

The operational semantics of a component can be captured by defining its executions. To execute a component, we first initialize all the variables to obtain an initial state. The execution then proceeds for a finite number of rounds. In each round, values for input variables are chosen. Then the code in the reaction description of the component is executed to determine its output and updated state.

Formally, an *execution* of a synchronous reactive component C of length k, where $k \geq 0$, consists of a finite sequence of the form

$$s_0 \xrightarrow{i_1/o_1} s_1 \xrightarrow{i_2/o_2} s_2 \xrightarrow{i_3/o_3} s_3 \cdots\cdots s_{k-1} \xrightarrow{i_k/o_k} s_k,$$

where

1. for $0 \leq j \leq k$, each s_j is a state of C, and for $1 \leq j \leq k$, each i_j is an input of C, and each o_j is an output of C;

2. s_0 is an initial state of C; and

3. for $1 \leq j \leq k$, $s_{j-1} \xrightarrow{i_j/o_j} s_j$ is a reaction C.

For instance, one possible execution of the `Delay` component of length 6 is:

$$0 \xrightarrow{1/0} 1 \xrightarrow{1/1} 1 \xrightarrow{0/1} 0 \xrightarrow{1/0} 1 \xrightarrow{1/1} 1 \xrightarrow{1/1} 1.$$

Exercise 2.1: Consider a modified version of the `Delay` component, called `OddDelay`, that has a Boolean input variable *in*, a Boolean output variable *out*, and two Boolean state variables x and y. Both the state variables are initialized to 0, and the reaction description is given by:

> **if** y **then** *out* $:= x$ **else** *out* $:= 0$;
> $x := in$;
> $y := \neg y$.

Describe in words the behavior of the component `OddDelay`. List a possible execution of the component if it is supplied with the sequence of inputs $0, 1, 1, 0, 1, 1$ for the first six rounds. ∎

2.1.6 Extended-State Machines

State machines are commonly used for describing the behavior in model-based design. Figure 2.2 describes the component `Switch` that models a light switch. The component has a single Boolean input variable *press*. In every round, the value 1 for the input variable indicates that the switch is pressed. Initially, the light is off, and when the switch is pressed, it is turned on. The light gets switched off either when the switch is pressed again or if 10 rounds elapse without the switch being pressed.

In the state-machine notation, there is an implicit state variable, called the *mode* of the state machine, ranging over an enumerated type. For the component `Switch`, the mode ranges over the enumerated type {off, on}. The different modes of the machine are drawn as circles. This visual representation highlights different modes of operation for the component. The sourceless incoming arrow for the mode off indicates that the initial value of the variable *mode* is off.

In *extended* state machines, the description using the modes of the machine is augmented with additional state variables. In our example, the component `Switch` uses an additional state variable x of type **int**. The initialization arrow into the initial mode is labeled with the declaration of these additional state variables along with their initial values. In our example, there is a single additional state variable x of type **int**, and it is initialized to the value 0.

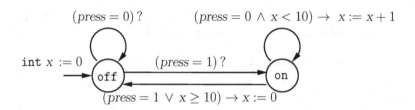

Figure 2.2: Description of `Switch` as an Extended-State Machine

In extended-state machines, reactions are specified using mode-switches. A mode-switch is depicted as an edge between two modes and has an associated guard-condition and a code fragment to update variables. If the guard-condition is the expression *Guard* and the code to update variables is *Update*, then the edge is annotated with *Guard* → *Update*. If the guard-condition always holds (that is, it equals the constant 1), then it is omitted, and the edge is annotated only with the code *Update*; and if the update code does not modify any variables, then it is omitted, and the edge is annotated only with the guard-condition *Guard*?

In our illustrative example, we have four mode-switches. The mode-switch from `off` to `on` is guarded with the condition $press = 1$ and does not change the value of x; the mode-switch from `off` to `off` is guarded with the condition $press = 0$ and does not change the value of x; the mode-switch from `on` to `on` is guarded with the conjunctive condition ($press = 0 \land x < 10$), and it increments the value of x; and the mode-switch from `on` to `off` is guarded with the disjunctive condition ($press = 1 \lor x \geq 10$), and it resets the value of x to 0.

When the mode is `off`, if the input is 0, then the guard-condition for the mode-switch from `off` to `off` is satisfied, and this mode-switch is executed. The new mode is the same as the old mode `off`, and since there is no explicit update code, the value of the state variable x stays unchanged. If the input is 1, then the guard-condition for the mode-switch from `off` to `on` is satisfied, and this mode-switch is executed. As a result, the updated value of the mode is `on`, with the value of x unchanged.

When the mode is `on`, if the input is 0 and the value of x is still below the timeout-threshold 10, the mode-switch from `on` to `on` is executed, leaving the mode unchanged while incrementing x. Thus, the component can stay in the mode `on` for at most 10 consecutive rounds. When either the input *press* has the value 1 or the value of x reaches 10, the guard-condition for the mode-switch from `on` to `off` is satisfied. Executing this mode-switch updates the mode to `off` and x to 0.

In this example, each state can be represented as a pair, where the first component belongs to the enumerated type {`off`, `on`} and the second component is an integer. The set $[\![Init]\!]$ of initial states contains the sole state (`off`, 0).

We can associate a set $[\![React]\!]$ of reactions to formally capture the meaning of the mode-switches. Given a state that assigns values to the mode and x and an input value for *press*, we can obtain a reaction by executing the update code of a mode-switch out of the current mode, provided the corresponding guard-condition is satisfied. For every integer n, we have reactions

$$(\text{off}, n) \xrightarrow{1/} (\text{on}, n); \; (\text{off}, n) \xrightarrow{0/} (\text{off}, n); \; (\text{on}, n) \xrightarrow{1/} (\text{off}, 0);$$

for every integer $n < 10$, we have the reaction $(\text{on}, n) \xrightarrow{0/} (\text{on}, n + 1)$; and for every integer $n \geq 10$, we have the reaction $(\text{on}, n) \xrightarrow{0/} (\text{off}, 0)$.

In this example, the component `Switch` has no output variables. In the presence of output variables, the update code associated with each mode-switch assigns values to all the output variables. Also, in our example, the guard-conditions of two mode-switches out of the same mode are disjoint, and thus there is no choice in terms of which mode-switch should be executed. In general, this assumption need not hold. In fact, the state-machine notation is a convenient way of specifying multiple choices as later examples will illustrate.

Exercise 2.2: Describe the component `OddDelay` from Exercise 2.1 as an extended-state machine with two modes. The mode of the state machine should capture the value of the state variable y, while the state variable x should be updated using assignments in the mode-switches. ∎

Exercise 2.3: We want to design a reactive component with three Boolean input variables x, y, and *reset* and a Boolean output variable z. The desired behavior is the following. The component waits until it has encountered a round in which the input variable x is high and a round in which the input variable y is high, and as soon as both of these have been encountered, it sets the output z to high. It repeats the behavior when, in a subsequent round, the input variable *reset* is high. By default the output z is low. For instance, if x is high in rounds 2,3,7,12, y is high in rounds 5,6,10, and *reset* is high in round 9, then z should be high in rounds 5 and 12. Design a synchronous reactive component that captures this behavior. You may want to use the extended-state machine notation. ∎

2.2 Properties of Components

2.2.1 Finite-State Components

In many embedded applications, it suffices to consider types with only finitely many values. The type `bool` and enumerated types are finitely valued, whereas the numerical types such as `nat`, `int`, and `real` are not finite. When all the variables of a component have finite types, the set Q_I of inputs, the set Q_O of outputs, and the set Q_S of states are all finite. This is the case for the

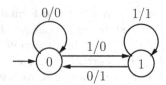

Figure 2.3: Mealy Machine Corresponding to the Component `Delay`

component `Delay`. Such components are called finite-state components and are amenable to powerful automated analysis.

FINITE-STATE COMPONENT

A synchronous reactive component C is said to be *finite-state* if the type of each of its input, output, and state variables is finite.

Note that the component `Switch` of figure 2.2 is not finite-state according to the definition due to the integer-valued state variable `x`. A close examination of this component reveals that, along every possible execution of the component, the value of x never exceeds 10. Thus, the only relevant range of values for x is from 0 to 10, and we can modify the description of `Switch` by changing the type of x to the range-type `int`$[0,10]$. The resulting component is a finite-state component. In general, we allow restricting the numerical types `int`, `nat`, and `real` to the corresponding range-types `int`$[low, high]$, `nat`$[low, high]$, and `real`$[low, high]$, where *low* and *high* are numerical constants of the corresponding types. The resulting restricted versions of `int` and `nat` are finite types.

For a finite-state component, its behavior can be illustrated by a labeled finite graph. The nodes of the graph are states of the component. If s is an initial state of the component, then there is a sourceless edge incident on s. If $s \xrightarrow{i/o} t$ is a reaction of the component, then there is an edge from node s to node t labeled with input i and output o. Such graphs are called *Mealy machines*. Executions of the component are simply paths through this graph starting at an initial state.

The Mealy machine representation of the `Delay` component is shown in figure 2.3.

Exercise 2.4: Consider the component `OddDelay` from exercise 2.1. Is the component finite-state? Draw the corresponding Mealy machine. ∎

2.2.2 Combinational Components

Consider the `Comparator` component shown in figure 2.4. The component has two input variables, in_1 and in_2, both of type `nat`. It has a Boolean output variable *out*. In each round, the component reads the inputs in_1 and in_2 and

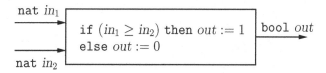

Figure 2.4: Combinational Component `Comparator`

sets the output variable *out* to 1 if the value of in_1 is greater than or equal to the value of in_2, and to 0 otherwise.

The component does not need to maintain any internal state and, hence, has no state variables. When there are no state variables, there is no initialization, and the reaction description assigns values to the output variables in terms of values of the input variables.

When the set S of state variables is empty, formally there is a unique valuation for S, and let us denote this unique state by $s_\emptyset$. This will also be the initial state. For every pair of natural numbers m and n, if $m \geq n$, `Comparator` has a reaction $s_\emptyset \xrightarrow{(m,n)/1} s_\emptyset$, and if $m < n$, then it has a reaction $s_\emptyset \xrightarrow{(m,n)/0} s_\emptyset$. A possible execution of the component is

$$s_\emptyset \xrightarrow{(2,3)/0} s_\emptyset \xrightarrow{(5,1)/1} s_\emptyset \xrightarrow{(40,40)/1} s_\emptyset.$$

Components such as `Comparator` without any state variables are called combinational.

COMBINATIONAL COMPONENT

A synchronous reactive component C is said to be *combinational* if the set of its state variables is empty.

Note that the component `Comparator` corresponds to the Boolean-valued expression $in_1 \geq in_2$. A component C that has the variable *out* as one of its input variables can instead take both in_1 and in_2 as input variables, and we can substitute every occurrence of *out* in the reaction description of C by the expression $in_1 \geq in_2$ without changing its behavior. Conversely, note that the values of the expressions used in the reaction description of a component can be explicitly modeled as combinational components. For example, to take logical conjunction of two Boolean variables x and y, we can simply use the expression $x \wedge y$ or construct a combinational component that takes two input variables x and y and produces an output that has value 1 precisely when both input variables are 1. Whether a desired expression is modeled as a combinational component is a design choice, which is influenced by the primitives supported by the language used to describe the reactions.

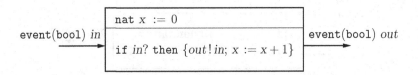

Figure 2.5: Event-Triggered Component `TriggeredCopy`

2.2.3   Event-Triggered Components *

In the synchronous execution model of reactive components, the notion of a *round* is global, and each component participates in every round. This may not be realistic in some scenarios, and we want to allow the possibility of a component to specify its own notion of a round. For example, a system may consist of multiple hardware components, each operating at a different clock frequency, and in this case, each component will participate in only those rounds in which its own clock signal is high. To model this behavior, we use input variables of type `event`. The basic type `event` is the enumerated type $\{\top, \bot\}$, where $\top$ denotes that the event is present and $\bot$ denotes that the event is absent. More generally, we allow the event type to be parameterized by another type: the event can be absent or can be present with a value belonging to the parameter type. For example, the type `event(bool)` has three values 0, 1, and $\bot$. For an event variable x, the Boolean expression x?, meaning x is present, stands for the expression $x \neq \bot$. Note that while input, output, and local variables can be of type `event`, a state variable cannot be of this type, as it is not meaningful to consider a state to be absent.

As an example, consider the component `TriggeredCopy`, shown in figure 2.5, that copies its input to output. The type of input variable *in* is `event(bool)`: in each round, the input can be absent and, if present, takes a Boolean value. Whenever the input is present, denoted by *in*?, it is copied to the output; when input is absent, so is the output. The type of the output variable *out*, thus, is also `event(bool)`. The assignment *out* := *in* is written as *out*!*in* to highlight that the event *out* is issued.

By default, an output event variable is absent, that is, if the event output variable *out* is not explicitly assigned a value during a round, then its value is assumed to be $\bot$. In the reaction description of the component `TriggeredCopy`, if the input event is absent, then the code does not assign any value to the output variable, and thus the component responds to an absent input with an absent output.

The component `TriggeredCopy` does maintain a state variable x: initially x is 0, whenever the input is present, x is incremented, and whenever the input is absent, x stays unchanged. Thus, the value of the state x shows the number of past rounds in which the input has been present. For every natural number n,

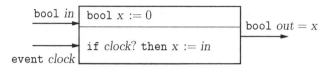

Figure 2.6: Event-Triggered Component `ClockedCopy`

the component has three reactions:

$$n \xrightarrow{\perp/\perp} n; \quad n \xrightarrow{0/0} n+1; \quad n \xrightarrow{1/1} n+1.$$

A sample execution of `TriggeredCopy` is

$$0 \xrightarrow{\perp/\perp} 0 \xrightarrow{0/0} 1 \xrightarrow{1/1} 2 \xrightarrow{\perp/\perp} 2 \xrightarrow{\perp/\perp} 2 \xrightarrow{1/1} 3.$$

We say that the input variable *in* is a *trigger* for the component `TriggeredCopy`, and the component is *event-triggered*. If the input is absent in a round, then the component is passive: the output is absent, and the state stays unchanged. Such a reaction is called a *stuttering* reaction. In the implementation, the component does not have to be "executed" to produce such a reaction.

As another example, consider the event-triggered component `ClockedCopy` shown in figure 2.6. It has a Boolean input variable *in* and an input event variable *clock* that acts as a trigger. Every time the clock event is present, the component updates its state variable x to the current value of the input variable *in*. Any changes to the input *in* during rounds in which the event *clock* is absent are ignored. The output *out* is a Boolean variable, and its value equals the updated state. Such an output variable is called a *latched* output. In this case, the component does not need to explicitly compute the value of *out*, and this is indicated by associating *out* with the state variable x in the declaration of the output.

The output variable *out* always has a value, even during rounds in which the trigger *clock* is absent, and equals the value of the input variable *in* from the most recent round in which the event *clock* was present. One possible execution of the component is shown below, where input is listed as a pair with the value of *in* followed by the value of *clock*:

$$0 \xrightarrow{(1,\perp)/0} 0 \xrightarrow{(1,\top)/1} 1 \xrightarrow{(0,\perp)/1} 1 \xrightarrow{(0,\perp)/1} 1 \xrightarrow{(0,\top)/0} 0.$$

Formally, for a synchronous reactive component C, an output variable y is said to be *latched* if there exists a state variable x such that in every reaction $s \xrightarrow{i/o} t$ of the component, the value of the output variable y is the updated value of the state variable x: $o(y) = t(x)$. In the implementation of a component, a latched output does not need to be explicitly stored or computed; the corresponding state variable needs to be made accessible to other components.

Now we can define the notion of an event-triggered component in its generality. Each of its output variables should be either latched or an event. When the triggering input events are absent, the state should stay unchanged (and, as a result, the latched outputs also stay unchanged), and event outputs should be absent.

EVENT-TRIGGERED COMPONENT

For a synchronous reactive component $C = (I, O, S, Init, React)$, a set $J \subseteq I$ of input variables is said to be a *trigger* if:

1. every input variable in J is of type `event`;

2. every output variable either is latched or is of type `event`; and

3. if i is an input with all events in J absent (that is, for all input variables $x \in J$, $i(x) = \bot$), then for all states s, if $s \xrightarrow{i/o} t$ is a reaction, then $s = t$ and $o(y) = \bot$ for every output variable y of event type.

A component C is said to be *event-triggered* if there exists a subset $J \subseteq I$ of its input variables such that J is a trigger for C.

Exercise 2.5: Design an event-triggered combinational component `ClockedMax` with two input variables x and y of type `nat` and an input event variable *clock*. The output variable z of the component should be of type `event(nat)` such that the value of z should be the maximum of the inputs x and y during rounds in which *clock* is present. ∎

Exercise 2.6: Design an event-triggered component `SecondToMinute` with the input event variable *second* and the output event variable *minute* such that *minute* is present every 60th time the event *second* is present. ∎

Exercise 2.7: Design a component `ClockedDelay` with a Boolean input variable x, an input event variable *clock*, and an output variable y of type `event(bool)` with the following behavior: if *clock* is present during rounds, say, $n_1 < n_2 < n_3 < \cdots$ then in round n_1, the output should be some default value, say 0; in round n_{j+1}, for each j, the output should equal the value of x in round n_j; and in the remaining rounds (that is, rounds during which the input event *clock* is absent), output should be absent. ∎

2.2.4   Nondeterministic Components

In our examples so far, the components are *deterministic*: for a given sequence of inputs, the component has a unique execution producing a unique sequence of outputs. Such deterministic behavior is ensured if the component has a single initial state, and in every state, for a given input, there is exactly one possible reaction.

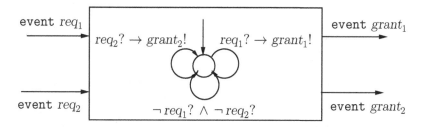

Figure 2.7: Nondeterministic Component `Arbiter`

DETERMINISTIC COMPONENT

A synchronous reactive component C is said to be *deterministic* if:

1. C has a single initial state, and

2. for every state s and every input i, there is precisely one output o and one state t such that $s \xrightarrow{i/o} t$ is a reaction of C.

The components `Delay`, `Switch`, `Comparator`, `TriggeredCopy`, and `ClockedCopy` are all deterministic. Determinism is a desirable property for components that are meant to be implemented.

Nondeterministic components, in contrast, can respond with different output sequences for the same input sequence. Such components are useful for modeling parts of the system that are not yet fully designed and for capturing constraints on the environment. As an example, consider the component `Arbiter` shown in figure 2.7. It has two input variables, req_1 and req_2, and two output variables $grant_1$ and $grant_2$, all of which are events. This component is designed to resolve contention among incoming requests. The dynamics of the component are described using the extended-state machine notation. The machine has only one mode, and thus no state needs to be maintained explicitly to record this mode. In each round, a mode-switch whose guard-condition is satisfied is chosen, and the corresponding update code is executed.

When only the request req_1 is present, the guard-condition of the mode-switch "$req_1? \rightarrow grant_1!$" is satisfied, and the event $grant_1$ is issued. Note that the event $grant_2$ is absent by default in this case. The case when only the request req_2 is present is symmetric. If both requests are absent, then the guard-condition of the mode-switch labeled "$\neg req_1? \wedge \neg req_2?$" is satisfied. For this mode-switch, there is no explicit update code, and thus both the output events are absent by default. However, if both input requests are present, then the guard-conditions $req_1?$ and $req_2?$ of two mode-switches are satisfied. In such a case, one of them gets executed. There are two possible reactions of the component: either $grant_1$ is present and $grant_2$ is absent or $grant_1$ is absent and $grant_2$ is present. Such a nondeterministic behavior captures what an arbiter should do, namely,

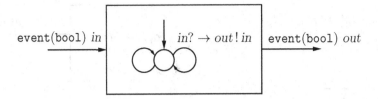

Figure 2.8: Nondeterministic Component `LossyCopy`

a grant output should be issued only when requested, and at most one grant output should be issued in any round, without constraining how the contention is resolved, leaving open the possibility of different implementations. Note that the component `Arbiter` is combinational and event-triggered.

As another example of a nondeterministic component, consider the combinational and event-triggered component `LossyCopy` shown in figure 2.8. It has an input event *in* and an output event *out*. The desired behavior of the component is that in each round, either the input is copied to the output or the output is absent. This is again described by a single-mode extended-state machine with two mode-switches: one with guard-condition *in?* and update code *out!in* and one with default guard-condition that always holds and default update code that does not assign any explicit value to *out*. When the input event is present, the guard-conditions of both the mode-switches are satisfied. In one case, the value of the input is issued on the output event; in the other case, no action is taken, and thus the output event is absent. When the input event is absent, only the mode-switch with default guard-condition is enabled, and the output event is absent. Such a component can be used to model potential loss of messages along a network link. One possible execution of the component is

$$s_\emptyset \xrightarrow{0/0} s_\emptyset \xrightarrow{1/\bot} s_\emptyset \xrightarrow{\bot/\bot} s_\emptyset \xrightarrow{1/1} s_\emptyset \xrightarrow{\bot/\bot} s_\emptyset \xrightarrow{0/\bot} s_\emptyset.$$

Exercise 2.8: Consider the `Delay` component of figure 2.1, and suppose we replace the reaction description by

$$out := x;$$
$$x := \mathsf{choose}(in, x)$$

Describe in words the behavior of the modified component. Draw the Mealy machine corresponding to the component. ∎

Exercise 2.9: For the nondeterministic component `Arbiter` of figure 2.7, the reactions are expressed using the extended-state machine notation. Write an equivalent description using straight-line update code. You can use a local variable whose value is assigned nondeterministically using the **choose** construct. ∎

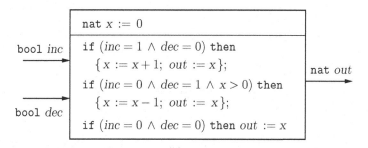

Figure 2.9: Component `Counter` with Input Assumptions

2.2.5 Input-Enabled Components

All the components we have seen so far have the following property: in every state and for every input, the component has *at least* one reaction. This property makes the components *input-enabled*. For a given sequence of inputs, an input-enabled component has at least one corresponding execution.

INPUT-ENABLED COMPONENT

For a synchronous reactive component C, an input i is said to be *enabled* in a state s if there exists an output o and a state t such that $s \xrightarrow{i/o} t$ is a reaction of C. The component C is said to be *input-enabled* if every input is enabled in every state.

There are design problems where it is useful to make assumptions about the inputs that the environment may supply. As an example, consider the component `Counter` of figure 2.9, which maintains a non-negative counter using a state variable x. The initial value of x is 0. The component has two Boolean input variables *inc* and *dec*: when the input *inc* is 1, the counter state is incremented by 1, and when the input *dec* is 1, the counter state is decremented by 1. The counter does not expect both input variables *inc* and *dec* to be 1 simultaneously, nor does it expect the counter to be decremented when the counter value is 0. We describe the reactions of `Counter` only for inputs satisfying this assumption as shown in figure 2.9. The output of the component is the updated value of x. When both input variables *inc* and *dec* are 1, and when *dec* is 1 with the state x equal to 0, the reaction description does not assign any value to the output, and thus there is no corresponding reaction.

In general, the input assumption can be a constraint on the sequence of inputs supplied to the component. When a component C with input assumptions is used as part of a larger system, we need to check that its input assumptions are indeed satisfied by the components that supply its inputs.

Note that, by definition, every deterministic component has exactly one reaction corresponding to a given state and input and, thus, is input-enabled.

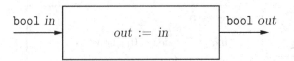

Figure 2.10: The Combinational Component `Relay`

Exercise 2.10: Design a nondeterministic component `CounterEnv` that supplies inputs to the counter of figure 2.9. The component `CounterEnv` has no inputs, and its outputs are the Boolean variables *inc* and *dec*. It should produce all possible combinations of outputs as long as the component `Counter` is willing to accept these as inputs: it should never set both *inc* and *dec* to 1 simultaneously, and it should ensure that the number of rounds with *dec* set to 1 never exceeds the number of rounds with *inc* set to 1. ∎

2.2.6   Task Graphs and Await Dependencies

Let us consider a component `Relay`, shown in figure 2.10, that has a Boolean input variable *in* and a Boolean output variable *out*. The component `Relay` is a combinational component without any state variables and, in each round, simply copies the input to the output.

Let us compare the components `Relay` and `Delay`. Observe that they have identical input/output variables. In a given round, the output of the component `Delay` does not depend on its input in that round, whereas the component `Relay` can produce its output only after reading the input for the current round. Intuitively, the output of `Relay` must *await* its input, whereas this is not necessary for `Delay`. This crucial *intra-round dependency* of output variables on input variables can impact how components can be composed.

In the current reaction description of the component `Delay`, given the input and the current state, the update code consisting of two assignments computes the output and updated state. This monolithic description hides the intra-round independence of the output on the input. To avoid this problem, we allow the reaction description to be split into multiple tasks. This is illustrated by revising the description of the component `Delay` to obtain the component `SplitDelay` shown in figure 2.11.

The reaction description for the component `SplitDelay` is split into execution of two tasks A_1 and A_2. The task A_1 computes the value of the variable *out* using the value of the state variable x, whereas the task A_2 updates the value of the state variable x using the value of the input variable *in*. In general, each task has an associated *read-set R* and *write-set W* of variables, and it assigns values to variables in the write-set given the values of variables in the read-set. Note that the write-set of a task should not include any input variables of the component. The read-/write-sets can also include local variables used in

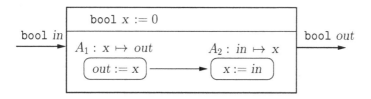

Figure 2.11: Component `SplitDelay` with Split Reaction

the reaction description. Since an output variable is used for communication with other components, it should be written by exactly one task. With this restriction, once the task A of a component C responsible for writing an output variable y executes, the value of y will no longer change within the current round and can be used by other components, even if some other tasks within the component C have not yet executed.

In our illustrations, we depict tasks as rectangular boxes with rounded corners. The declaration

$$A : x_1, x_2, \ldots x_m \mapsto y_1, y_2, \ldots y_n$$

indicates that the task A has the read-set $R = \{x_1, x_2, \ldots x_m\}$ and the write-set $W = \{y_1, y_2, \ldots y_n\}$. The *update description* of a task describes its computation as a sequence of assignment and conditional statements or, alternatively, using the extended-state machine notation. The update description can be nondeterministic: the assignments may use the **choose** construct, and in the extended-state machine, guards of multiple mode-switches out of the same mode may be simultaneously satisfied. Thus, the mathematical semantics of the update description *Update* of a task is a relation $[\![Update]\!]$ between the values of variables in the read-set and the values of variables in the write-set; that is, $[\![Update]\!]$ contains pairs of the form (s, t) with $s \in Q_R$ and $t \in Q_W$.

In example of figure 2.11, for the task A_1, the read-set is $\{x\}$, the write-set is $\{out\}$, and the update is described by the assignment $out := x$; and for the task A_2, the read-set is $\{in\}$, the write-set is $\{x\}$, and the update is described by the assignment $x := in$.

When the reaction description is split into multiple tasks, we need to specify constraints on the order in which the tasks should be executed. In our example of `SplitDelay`, the task A_1 must be executed before the task A_2; executing the assignment statements corresponding to these two tasks in this order is necessary for the desired behavior. We write $A_1 \prec A_2$ to express the precedence constraint that the task A_1 should be executed before the task A_2. In our illustrations, the precedence constraint $A_1 \prec A_2$ is captured by an arrow from the task A_1 to the task A_2. Thus, in the *task graph* description of the component, nodes correspond to tasks (with associated read-sets, write-sets, and update descriptions), and edges correspond to precedence constraints on the order of execution of the tasks within a round.

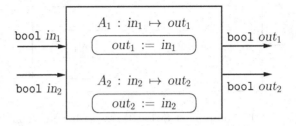

Figure 2.12: Component `ParallelRelay`: out_1 awaits in_1 and out_2 awaits in_2

Given the precedence relation $\prec$ over tasks, the relation $\prec^+$ denotes the *transitive closure* of the relation $\prec$: for two tasks A and A', $A \prec^+ A'$ holds if there is a path from A to A' in the task graph. In other words, if there is a chain of precedence constraints $A_1 \prec A_2 \prec \cdots \prec A_n$, for $n > 1$, then $A_1 \prec^+ A_n$. If $A \prec^+ A'$ holds, then the task A must execute before the task A'. Thus, the relation $\prec^+$ captures all the constraints on the execution order implied by the precedence constraints. We require that the precedence relation $\prec$ is *acyclic*: there is no task A such that $A \prec^+ A$, that is, the task graph does not contain any cycles. In particular, if $A_1 \prec A_2$ is a precedence constraint, then we cannot have the constraint $A_2 \prec A_1$.

The precedence constraints captured by the relation $\prec$ lead to *await* dependencies between output and input variables. Output variables written by a task must await the input variables this task reads, and if $A_1 \prec^+ A_2$, then the output variables written by A_2 must await the input variables read by A_1. We write $y \succ x$ if the output variable y awaits the input variable x according to the precedence constraints $\prec$ over the tasks.

In our example component `SplitDelay`, by the above definitions, the output variable *out* does not await the input variable *in*. By default, when the reaction description of a component is not explicitly split into tasks, it can be viewed as a single task with its read-set containing all the input and state variables, its write-set containing all the output and state variables, and the update description same as the reaction description. In such a case, each output variable awaits each input variable. Then for both `Relay` and `Delay` components, the output variable *out* awaits the input variable *in*.

An output variable may await only some of the input variables, and different output variables may await different input variables. This can be captured if we allow the precedence relation $\prec$ to express ordering constraints among tasks only partially. To illustrate this point, consider the combinational component `ParallelRelay` (see figure 2.12) with two input variables, in_1 and in_2, and two output variables, out_1 and out_2. In each round, the component copies the input variable in_1 to output variable out_1 and copies the input variable in_2 to output variable out_2. We wish to express that out_1 awaits in_1 but not in_2, and out_2

awaits in_2 but not in_1. This is achieved by splitting the reaction description into two tasks, A_1 and A_2. The task A_1 reads in_1 and writes out_1, whereas the task A_2 reads in_2 and writes out_2. There is no precedence constraint between the two tasks, and thus the task graph has no edges. This means that these two tasks are independent and can be executed in any order (and even concurrently if implemented on parallel hardware).

A *task schedule* is a linear ordering of all the tasks that is consistent with the precedence relation. For `SplitDelay`, we have only one task schedule A_1, A_2, whereas for `ParallelRelay`, we have two possible task schedules, A_1, A_2 and A_2, A_1. In general, for a task graph with k tasks, a schedule is an ordering $A_1, A_2, \ldots A_k$ of all the tasks, such that if there is a precedence constraint from the task A to task A', then A must appear before A' in the schedule.

Thus, the ordering constraints expressed by $\prec$ can allow multiple schedules. We can allow any binary relation $\prec$ over tasks as the precedence relation as long as it obeys the following rules. As already discussed, the precedence relation should be acyclic, and this ensures that there is at least one possible way of ordering all the tasks to get a schedule. Second, if a task A_2 is reading an output or a local variable y, then the precedence constraints should enforce that the value of y is already computed when A_2 executes in every possible schedule. This is ensured if there exists a task A_1 that writes y, such that $A_1 \prec^+ A_2$ holds. Third, we want to ensure that two tasks that are independent according to the precedence constraints can be executed in either order without affecting the result of the execution. For example, in the component `SplitDelay`, the task A_1 reads x, and the task A_2 writes x. If these two tasks were unordered (that is, if the precedence edge $A_1 \prec A_2$ was missing), then it should be considered a syntactic error as the result depends on which of the two tasks executes first. In general, two tasks have a *write-conflict* if there is a variable that belongs to the write-set of one of the tasks and also belongs to either the read-set or the write-set of the other task. The tasks A_1 and A_2 have a write-conflict in `SplitDelay` but not in `ParallelRelay`. When two tasks have a write-conflict, the order in which they execute matters, and hence the precedence relation should not leave their ordering unconstrained.

As an illustrative example of the general specification of the tasks, consider the task graph shown in figure 2.13 for a component with state variables x_1 and x_2, input variables in_1 and in_2, and output variables out_1, out_2, and out_3. The reaction description also uses a local variable y. Each output variable is written by exactly one task: out_1 by task A_3, out_2 by task A_2, and out_3 by task A_4. The precedence relation given by $A_1 \prec A_3$, $A_1 \prec A_4$, and $A_2 \prec A_4$ is acyclic. The task A_4 reads the output out_2, and this is legal since it has a precedence-edge from the task A_2 that writes out_2. Similarly, the local variable y is guaranteed to be written by the task A_1 before it is used by A_4. The task A_2 is not ordered with respect to the tasks A_1 and A_3 according to the precedence constraints, and this means that the task A_2 should have no write-conflicts with both A_1 and A_3. This is indeed the case. Finally, verify that for each of the state variables x_1

and x_2, the task that writes to the variable is ordered with respect to the tasks that read this variable. For instance, the task A_2 will read the "old" value of x_2, whereas the task A_3 will read the value of x_1 updated by A_1 and rewrite it. The await dependencies implied by the precedence constraints are: the output variable out_1 awaits the input in_1, the output variable out_2 does not await any inputs, and the output variable out_3 awaits both the input variables in_1 and in_2. Thus, $out_1 \succ in_1$, $out_3 \succ in_1$, and $out_3 \succ in_2$.

The definition and rules for splitting the update into multiple tasks, and the await dependencies they induce among output and input variables, are summarized below.

TASK GRAPHS AND AWAIT DEPENDENCIES

For a synchronous reactive component C with input variables I, output variables O, and state variables S, a task-graph description of the reactions using a set L of local variables consists of a set of tasks and a binary precedence relation $\prec$ over these tasks. Each task A has a read-set $R \subseteq I \cup S \cup O \cup L$, a write-set $W \subseteq O \cup S \cup L$, and an update description $Update$ with $[\![Update]\!] \subseteq Q_R \times Q_W$ such that:

1. The precedence relation $\prec$ is acyclic.

2. Each output variable belongs to the write-set of exactly one task.

3. If an output or a local variable y belongs to the read-set of a task A, then there exists a task A' such that y is in the write-set of A' and $A' \prec^+ A$.

4. If a state or a local variable x belongs to the write-set of a task A and also to either the read-set or write-set of a different task A', then either $A \prec^+ A'$ or $A' \prec^+ A$.

For an output variable y and an input variable x, y *awaits* x (also written $y \succ x$), precisely when for the unique task A, such that y belongs to the write-set of A, either x belongs to the read-set of A, or there exists a task A' such that $A' \prec^+ A$ and x belongs to the read-set of A'.

To execute a component with a task-graph description of reactions, we choose a schedule that orders all the tasks in a manner consistent with the precedence relation. The tasks are then executed one by one in this order. Executing a task A means assigning values to the variables in its write-set based on the values of the variables in its read-set. Note that our consistency requirements make sure that an output variable gets assigned a value exactly once, and if it is read by a task A, then the output value has already been assigned when A is executed. Also, if a task A reads a state variable x, then either A reads the value of x at the beginning of the round (this happens if there is no task with x in its write-set that precedes A), or there is a unique task A' with x in its write-set such that A always reads the value written by A' (irrespective of the schedule

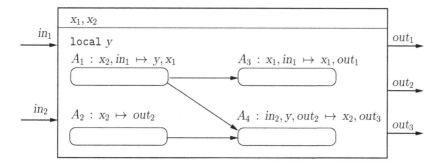

Figure 2.13: Illustrative Task Graph

chosen) since the precedence relation totally orders A with respect to all the tasks that write to x.

In our example of figure 2.13, there are five possible schedules:

A_1, A_2, A_3, A_4; A_1, A_2, A_4, A_3; A_1, A_3, A_2, A_4; A_2, A_1, A_3, A_4; A_2, A_1, A_4, A_3.

The set of possible reactions of the component will depend on the update descriptions of the four tasks but is independent of the schedule.

Properties such as determinism and input-enabledness can be naturally defined for tasks so that they imply the corresponding properties of the component:

- **Deterministic Tasks:** A task A with read-set R, write-set W, and update description *Update* is said to be *deterministic* if for every valuation s over R, there exists a unique valuation t over W such that $(s, t) \in [\![Update]\!]$. Thus, given values for the read variables, a deterministic task assigns unique values to the variables it writes. If a component has a single initial state and all the tasks in the task-graph description of reactions are deterministic, then the component must be deterministic. This is because the requirements on what constitutes a legal precedence relation ensure that the schedule does not affect the result of executing tasks.

- **Input-Enabled Tasks:** An update task A with read-set R, write-set W, and update description *Update* is said to be *input-enabled* if for every valuation s over R, there exists at least one valuation t over W such that $(s, t) \in [\![Update]\!]$. Thus, given values for the read variables, an input-enabled task produces at least one result. Now consider a component with a task-graph description of its reactions so that all the tasks are input-enabled. Given a state and an input, we can execute all the update tasks in an order consistent with the precedence constraints. Since each task is input-enabled, there is a way to progress at every step, and thus the component can produce at least one reaction. Thus, such a component is input-enabled.

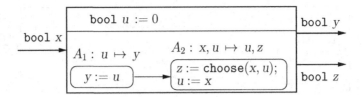

Figure 2.14: Component Specified Using Task Graph

Exercise 2.11: Consider the component from exercise 2.3. Split the reaction description into two tasks so that the output z awaits the inputs x and y but not the input *reset*. ∎

Exercise 2.12: Consider a synchronous reactive component C with an input variable x and output variables y and z. The component has two tasks, A_1 and A_2, such that the output y belongs to the write-set of the task A_1, and the output z belongs to the write-set of the task A_2. If we know that the output y awaits the input x, but the output z does not await x, then what can we conclude regarding the precedence constraints between the tasks A_1 and A_2? ∎

Exercise 2.13: Consider the synchronous reactive component shown in figure 2.14. List all the possible reactions of the component. Does the output y await x? Does the output z await x? ∎

Exercise 2.14: Design a synchronous reactive component `ComputeAverage` with an integer input variable x, an input event variable *clock*, and a real-valued output variable y with the following behavior: in the first round, the output y is 0; in a subsequent round i, let $j < i$ be the most recent round before round i in which the input event *clock* is present (if *clock* is absent in all rounds before i, then let $j = 0$), the output should be the *average* of input values for x in rounds j, $j + 1$, upto $i - 1$. The following is a sample behavior of the desired component:

Clock	$\perp$	$\perp$	$\top$	$\perp$	$\perp$	$\perp$	$\top$	$\perp$
x	5	2	−3	1	6	5	−2	11
y	0	5	3.5	−3	−1	1.33	2.25	−2

The component should be designed so that the output y does not await any of the input variables. ∎

2.3   Composing Components

2.3.1   Block Diagrams

Suppose we want to design a reactive component with a Boolean input variable *in* and a Boolean output variable *out*, such that in the first two rounds the

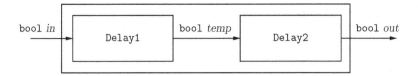

Figure 2.15: Block Diagram for `DoubleDelay` from Two `Delay` Components

output is 0 and in every subsequent round n, the output equals the input in round $n-2$. Instead of designing this component from scratch, we would like to reuse the component `Delay`. Composing two `Delay` components in series gives the desired component. The resulting design of the component `DoubleDelay` is shown in figure 2.15. The design of the component should be obvious from the block diagram, and given the intuitive appeal of such diagrammatic descriptions, almost all tools for high-level embedded systems design support such diagrams. A careful examination of the block diagram reveals that there are three operations on components in such a diagram:

- **Instantiation:** The components `Delay1` and `Delay2` are both instances of the component `Delay`. Such instances are obtained by renaming the input/output variables. For example, the component `Delay1` is exactly like the component `Delay` except its output variable is called *temp* instead of *out*.

- **Parallel Composition:** The two components `Delay1` and `Delay2` run in parallel. The block diagram shows that the output of the component `Delay1` is the same as the input of the component `Delay2`, and this achieves communication between the two components. The communication is synchronous. In each round, the component `Delay1` reads its input *in*, produces output *temp*, and updates its internal state to record the current value of *in*. In the same round, the component `Delay2` reads its input *temp* — as supplied by the component `Delay1`, produces its output *out*, and updates its internal state to record the current value of *temp*.

- **Output Hiding:** For the component `DoubleDelay`, the relevant output variable is *out*, and the variable *temp* is only an auxiliary variable that is used in implementing `DoubleDelay`. The block diagram shows that the variable *temp* is local and not exported to the outside world.

The component `DoubleDelay` is textually defined as

$$(\texttt{Delay}[\, out \mapsto temp\,] \;\|\; \texttt{Delay}[\, in \mapsto temp\,]) \setminus temp.$$

We proceed to discuss the three operations in the above expression, namely, parallel composition $\|$, renaming $\mapsto$, and hiding $\setminus$, in more details.

2.3.2   Input/Output Variable Renaming

Before composing and connecting components, we may need to rename variables so that there are no name conflicts among state variables of different components, and common names for input/output variables indicate desired input/output connections. It is common practice to assume that the renaming of state variables is *implicit* and performed mechanically without burdening the designer. For instance, in figure 2.15, we can assume that the state variable of the component Delay1 is called x_1 instead of x, and the state variable of the component Delay2 is called x_2. The renaming of input/output variables needs to be defined *explicitly* since it establishes the intended communication pattern.

Let $C = (I, O, S, Init, React)$ be a synchronous reactive component, x be an input or an output variable, and y be a *fresh* variable (that is, y is not a state, input, or output variable of C), such that the types of x and y are the same. Then the component obtained by *renaming x to y in C*, denoted $C[x \mapsto y]$, is the synchronous reactive component obtained by substituting the variable name x by y in the description of C.

With this notation, the component Delay1 is defined as $\mathtt{Delay}[out \mapsto temp]$. For the component Delay1, the set of input variables is $\{in\}$, the set of output variables is $\{temp\}$, the set of state variables is $\{x_1\}$, the initialization is $x_1 := 0$, and the reaction description is $temp := x_1; x_1 := in$. Similarly, the component Delay2 is defined as $\mathtt{Delay}[in \mapsto temp]$.

Observe that variable renaming does not change properties of a component. For instance, if a component is deterministic, so is its renamed instance, and if a component is event-triggered, so is its renamed instance.

2.3.3   Parallel Composition

The parallel composition operation combines two components into a single component whose behavior captures the synchronous interaction between the two components running concurrently.

Compatibility in Variable Names

Consider $C_1 = (I_1, O_1, S_1, Init_1, React_1)$ and $C_2 = (I_2, O_2, S_2, Init_2, React_2)$. Before we can compose these two components, we need to check for *compatibility* in their variable declarations. First, there should be no name conflicts concerning state variables. If x is a state variable of C_1, then no variable of C_2 should be called x. That is, the set S_1 should be disjoint from each of the sets I_2, O_2, and S_2; symmetrically, the set S_2 should be disjoint from each of I_1, O_1, and S_1. Note that the names of state variables are really private to a component. We can always rename them to avoid name conflicts before taking the composition. For instance, the variable name may be prefixed by the name of the component instance. Henceforth, we will assume that names of state variables are chosen according to a scheme that avoids name conflicts. Similarly, if the reaction description uses local variables, we will assume that the names of these local variables are unique and do not conflict with the names of other variables.

Second, a variable can be an input variable to both the components, and an output variable of one component can be an input variable to the other, but a variable cannot be an output variable of both the components. That is, the sets O_1 and O_2 should be disjoint. A consequence of this requirement is that only one component is responsible for controlling the value of any given variable.

Product Variables

When two components C_1 and C_2 are compatible, we want to define their parallel composition, denoted $C_1 \| C_2$, to be another synchronous reactive component C. We will also refer to the composition C as the *synchronous product* of the components C_1 and C_2. We proceed to describe how to construct the input variables, output variables, state variables, initialization, and reaction description of the product C.

Each state variable of a component is a state variable of the product. That is, the set S of state variables of C is the union $S_1 \cup S_2$. Each output variable of a component is an output variable of the product. That is, the set O of output variables of C is the union $O_1 \cup O_2$. Each input variable of a component is an input variable of the product, provided it is not an output variable of the other component. That is, the set I of input variables of C is the set $(I_1 \cup I_2) \setminus O$, denoting the difference of the two sets $I_1 \cup I_2$ and O.

For example, the composition of the components `Delay1` and `Delay2` gives the component with state variables $\{x_1, x_2\}$, output variables $\{temp, out\}$, and input variables $\{in\}$.

Product States

A state of the product C assigns values to variables in S_1 as well as variables in S_2. The initial states of C are obtained by choosing the values for variables in S_1 according to the initialization $Init_1$ of the component C_1 and choosing the values for variables in S_2 according to the initialization $Init_2$ of the component C_2. If the two initializations $Init_1$ and $Init_2$ are given as sequences of assignment statements, then the initialization $Init$ for the product can be defined to be $Init_1; Init_2$ or, equivalently, $Init_2; Init_1$ since there can be no write-conflicts in the two blocks of initial assignments. In the example of composing the components `Delay1` and `Delay2`, the sole initial state of the product assigns the value 0 to both the state variables x_1 and x_2, and this can be described by the initialization $x_1 := 0; x_2 := 0$.

Reaction Description of the Product

Let us consider how to obtain the reaction description and the corresponding set of reactions of the product. If the reaction descriptions of the two components

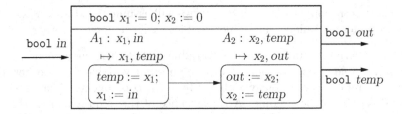

Figure 2.16: Parallel Composition of Delay1 and Delay2

C_1 and C_2 use local variables L_1 and L_2, respectively, then the set of local variables for the reaction description of the product is $L_1 \cup L_2$.

If there is no communication between the two components, and this is the case when outputs of one component are not inputs to the other, then the two components can be executed independently. To obtain the reactions of the product, we can execute the update code of one followed by the other, and the order would not matter. However, when outputs of one component are inputs to the other, we have ordering constraints on how components should execute within a round. If the input/output connections are only one way, as is the case in the example of composing the components Delay1 and Delay2, where the output of Delay1 is an input to Delay2 but not vice versa, then we can execute the updates of the components in the order suggested by the connections: we can first execute the component Delay1 and then execute the component Delay2. In other words, the reaction description for the product consists of a task graph with two tasks, the task A_1 corresponding to the reaction description of Delay1, the task A_2 corresponding to the reaction description of Delay2, and a precedence edge from A_1 to A_2. The product is shown in figure 2.16. The reactions of the product are listed below, where in each state we list the values of x_1 and x_2, in that order, and in each output, we list the values of $temp$ and out, in that order:

$$(0,0) \xrightarrow{0/(0,0)} (0,0); \quad (0,0) \xrightarrow{1/(0,0)} (1,0); \quad (0,1) \xrightarrow{0/(0,1)} (0,0); \quad (0,1) \xrightarrow{1/(0,1)} (1,0);$$
$$(1,0) \xrightarrow{0/(1,0)} (0,1); \quad (1,0) \xrightarrow{1/(1,0)} (1,1); \quad (1,1) \xrightarrow{0/(1,1)} (0,1); \quad (1,1) \xrightarrow{1/(1,1)} (1,1).$$

Composing Task Graphs

As another example of parallel composition, consider the composition shown in figure 2.17, in which the output out of the component SplitDelay is connected to the input of the component Inverter and vice versa. The component Inverter is a combinational component that sets its output to the negation of its input. This form of cyclic composition is called *feedback composition*. The result of the composition is the product component shown in figure 2.18. The product has no input variables, two output variables in and out, and one state

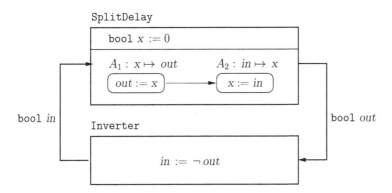

Figure 2.17: Feedback Composition of a `SplitDelay` with an `Inverter`

variable x. The task-graph description of the update for `SplitDelay` suggests an execution schedule in each round for the product. In each round, we first execute the task A_1 of `SplitDelay` to assign a value to the variable *out*. Now, the component `Inverter` can be executed as its input is available, and this assigns a value to the variable *in*. Subsequently, the task A_2 of `SplitDelay` can be executed using this value. In essence, we are constructing the task graph of the product by merging the task graphs of the two components. In this case, the component `Inverter` has a single task A, which reads *out* and writes *in*. We retain the original precedence constraints (the edge from A_1 to A_2 in the task graph of `SplitDelay`) and add additional precedence edges to reflect variable dependencies. The general rule for these additional cross-component edges is:

> If a task A belonging to one component reads a variable y, which is an output variable of the other component, then add a precedence edge from the unique task that writes y to the task A.

This rule gives us the edge from the task A_1 of `SplitDelay` to the task A of `Inverter` as the latter reads the variable *out* computed by A_1, and the edge from the task A of `Inverter` to the task A_2 of `SplitDelay` as the latter reads the variable *in* computed by A. In the first round, *out* is 0 and *in* is 1, and in every subsequent round, both of these values toggle. That is, the sequence of outputs produced by the product, listing the value of *in* first and *out* second, is $10, 01, 10, 01, 10, \ldots$

Now we can propose a precise definition for the reaction description of the product. We assume that the reaction descriptions for both the components C_1 and C_2 are given as task graphs. Recall that when the reaction description is not explicitly split into tasks, we interpret it as a single task that reads all the state and input variables and writes all the state and output variables, thus leading to await dependency of each of the output variables on all of the input variables. The set of tasks in the product is the union of the tasks of the two components. The precedence relation $\prec$ of the product is the union of the

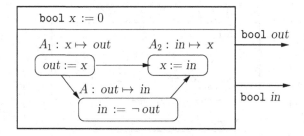

Figure 2.18: Parallel Composition of SplitDelay and Inverter

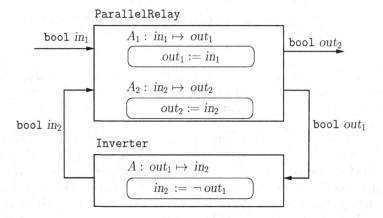

Figure 2.19: Parallel Composition of ParallelDelay and Inverter

precedence relations $\prec_1$ and $\prec_2$ of the component task graphs, together with the cross-component edges according to the rule above.

As another example, consider the composition of the components ParallelRelay and Inverter, shown in figure 2.19. The two tasks A_1 and A_2 are independent in ParallelRelay. Since the component Inverter reads out_1 and writes in_2, we get cross-component edges from A_1 to A and from A to A_2. This implies a new transitive precedence constraint: the task A_1 must be executed before A_2 in the product.

Acyclicity of Await Dependencies

The cross-component precedence edges can lead to a cycle among precedence constraints. This problem can be traced to cycles in the input/output await dependencies. Observe that in the feedback composition of the components SplitDelay and Inverter of figure 2.17, for the Inverter component, the variable in awaits the variable out, but for the SplitDelay component, there is no await dependency between the variables out and in. The absence of mutual

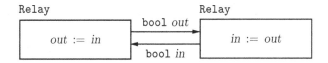

Figure 2.20: Ill-Formed Combinational Loop with Two `Relay` Components

await dependency is a key for well-formed behavior of the product. Composing components with mutually cyclic await dependencies can lead to unexpected behaviors, even when the individual components are deterministic. Let us illustrate two kinds of basic problems using two examples.

Figure 2.20 shows composition of two `Relay` components: the left component copies its input *in* to its output *out*, whereas the right component copies its input *out* to its output *in*. For one component, the variable *out* awaits the variable *in*, whereas for the other, the variable *in* awaits the variable *out*, and it is not possible to order the updates of the components in a consistent manner. If we just consider the set of reactions of the two components and mathematically compose these sets to obtain reactions that are consistent with the descriptions of both, then in each round, the product can produce two outputs: one possibility is that both the variables *in* and *out* are set to 0, and the other possibility is that both the variables are set to 1. Thus, if we were to allow composition of these two components, then we would obtain a nondeterministic component by composing deterministic ones.

A converse problem to the one of multiple possible consistent reactions arises in composition of an inverter component with a relay component shown in figure 2.21. The left component `Inverter` sets its output to the negation of its input, and the right component `Relay` copies its input to its output. For the left component, the variable *out* awaits the variable *in*, whereas for the right component, the variable *in* awaits the variable *out* causing cyclic await dependencies. In this case, there is no assignment of values to the two variables that is consistent with the reactions of the two components, and thus the product would have no behaviors.

Thus, it is imperative to detect cyclic await dependencies and rule out such ill-formed compositions. Let $\succ_1$ and $\succ_2$ represent the await dependencies of the two components. For instance, in figure 2.20, the await dependency for the top component is *out* $\succ_1$ *in* and for the bottom component is *in* $\succ_2$ *out*. The union of the two relations gives the cycle *out* $\succ_1$ *in* $\succ_2$ *out*. We allow two components to be composed only when the union of the two await-dependency relations $\succ_1 \cup \succ_2$ is acyclic, that is, there do not exist common input/output variables $x_1, x_2, \ldots x_n$, with $x_n = x_1$, such that for each $1 \leq j < n$, x_{j+1} awaits x_j according to either one of the two await-dependency relations. This condition can be checked automatically by the compiler for the modeling language.

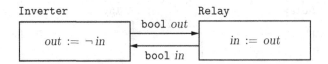

Figure 2.21: Ill-Formed Combinational Loop with a `Relay` and an `Inverter`

The compatibility conditions for two components to be composable are summarized in the definition below.

COMPONENT COMPATIBILITY

The components C_1 with input variables I_1, output variables O_1, and input/output await-dependency relation $\succ_1 \subseteq O_1 \times I_1$, and C_2 with input variables I_2, output variables O_2, and input/output await-dependency relation $\succ_2 \subseteq O_2 \times I_2$, are said to be *compatible* if (1) the sets O_1 and O_2 are disjoint, and (2) the relation $(\succ_1 \cup \succ_2)$ is acyclic.

We can take parallel composition of two components only when they are compatible by the above definition. Observe that by this definition, in figure 2.18, if we replace the component `SplitDelay` with the component `Delay`, then the composition is not allowed as the compatibility check would fail. This is because the component `Delay` has a single task, and thus its output *out* awaits its input *in*. Thus, our approach to ensuring well-behaved composition is conservative as it relies on the syntactic decomposition of the reaction description into tasks given by the designer.

Interfaces

One appealing feature of the compatibility check is that it refers only to input/output variables and their await dependencies. We can think of the inputs I, outputs O, and await-dependency relation $\succ \subseteq O \times I$ as an *interface* for the component. To form block diagrams consisting of multiple components, the designer can focus only on the interfaces of components to ensure compatibility and consistent usage and does not need to know internal details such as state variables and task graphs.

As an example of compatibility check using interfaces, consider the components shown in figure 2.22. The interface of the component C_1 corresponds to the task graph illustrated in figure 2.13. The interface simply shows the input variables in_1 and in_2, the output variables out_1, out_2, and out_3, and the dependencies that the output out_1 awaits in_1 and the output out_3 awaits both in_1 and in_2. The block-diagram connects this component C_1 with another component C_2. The interface of C_2 shows its input variables to be out_1, out_2, and out_3, its output variables to be in_1 and in_2, and the dependencies that in_1 awaits out_2 and in_2 awaits both out_1 and out_2. Verify that there is no cycle in the combined

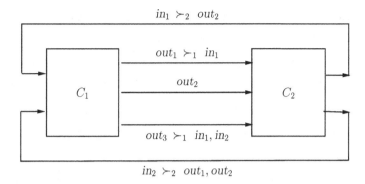

$in_1 \succ_2 out_2$

$out_1 \succ_1 in_1$

out_2

C_1 C_2

$out_3 \succ_1 in_1, in_2$

$in_2 \succ_2 out_1, out_2$

・ Figure 2.22: Composing Interfaces

dependencies and the interfaces are compatible. This suffices for us to conclude that the composition illustrated in figure 2.22 is well defined.

The next proposition asserts that indeed absence of cycles in the union of the await-dependency relations over input/output variables implies absence of cycles in the precedence constraints in the task graph of the product component.

Proposition 2.1 [Await Compatibility Implies Acyclic Product Task Graph]
Let C_1 and C_2 be compatible reactive components. Then the task graph over the set of tasks of C_1 and C_2 obtained by retaining the precedence edges in the individual components and adding cross-component edges from a task A_1 of one component to a task A_2 of another component whenever A_1 writes a variable read by A_2, is acyclic.

Proof. Consider two compatible components, C_1 and C_2. Let $\prec_1$ and $\prec_2$ be their respective precedence relations over their task sets, and let $\succ_1$ and $\succ_2$ be the corresponding input/output await dependencies. Consider the combined task graph over the tasks of both the components obtained by retaining edges of the individual precedence relations and cross-component edges from a task A_1 of one component to a task A_2 of another component if A_2 reads a variable written by A_1. We will prove that if this task graph contains a cycle, then the union relation $(\succ_1 \cup \succ_2)$ over input and output variables also contains a cycle, thereby contradicting the assumption that the two components are compatible.

Consider a cycle in the combined task graph. This type of cycle must alternate between stretches of tasks, such that each stretch contains one or more tasks of a single component, and there is a cross-component edge from the last task of one stretch to the first task of the next stretch. Let $(A_1, B_1), (A_2, B_2), \ldots (A_k, B_k)$ be the pairs of tasks connected by cross-component edges in the order in which these cross-component edges appear in the cycle. For each j, tasks A_j and B_j belong to different components. Let x_j be the variable written by task A_j and read by task B_j. Thus, x_j must be an output variable of the component to which the task A_j belongs and an input variable of the other component to which the task B_j belongs. From each task B_j, there is a stretch of the cycle to the task A_{j+1} within the same component (define $A_{k+1} = A_1$ for the cycle to wrap around). Consider the case when the task B_j belongs to the component C_1. Then either $B_j = A_{j+1}$ or $B_j \prec_1^+ A_{j+1}$. The variable x_j is an input to C_1 belonging to the read-set of B_j, and the variable x_{j+1} is an output of C_1 belonging to the write-set of A_{j+1}. Thus, the component C_1 cannot produce the output variable x_{j+1} before the input variable x_j is available. By definition of await dependencies, $x_{j+1} \succ_1 x_j$. The case when the task B_j belongs to the component C_2 is symmetric and leads to $x_{j+1} \succ_2 x_j$. Note that in this argument, the task A_{k+1} is the same as A_1, and thus the variable x_{k+1} is the same as x_1. This gives a cycle of await dependencies, alternating between $\succ_1$ and $\succ_2$, among the sequence of input/output variables $x_1, x_2, \ldots x_k, x_1$.

Thus, we have established that the existence of a cycle in the task graph of the product implies the existence of a cycle in the combined await dependencies and, thus, incompatibility of the two components. ∎

The following definition summarizes the parallel composition operation.

COMPONENT COMPOSITION

Let $C_1 = (I_1, O_1, S_1, Init_1, React_1)$ and $C_2 = (I_2, O_2, S_2, Init_2, React_2)$ be compatible synchronous reactive components. Suppose the reaction description $React_1$ is given using local variables L_1 by a task graph with the set $\mathcal{A}_1$ of tasks and the precedence relation $\prec_1$, and the reaction description $React_2$ is given using local variables L_2 by a task graph with the set $\mathcal{A}_2$ of tasks and the precedence relation $\prec_2$. Then the *parallel composition* $C_1 \| C_2$ is a synchronous reactive component C such that:

- the set S of state variables is $S_1 \cup S_2$;

- the set O of output variables is $O_1 \cup O_2$;

- the set I of input variables is $(I_1 \cup I_2) \setminus O$;

- the initialization for a state variable x is given by $Init_1$ for $x \in S_1$ and by $Init_2$ for $x \in S_2$; and

- the reaction description of C uses the local variables $L_1 \cup L_2$ and is given by the task graph such that (1) the set of tasks is $\mathcal{A}_1 \cup \mathcal{A}_2$, and (2) the precedence relation is the union of $\prec_1$ and $\prec_2$ and task pairs (A_1, A_2), such that A_1 and A_2 are tasks of different components with some variable occurring in both the write-set of A_1 and the read-set of A_2.

Properties of Parallel Composition

Let C_1 and C_2 be compatible components. Then by the above definition, the product $C_1 \| C_2$ is the same as the product $C_2 \| C_1$. Thus, the parallel composition operation is *commutative*.

The parallel composition is also associative. Suppose two components, C_1 and C_2, are compatible, and their product, $C_1 \parallel C_2$, is compatible with a third component, C_3. Then compatibility also holds for components C_2 and C_3 and for C_1 and $C_2 \parallel C_3$. Furthermore, $(C_1 \parallel C_2) \parallel C_3$ is the same as $C_1 \parallel (C_2 \parallel C_3)$. Thus, if we want to compose multiple components, then we can compose two, compose the result with a third one, and so on, and we get the same final result irrespective of the order of composition. At some step, we may discover incompatibility due to either common outputs or cyclic await dependencies, and we may not be able to compose all the components, but this failure does not depend on the order in which the components are composed.

If both the components C_1 and C_2 are finite-state, then so is the product $C_1 \parallel C_2$. If C_1 has n_1 states and C_2 has n_2 states, then $C_1 \parallel C_2$ has $n_1 * n_2$ states. For example, in the composition of the components Delay1 and Delay2, each component has two states, and the product has four states. If we were to compose n instances of the component Delay in a chain to construct a component that outputs, in each round, the value of the input n rounds earlier, then it will have 2^n states. The fact that the number of states grows *exponentially* with the number of components is sometimes referred to as the *state-space explosion problem*, and it poses a challenge to analysis tools in terms of scalability.

Note that when all the tasks of two compatible components C_1 and C_2 are deterministic, then the product $C_1 \parallel C_2$ is guaranteed to be deterministic. Similarly, if all the tasks of two compatible components C_1 and C_2 are input-enabled, then the product $C_1 \parallel C_2$ is guaranteed to be input-enabled.

2.3.4 Output Hiding

The final operation needed to define the semantics of block diagrams is hiding of output variables. If y is an output variable of a component C, then the result of *hiding* y in C, denoted $C \setminus y$, gives a component that behaves the same way as the component C, but y is no longer an output that is observable outside. This is achieved by removing y from the set of output variables and declaring y to be a local variable in the reaction description.

Let us revisit the component Delay1 $\parallel$ Delay2 (see figure 2.16). If we hide the intermediate output *temp*, then we get the desired product component DoubleDelay: the set of state variables is $\{x_1, x_2\}$, the set of output variables is $\{out\}$, and the set of input variables is $\{in\}$. The resulting component is shown in figure 2.23. Note that the initial state of the component DoubleDelay is $(0, 0)$, and its reactions are:

$$(0,0) \xrightarrow{0/0} (0,0); \quad (0,0) \xrightarrow{1/0} (1,0); \quad (0,1) \xrightarrow{0/1} (0,0); \quad (0,1) \xrightarrow{1/1} (1,0);$$
$$(1,0) \xrightarrow{0/0} (0,1); \quad (1,0) \xrightarrow{1/0} (1,1); \quad (1,1) \xrightarrow{0/1} (0,1); \quad (1,1) \xrightarrow{1/1} (1,1).$$

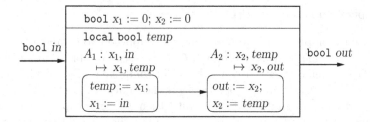

Figure 2.23: The Component `DoubleDelay`

Hiding preserves all the following properties of components: being finite-state, combinational, deterministic, input-enabled, and event-triggered.

When we want to hide multiple output variables, the order in which we apply the hiding operator does not matter. If x and y are two output variables of a component C, then the components $(C \setminus x) \setminus y$ and $(C \setminus y) \setminus x$ are exactly the same, and we can use $C \setminus \{x, y\}$ as an abbreviation to indicate hiding of both the output variables.

Exercise 2.15: Consider the components `ClockedDelay` from exercise 2.7 and `Comparator` of figure 2.4. The component `ClockDelayComparator` is defined as follows:

$$(\text{Comparator}[out \mapsto x] \parallel \text{ClockedDelay}) \setminus x$$

Describe the input-output behavior of the component `ClockDelayComparator`. ∎

Exercise 2.16: Consider the component `DoubleSplitDelay` defined as

$$(\text{SplitDelay}[out \mapsto temp] \parallel \text{SplitDelay}[in \mapsto temp]) \setminus temp$$

This component is similar to the component `DoubleDelay` except we use instances of the component `SplitDelay` instead of `Delay`. Show the "compiled" version of `DoubleSplitDelay`, that is, list its state, input, output, and local variables, tasks, and precedence constraints. What are the await dependencies among output and input variables for `DoubleSplitDelay`? ∎

Exercise 2.17: Recall the event-triggered component `SecondToMinute` from exercise 2.6 with the input event variable *second* and the output event variable *minute* such that *minute* is present every 60th time the event *second* is present. Now suppose we want to design an event-triggered component `SecondToHour` with an input event variable *second* and an output event variable *hour*, such that the output event *hour* is present every 3600th time the event *second* is present. Show how to construct the desired component `SecondToHour` from the component `SecondToMinute` using the operations of parallel composition, instantiation, and output hiding. ∎

2.4   Synchronous Designs

Before we consider some illustrative design problems in the synchronous model, let us recap the salient features and assumptions of the model.

In the classical functional model of computation, the component reads its input and then computes, producing the output on termination. The desired behavior of the component is described as a function from inputs to outputs. Reactive components, in contrast, interact with their environment via inputs and outputs in an ongoing manner. In principle, the component never terminates. The desired behavior is described by the *sequence* of outputs that the component should produce in response to a given sequence of inputs.

In *synchronous* reactive computation, the computation proceeds in a well-defined sequence of rounds. All the components, along with the environment that is supplying the inputs, agree on what constitutes a round. Event-triggered modeling can be used to describe the situation where a component may not be interested in every round and actively participates only in those rounds in which one of the trigger events is present. The key assumption of the synchronous model is that the computation of all the tasks within a round and all the inter-task communication necessary to determine the values of all the variables logically happens *instantaneously*. The external inputs do not change during a round, and when the inputs do change, a new round is initiated with all the tasks ready to process the new inputs. This assumption is called the *synchrony hypothesis*. This idealized assumption leads to simplicity and predictability of designs.

The computation of a component within a round can be split into multiple tasks. The precedence constraints among tasks capture read/write dependencies among its variables and lead to await dependencies among output and input variables. While composing components, absence of mutually cyclic await dependencies, a condition that can be checked at design time, ensures well-behaved execution of the product. During a round, the order in which tasks execute does not affect the resulting reaction. Nondeterminism, that is, multiple reactions in response to the same input, needs to be explicitly programmed within the description of a task and is not an artifact of the interaction model. In particular, for deterministic components, the behavior is repeatable: if we execute the component again with the same sequence of inputs, then we will observe the same sequence of outputs. This is valuable in debugging and analysis of complex designs.

During implementation, one needs to ensure that the implementation faithfully implements the synchronous semantics. This is the case, for instance, if the upper bound on the time needed to compute a reaction, which may require inter-component communication, is less than the minimum delay between changes to the input. Real-time scheduling theory, to be studied in chapter 8, offers ways of checking this.

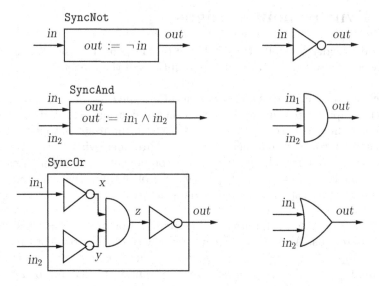

Figure 2.24: Synchronous Not, And, and Or Gates

2.4.1 Synchronous Circuits

Synchronous circuits are built from logic gates and memory cells that are driven by a sequence of clock ticks. Each logic gate computes a Boolean value once per clock cycle, and each memory cell stores a Boolean value from one clock cycle to the next. The design of synchronous circuits offers an excellent case study of how to build complex systems by putting together simpler components in a hierarchical manner. We can construct synchronous circuits from three basic building blocks: as basic logic gates we use the Not and And gates, and as basic memory cell we use a latch component that models a set-reset flip-flop. These building blocks are then combined to obtain circuits by applying the three operations of parallel composition, variable renaming, and output hiding.

Combinational Circuits

Figure 2.24 defines three deterministic and combinational synchronous reactive components for modeling Not, And, and Or gates. In the description of synchronous circuits, all variables are implicitly assumed to be of type `bool`.

The component `SyncNot` is the same as the component `Inverter` of figure 2.17 and models a Not gate, which takes a Boolean input variable *in* and produces a Boolean output *out*. The reaction description sets the output to the logical negation of the input value. Note that the output *out* awaits the input *in*. The component `SyncAnd` models an And gate in a similar fashion. The component takes two Boolean input variables in_1 and in_2 and produces a Boolean output

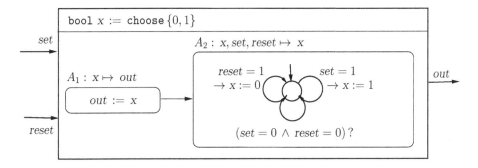

Figure 2.25: Synchronous `Latch` Component

out. The output is set to the logical conjunction of the two inputs and awaits both the input variables.

From `SyncNot` and `SyncAnd` gates we can build all combinational circuits. For example, by de Morgan's law, an Or gate can be defined by composing an And gate with three Not gates that negate both inputs and the output of the And gate. The block diagram is shown in figure 2.24. Note that instances of the components `SyncNot` and `SyncAnd` are shown by the corresponding symbolic representations commonly used in circuit diagrams. The resulting component `SyncOr` has two Boolean input variables in_1 and in_2 and produces a Boolean output *out*. The local variables x, y, and z of `SyncOr` represent internal wires that connect the four component gates. The component `SyncOr` is deterministic and combinational, and its output awaits both its input variables. The component `SyncOr` can be equivalently described using the operations of instantiation, parallel composition, and hiding:

$$
\begin{aligned}
\texttt{SyncNot1} &= \texttt{SyncNot}[in \mapsto in_1][out \mapsto x], \\
\texttt{SyncNot2} &= \texttt{SyncNot}[in \mapsto in_2][out \mapsto y], \\
\texttt{SyncNot3} &= \texttt{SyncNot}[in \mapsto z], \\
\texttt{SyncAnd1} &= \texttt{SyncAnd}[in_1 \mapsto x][in_2 \mapsto y][out \mapsto z], \\
\texttt{SyncOr} &= (\texttt{SyncNot1} \,\|\, \texttt{SyncNot2} \,\|\, \texttt{SyncAnd1} \,\|\, \texttt{SyncNot3}) \setminus \{x, y, z\}.
\end{aligned}
$$

Sequential Circuits

The combinational circuits are stateless. To model sequential circuits, we need a component that can store a value in its state from one round to the next. Figure 2.25 defines a nondeterministic component for modeling a unit-delay latch. The latch takes two Boolean input variables *set* and *reset* and produces a Boolean output *out*. The latch has a Boolean state, which is represented by the state variable x. The initial value of the state is unconstrained, and this is expressed using the `choose` construct in the initialization. In every round, the latch first issues its state as output and then waits for the input values

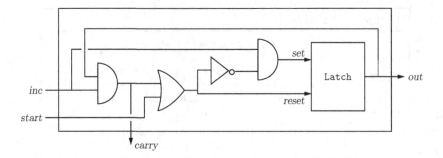

Figure 2.26: Synchronous Component `1BitCounter`

to compute its next state. For this purpose, the reaction description is split into two tasks: the task A_1 for computing the output *out* and the task A_2 for updating x. The update of the state variable x is described using a single-mode extended-state machine with three mode-switches. If the value of *set* is 1, then the latch can change its state to 1 using the mode-switch "*set* = 1 → $x := 1$"; and if the value of *reset* is 1, then the latch can change its state to 0 using the mode-switch "*reset* = 1 → $x := 0$". If both input variables have value 1, then the guards of both mode-switches are satisfied, and one of them is executed nondeterministically; thus, the next state of the latch may be either 0 or 1. If both input variables have value 0, then the state stays unchanged using the mode-switch "*set* = 0 ∧ *reset* = 0".

Note that the component `Latch` is nondeterministic and finite-state, and its output does not await either of its input variables. The fact that the output of the latch is available before the values of its inputs are known is essential for composing latches with logic gates, which in every round (clock cycle) provide the latch inputs dependent on the latch outputs.

Binary Counter

As an example of a sequential circuit, we design a three-bit binary counter. The counter has two Boolean input variables *start* and *inc*, for starting and incrementing the counter, respectively. The counter value ranges from 0 to 7 and is represented by three bits. We do not make any assumption about the initial counter value. When the input *start* is 1, the counter value is reset to 0 independent of the value of the other input *inc*. Otherwise when the input *inc* is 1, the counter value increases by 1. If the counter value is 7, then an increment changes the counter value to 0. In every round, the counter issues its value as output—the low bit on the output variable out_0, the middle bit on the output variable out_1, and the high bit on the output variable out_2.

Figure 2.26 shows a possible design of a one-bit counter. It uses one `Latch` component to store one bit of state, and its logic is implemented using two And

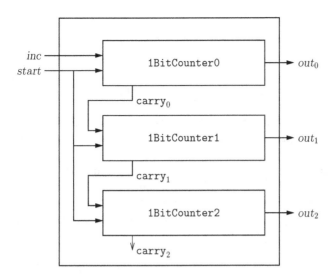

Figure 2.27: Synchronous Component `3BitCounter`

gates, one Not gate, and one Or gate. Verify that there are no awaits cycles and, thus, the components are compatible. The value of the output *out* equals the state of `Latch` at the beginning of the round. Let us consider all possible cases to understand how this circuit works.

Suppose the state of the latch (that is, the value of the counter) is 0. Then the output *carry*, which indicates overflow in the counter value, is 0. The value of the local variable *reset* equals the input *start*, and the value of the local variable *set* equals the conjunction $inc \wedge \neg start$. Observe that both *set* and *reset* cannot be 1 simultaneously. When *start* is 1, only *reset* is 1, and in this case, the latch state is reset to 0. When *start* is 0, *set* equals *inc*: if *inc* is 1, then the latch state is updated to 1; otherwise, the latch state stays 0.

Suppose the state of the latch is 1 and the input *inc* is 0. Then again, the value of the output *carry* is 0. The value of the local variable *reset* equals the input *start*, and the value of the local variable *set* equals 0. When *start* is 1, the latch state is reset to 0; otherwise, the latch state stays 1.

Finally, suppose the state of the latch is 1 and the input *inc* is 1. In this case, the value of the output *carry* is 1, indicating overflow. The variable *reset* equals 1 and the variable *set* equals 0. Thus, no matter what the value of *start* is, the latch state is updated to 0.

Figure 2.27 shows the block diagram for connecting three instances of the 1-bit counter to implement a 3-bit counter in a natural way. The input variable *start* acts as the command to reset to all the three instances. The input variable *inc* acts as the command to increment only to the 1-bit counter `1BitCounter0` corresponding to the least significant bit. The *carry* output of `1BitCounter0` is used as the command to increment the next significant bit stored in `1BitCounter1`, whose carry overflow output acts as the command to increment the most significant bit.

Exercise 2.18: An XOR (Exclusive-Or) gate has two Boolean inputs in_1 and in_2, and a boolean output *out*. The output is 1 when exactly one of its two inputs are 1 and is 0 otherwise. Define the combinational component `SyncXor` to capture this desired functionality by composing And, Or, and Not gates. ∎

Exercise 2.19: A *parity* circuit has n boolean input variables $in_1, in_2, \ldots in_n$ and a boolean output *out*. The value of the output should be 1 if an odd number of input variables have the value 1 and should be 0 otherwise. Construct the component `Parity`$_n$ that computes the parity of n input variables by composing instances of the component `SyncXor` defined in exercise 2.18. ∎

Exercise 2.20: Design a 1-bit synchronous adder `1BitAdder` by composing instances of And, Or, Not, and Xor gates. The component `1BitAdder` has three input variables x, y, and *carry-in* and two output variables z and *carry-out*. In each round, the value encoded by the two output bits z and *carry-out*, where z is the least significant bit, should equal the sum of the values of three input variables. Then, design a 3-bit synchronous adder `3BitAdder` by composing three instances of the component `1BitAdder`. The component `3BitAdder` has input variables x_0, x_1, x_2, y_0, y_1, y_2, and *carry-in* and has output variables z_0, z_1, z_2, and *carry-out*. In each round, the 4-bit number encoded by the output variables z_0, z_1, z_2, and *carry-out* should equal the sum of the 3-bit number encoded by the input variables x_0, x_1, and x_2, the 3-bit number encoded by the input variables y_0, y_1, and y_2, and the input value of *carry-in*. ∎

2.4.2   Cruise Control System

We will illustrate concepts of top-down component-based design using a simplified design of a cruise-control system for a car.

Top-Level Specification

The inputs and outputs of the system are shown in figure 2.28. The driver interacts with the cruise-controller with three buttons: one to turn the cruise controller on and off, one to increment the desired speed, and one to decrement the desired speed. These are modeled by three input event variables *cruise*, *inc*, and *dec*. Presence of the event *cruise* should toggle the controller between on and off modes. When it is turned on, the desired cruising speed should be set to the current speed, and the events *inc* and *dec*, when present, should cause the desired cruising speed to increment and decrement, respectively. We should ensure that this value stays within a reasonable cruising range, given by a minimum value, denoted `minSpeed`, and a maximum value, denoted `maxSpeed`.

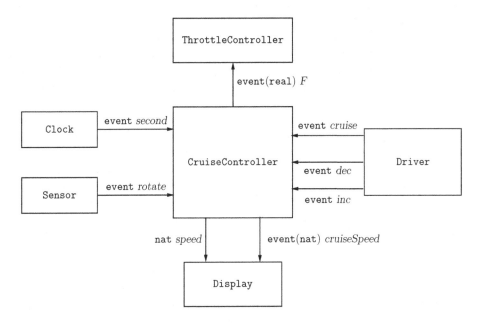

Figure 2.28: Inputs and Outputs of the Cruise-Control System

The cruise controller needs to measure the current speed to make its decisions. This is achieved using two input events: *rotate* and *second*. Whenever the wheel completes a rotation, a sensor associated with the wheel-shaft issues the input event *rotate*, and every second a system-wide clock issues the input event *second*. Thus, the controller can count the number of rotations every second and compute the current speed.

The controller should send information to the display regarding the current settings. This is modeled by output variables *speed*, denoting the current speed, and *cruiseSpeed*. The value of *cruiseSpeed* is absent if the cruise control is turned off and, when on, equals the current cruising speed set by the driver.

Finally, the output *F* is sent to the throttle control system and corresponds to the force needed to adjust the throttle to regulate the current speed so as to track the desired cruising speed.

Decomposing into Subsystems

As a next step in the design, we decompose the controller into three subsystems: the component **MeasureSpeed** to compute the current speed based on the inputs *rotate* and *second*, the component **SetSpeed** to keep track of the desired cruise settings based on the inputs from the driver and the current speed, and the component **ControlSpeed** to process the differential between the current speed and the desired speed in order to compute the output force. The interconnections among these subcomponents are shown in figure 2.29. The design of the

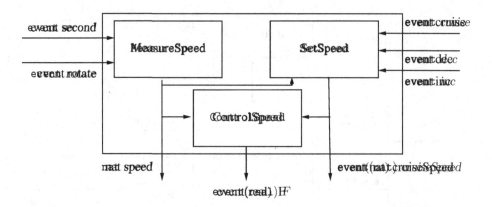

Figure 2.29: Components of the Cruise-Control System `CruiseController`

component `ControlSpeed` requires understanding the dynamics of the car and control theory, a topic to be discussed in chapter 6 on dynamical systems. We proceed to design the other two components.

Tracking Speed

The task for the component `MeasureSpeed` is to output the current speed of the car based on the two input event variables, *rotate* and *second*. The component is shown in figure 2.30. The component has a state variable *count* that counts the number of times the event *rotate* has occurred since the most recent occurrence of the event *second*. The initial value of *count* is 0. The state variable *s* remembers the current speed: it is 0 initially, and in every round during which the event *second* is present, the current value of *count* is used to update *s*.

More precisely, the rules for updating the state are as follows. If *rotate* event is present, then the component increments *count*. If *second* event is present, then the current value of *count* indicates the number of rotations of the wheel during a time interval of a second. To compute the speed, this value is multiplied by a constant, denoted k, which depends on the circumference of the wheel and rounded to the nearest integer (the function `round-off` returns the integer nearest to its argument). In this case, the value of *count* is reset to 0. Note that if both input events are absent, then the state stays unchanged. If both events are present, then *count* is first incremented, then used to compute the value of *s*, and then reset to 0.

The component has an output variable *speed*: in each round, the output is set to the updated value of *s*. Thus, it is a latched output. The display as well as the speed controller can access this output in any round. The component `MeasureSpeed` is deterministic and event-triggered.

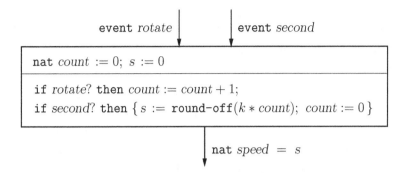

Figure 2.30: Component `MeasureSpeed`

Tracking Cruise Settings

Now consider the component `SetSpeed` shown in figure 2.31. The output variable *cruiseSpeed* is an event variable that is either absent (when the controller is off) or indicates the current desired speed. The component maintains two state variables: a Boolean variable *on* that keeps track of whether the controller is switched on and the current desired speed *s*.

Since there are three input events for the component, each of which can be present or absent, we should process all their combinations. In our design, we avoid this blow-up by considering these events in a priority order: first *cruise*, then *dec*, and then *inc*. If the driver presses two or more buttons simultaneously, then the effect will be equivalent to pressing a single button, with the highest priority among those pressed simultaneously. Alternatively, we could make an assumption about the environment that at any instant, at most one of the input events can be present.

The component updates the state variable *on* according to the following rule: every time the event *cruise* occurs, the variable *on* is toggled. The rule for updating the desired speed *s* is as follows. If the event *cruise* is present, then the variable *s* is set to the current speed, provided the current speed is within the legal cruising range from `minSpeed` to `maxSpeed`. Otherwise, if the event *dec* is present, then the variable *s* is decremented, provided its value is above the minimum threshold and the controller is on. Finally, if the event *inc* is present, then the variable *s* is incremented, provided its value is below the maximum threshold and the controller is on. If none of the rules applies, then the desired speed stays unchanged. After updating the state, the component decides on its output based on the following rule: if the updated value of *on* is 1, then the output *cruiseSpeed* is set to the updated value of *s* or else it is absent.

Note that the component `SetSpeed` is deterministic and event-triggered. Its output variable awaits all the four input variables.

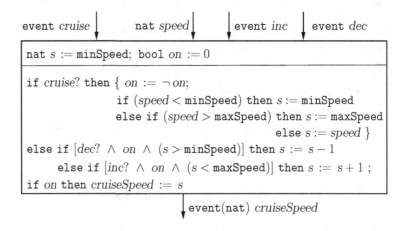

Figure 2.31: Component `SetSpeed`

Exercise 2.21: Consider the design of the component `SetSpeed` of figure 2.31. Suppose we want to add another input control for the driver, *pause*, with the following desired behavior. When the cruise controller is on, if the driver presses *pause*, then the controller is temporarily turned off. In the resulting paused state, the output *cruiseSpeed* should be absent, and the events *inc* and *dec* should be ignored. Pressing *pause* again in this paused state should resume the operation of the cruise controller, restoring the desired speed on pausing. Pressing *cruise* in the paused state should switch the system off, and when the controller is off, pressing *pause* should have no effect. Redesign the component `SetSpeed` with this additional input event *pause* to capture the above specification. ∎

2.4.3   Synchronous Networks *

In a synchronous network, communication happens in a sequence of time slots. The network topology determines the one-hop directed connectivity among network nodes. In each time slot, a node sends a message to all its neighbors connected by outgoing edges and receives messages from all its neighbors connected by incoming edges. We can model such networks as synchronous reactive components.

Modeling a Network Node

The design of an individual node should be independent of the network topology so that instances of the node can be connected in different ways to form different networks. For this purpose, each network node is modeled as a component `NetwkNode` with an input variable *in* and an output variable *out* as shown in figure 2.32. If the type of messages that a node sends in each round is `msg`, then

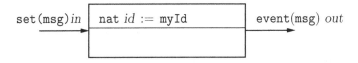

Figure 2.32: Schematic of a Synchronous Network Node `NetwkNode`

the type of the output variable *out* is `event`(msg) since in each round, a node may or may not send a message. The type of the input variable *in* is `set`(msg), and a value of this type is a set of messages of type `msg`. We want to design the component so that there is no await dependency between the output *out* and the input *in*: during each round, the component decides on its output message based on its state and then updates the state in response to the input that contains the set of messages it receives.

The description of the component `NetwkNode` is parameterized by an identifier `myId`. To form a desired network of components, we create as many instances of the component `NetwkNode` as needed. Each instance is given a unique identifier, which is used to instantiate `myId` and rename the input and output variables to avoid name conflicts.

Modeling the Interconnections

The communication network is modeled as a combinational component `Network`. It has one input and one output variable for each instance of `NetwkNode`.

As a concrete example, consider the communication network shown in figure 2.33 over four nodes with identifiers 1, 3, 5, and 8. The edges show connectivity: for example, node 3 has two outgoing edges connecting it to nodes 5 and 8 and two incoming edges connecting from nodes 1 and 5. In a single round, if node 3 chooses to send a message, then it will be delivered to both nodes 5 and 8, and the set of messages it receives contains messages sent by nodes 1 and 5 in this round.

Figure 2.34 shows the composition of components. There are four instances of the component `NetwkNode` corresponding to the four nodes. The network is captured by `Network` with input variables out_1, out_3, out_5, and out_8, each of type `event`(msg), each of which is connected to the output of the corresponding node component. It has output variables in_1, in_3, in_5, and in_8, each of type `set`(msg), connected to the input of the corresponding node component. In each round, the network reads the messages from all its input variables out_n, and for each node n, it collects the messages that are present on the incoming links for node n and delivers the corresponding set of messages by updating the output variable in_n. The reaction description in figure 2.34 first sets all the output sets in_n to empty sets. Then it checks all the input events one by one, adding it to the appropriate output sets. For instance, if the input message

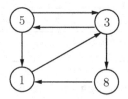

Figure 2.33: Example Communication Network with Four Nodes

out_3 is present, since the node 3 has outgoing links to nodes 5 and 8, then the message out_3 is added to the output sets in_5 and in_8.

More generally, let P be a set of node identifiers and let $E \subset P \times P$ denote the directed one-hop connectivity edges among nodes. Then for each $n \in P$, let NetwkNode$_n$ be the instance of NetwkNode obtained by instantiating myID to n and renaming each input and output variable x to x_n. The component Network$_{P,E}$ is a deterministic combinational component with the set $\{out_n \mid n \in P\}$ of input variables and the set $\{in_n \mid n \in P\}$ of output variables. In each round, for each $n \in P$, the output variable in_n equals the set containing the input values out_m, such that (1) the set E of network connections has an edge from node m to node n, and (2) the event out_m is present. The desired system is the parallel composition of all the components NetwkNode$_n$, for $n \in P$, and the interconnection network component Network$_{P,E}$.

Leader Election

To illustrate the design of algorithms for synchronous networks, let us consider the classical coordination problem of *leader election*: the nodes should exchange messages to decide on a unique leader. More precisely, let us assume that each node component has an output variable *status* that ranges over the enumerated type {unknown, leader, follower}. The nodes exchange messages updating the *status* so that (1) eventually every component sets the *status* output to be either leader or follower, and (2) exactly one component changes the *status* output to the value leader.

Since each node has a unique identifier, it is natural to use these identifiers for choosing the leader, say, the one with the highest value of the identifier. At the beginning, a node does not know which other nodes are part of the network, and the purpose of exchanging messages is to identify this highest identifier. We want the algorithm to work in as many networks as possible. Consider the algorithm shown in figure 2.35, which relies on two assumptions:

1. The network is *strongly connected*: for every pair of nodes m and n, there is a directed path from node m to node n.

2. Each node knows an upper bound N on the total number of nodes in the network.

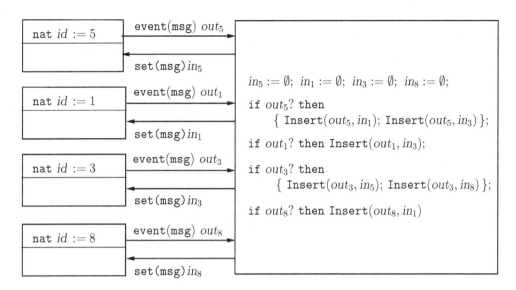

Figure 2.34: The Synchronous Network Component `Network`

The first condition is needed for information to flow from one node to another, and the second is used for termination.

In the algorithm of figure 2.35, called the *flooding* algorithm, a node maintains a state variable id that equals the highest identifier it knows so far. Initially, the value of id equals the node's own unique identifier. In each round, the node outputs this identifier to its neighbors, and if it receives any identifier higher than the current value of id, it updates this value. The reaction description is split into two tasks: the task A_1 computes the output out and updates the state variable r, and the task A_2 updates the state variable id and computes the output $status$. Note that the first task does not need the input, and thus only the output variable $status$ awaits the input variable in.

If the total number of nodes in a strongly connected network is N, then between each pair of nodes, there is a path with at most $N-1$ hops. Hence, after $N-1$ rounds, each node can be sure that its identifier has had a chance to propagate to every other node. More precisely, if a node's unique identifier is n, and if the shortest path from this node to another node m is of length j, then after j rounds, the value of id variable of node m will be n or higher. As a result, after $N-1$ rounds, the value of id variable of each node will be equal to the highest identifier in the network. At this point, each node can decide: if the value of its id variable equals its original identifier, then it is the leader, otherwise it is a follower.

Consider the four-node network shown in figure 2.34 so that each component is the instantiated version of the leader election component `SyncLENode` of fig-

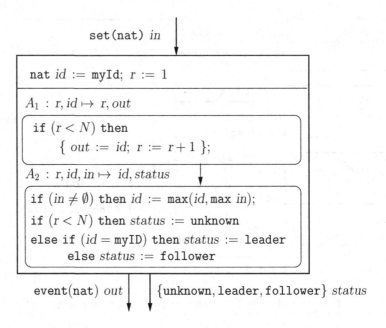

set(nat) *in*

nat *id* := myId; *r* := 1

A_1 : *r*, *id* ↦ *r*, *out*

if $(r < N)$ then
{ *out* := *id*; *r* := *r* + 1 };

A_2 : *r*, *id*, *in* ↦ *id*, *status*

if $(in \neq \emptyset)$ then *id* := max(*id*, max *in*);
if $(r < N)$ then *status* := unknown
else if $(id = myID)$ then *status* := leader
else *status* := follower

event(nat) *out*   {unknown, leader, follower} *status*

Figure 2.35: Component SyncLENode for Synchronous Leader Election

ure 2.35. Here is how their executions proceed:

1. In round 1, each of the nodes 1, 3, 5, and 8 output their original identifiers. The node 1 receives $\{5, 8\}$ and updates its *id* variable to 8; the node 3 receives $\{1, 5\}$ and updates its *id* variable to 5; the node 5 receives $\{3\}$, and its *id* variable remains 5; the node 8 receives $\{3\}$, and its *id* variable remains 8. All the nodes set the output *status* to the value unknown.

2. In round 2, nodes 1 and 8 output 8, and nodes 3 and 5 output 5. As a result, the *id* variable of node 3 gets updated to 8, and other nodes do not change their respective *id* variables. Again, all the nodes set the output *status* to the value unknown.

3. In round 3, nodes 1, 3, and 8 output 8, and node 5 outputs 5. The node 5 updates its *id* variable to 8. The updated value of the round-counting variable *r* equals $N = 4$; as a result, all the nodes decide based on the updated values of their respective *id* variables: the node 8 sets its output *status* to the value leader, and the rest set their output *status* to the value follower.

Observe that if the *diameter* of the network is D, that is, between every pair of nodes there is a path of length D or less, then after D rounds, the value of *id* variable of each node will be equal to the highest identifier in the network. An upper bound on D is $N - 1$, but D can be much less than this upper bound.

If the diameter D is known to the nodes in advance, then a node can decide at the end of round D.

Exercise 2.22: Consider the leader election algorithm in synchronous networks (figure 2.35). Argue that if the value of *id* does not change in a given round, then there is no need to send it in the following round (that is, the output *out* can be absent in the next round). This can reduce the number of messages sent. Modify the description of the component SyncLENode to implement this change. ∎

Exercise 2.23 *: In a strongly connected network, for each network node n, let D_n be the smallest integer j such that for every node m, there is a directed path of at most j links from m to n. For example, if the network is a complete graph (for every pair of nodes m and n, there is a link from m to n), D_n is 1 for every node n; if the network is unidirectional ring connecting all nodes in a single cycle, D_n is $N - 1$ for every node n, where N is the total number nodes; and in the network of figure 2.33, $D_5 = 3$, $D_3 = 2$, $D_1 = 2$, and $D_8 = 2$. Design an algorithm for synchronous networks so that each node n can figure out the value of D_n. As in the case of leader election, assume that each node has a unique identifier, and it knows the bound N on the total number of nodes. The algorithm should work for any network as long as it is strongly connected. Explain how your algorithm works. ∎

Bibliographic Notes

The term *reactive computation*, as opposed to the classical functional computation, was introduced in [HP85]. Since the 1980s, a number of formal models of synchronous reactive computation have been introduced and studied. Prominent examples include ESTEREL [BG88], LUSTRE [CPHP87], and STATE-CHARTS [Har87]. All of these have resulted in industrial-strength programming environments; see [BCE+03] for a survey of synchronous languages.

In theory of concurrency, a rich variety of formal models of reactive and concurrent computation with alternative forms of interaction among components has been studied. Example formalisms include CSP [Hoa85], CCS [Mil89], UNITY [CM88], data-flow networks [Kah74, LP95], I/O automata [Lyn96], TLA [Lam02], and BIP [Sif13].

The model of synchronous reactive components studied in this chapter is a simplified version of *Reactive Modules* [AH99b]. The presentation of the model follows the outline in [AH99a]. The description of the leader election algorithm is based on [Lyn96], which contains rigorous descriptions of algorithms for a large variety of distributed coordination problems in synchronous networks.

3

Safety Requirements

A reactive component interacts with the environment via inputs and outputs. A requirement for a component is a specification of acceptable or desired sequences of outputs in response to inputs. Design of high-assurance systems demands that requirements should be stated explicitly and as precisely as possible. Requirements can be classified in two broad categories: *safety* requirements assert that "nothing bad ever happens," and *liveness* requirements assert that "something good eventually happens." For instance, in the leader election problem of section 2.4.3, the main safety requirement is that no two nodes should ever declare themselves to be the leaders, and the main liveness requirement is that some node should eventually declare itself to be the leader, and the remaining nodes should eventually declare themselves to be the followers. Given a specific solution to the leader election problem, such as the one described in figure 2.35, the *verification* problem is to check whether the given implementation meets these requirements. For safety requirements, violation of a requirement can be demonstrated by a finite execution that illustrates the undesirable behavior. Typically, such requirements are captured by composing the system with a *monitor* that observes the inputs and outputs of the system and enters an *error* state if an undesirable behavior is detected. The safety verification problem then reduces to checking whether there is some execution of the system that leads the monitor to an error state. In this chapter, we will first study how reachability problems can be used to formalize safety requirements, and then we will explore verification techniques for establishing correctness of systems with respect to safety requirements.

3.1 Safety Specifications

3.1.1 Invariants of Transition Systems

A safety requirement for a system classifies its states into *safe* and *unsafe* and asserts that an unsafe state is never encountered during an execution of the system. Since the concept of such requirements and the tools for establishing correctness of systems with respect to such requirements do not specifically rely on the synchronous nature of interaction among reactive components, let us study them in a more general context of *transition systems*.

Transition Systems

A transition system is specified using variables whose valuations describe possible states of the system. The initialization describes the initial values for each of the system variables. Transitions of the system describe how the state evolves and are typically specified using a sequence of assignments and conditional statements that update the state variables, possibly using additional local variables. Following the convention analogous to the definition of synchronous reactive components, we use *Init* to denote the syntactic description of the initialization, with an associated semantics $[\![Init]\!]$ denoting the corresponding set of initial states. Similarly, *Trans* denotes the syntactic description of the transitions, and the associated semantics $[\![Trans]\!]$ is a set of pairs of states.

TRANSITION SYSTEM

A *transition system* T has:

- a finite set S of typed state variables defining the set Q_S of states,

- an initialization *Init* defining the set $[\![Init]\!] \subseteq Q$ of initial states, and

- a transition description *Trans* defining the set $[\![Trans]\!] \subseteq Q_S \times Q_S$ of transitions between states.

Synchronous Reactive Components as Transition Systems

With each synchronous reactive component $C = (I, O, S, Init, React)$, there is a naturally associated transition system: the set of state variables is S; the initialization is *Init*; and the transition description *Trans* is obtained from the reaction description *React* by declaring the input and output variables to be local variables, where the input variables are assigned nondeterministically chosen values. Consequently, the set of transitions contains pairs of states (s, t) such that $s \xrightarrow{i/o} t$ is a reaction for some input i and some output o.

For example, consider the component `TriggeredCopy` of figure 2.5. In the corresponding transition system, the set of state variables is $\{x\}$. The initialization is given by the assignment $x := 0$. The transition description is obtained by declaring the variables *in* and *out* as local and letting the input take every possible value using the `choose` construct:

```
local event(bool) in, out;
    in := choose {0, 1, ⊥};
    if in? then {out! in; x := x + 1 }.
```

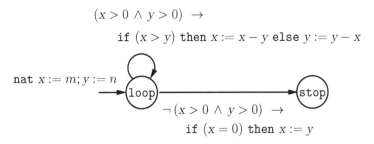

Figure 3.1: Euclid's GCD Program

In any given state, if the input event is present, then the value of x is incremented, and if the input event is absent, then the value of x stays unchanged. Thus, the corresponding transition system has transitions (n, n) and $(n, n + 1)$ for every natural number n.

Programs as Transition Systems

Sequential programs can also be modeled as transition systems. Consider the classical Euclid's algorithm for computing the *greatest common divisor* (GCD) of two natural numbers. Given two input numbers m and n, the algorithm computes their GCD using two variables x and y, both of type **nat**. The program executes the following code:

```
x := m;  y := n;
while (x > 0 ∧ y > 0)
    if (x > y) then x := x − y else y := y − x;
if (x = 0) then x := y.
```

The variable x contains the desired answer when the program terminates.

The program can be modeled as an extended-state machine shown in figure 3.1. In the initial mode, denoted **loop**, the system repeatedly decreases either x or y as long as the condition $(x > 0 \wedge y > 0)$ is true, and when the condition is false, it switches to the terminal mode **stop** and changes the variable x as needed so that it contains the desired answer. Note that the modes correspond to *program locations*, and such a representation of a program by an extended-state machine is sometimes called the *control-flow-graph* of the program.

We can use the extended-state machine representation of the program to associate a transition system with it. For given input numbers m and n, the behavior of the GCD program is captured by the transition system $\text{GCD}(m, n)$, whose description is parameterized by the numbers m and n. The state variables are x and y of type **nat** and the mode ranging over $\{\text{loop}, \text{stop}\}$. The sole initial state is (m, n, loop). Consider a state s of the form (j, k, loop). If $j > k > 0$, then the state s has a transition to state $(j - k, k, \text{loop})$; if $k \geq j > 0$, then the state s has a transition to state $(j, k - j, \texttt{loop})$; if $j = 0$, then the state s has a transition to state $(k, k, \texttt{stop})$; and if $j > 0$ and $k = 0$, then the state s has a transition to state $(j, k, \texttt{stop})$. A state in which the mode equals $\texttt{stop}$ has no outgoing transitions.

Reachable States

An execution of a transition system starts in an initial state and proceeds by following the transitions specified by *Trans*. States encountered during executions are *reachable* states of the system.

REACHABLE STATES OF TRANSITION SYSTEM

An *execution* of a transition system T consists of a finite sequence of the form $s_0, s_1, \ldots s_k$, such that:

1. for $0 \leq j \leq k$, each s_j is a state of T,

2. s_0 is an initial state of T, and

3. for $1 \leq j \leq k$, (s_{j-1}, s_j) is a transition of T.

For such an execution, the state s_k is said to be a *reachable* state of T.

For example, for $m = 6$ and $n = 4$, the transition system $\texttt{GCD}(6, 4)$ has the following execution

$$(6, 4, \texttt{loop}) \rightarrow (2, 4, \texttt{loop}) \rightarrow (2, 2, \texttt{loop}) \rightarrow (2, 0, \texttt{loop}) \rightarrow (2, 0, \texttt{stop}).$$

All the reachable states are the states appearing in this execution.

Invariants

For a transition system T, a *property* is a Boolean-valued expression over the state variables of T. A state q of T *satisfies* the property φ if φ evaluates to 1 when all variables are assigned values according to the valuation q. The set of all states that satisfy the property φ is denoted $[\![\varphi]\!]$.

Let us revisit the program computing the GCD of two natural numbers. Consider the following property of the transition system $\texttt{GCD}(m, n)$ (see figure 3.1):

$$\varphi_{gcd} : \texttt{gcd}(m, n) = \texttt{gcd}(x, y),$$

where $\texttt{gcd}$ represents the mathematical function that returns the greatest common divisor of its two arguments. The expression represents exactly those states in which the $\texttt{gcd}$ of the values of x and y in that state equals the $\texttt{gcd}$ of the parameters m and n (that is, the inputs to the GCD program).

A property is said to be an *invariant* of the transition system if all the reachable states of the system satisfy the property. For our example GCD program, the

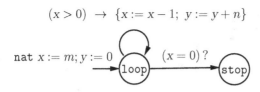

Figure 3.2: Program for Multiplication

property φ_{gcd} is indeed an invariant of the transition system $\mathrm{GCD}(m,n)$. This invariant captures the core logic of the program: during the execution of the program, even though the values of the state variables x and y change, their gcd stays the same. Since $\gcd(p, 0)$ equals p, for every natural number p, it follows that when the variable x has the value 0, $\gcd(x, y)$ is the same as y. By a symmetric argument, when the variable y has the value 0, $\gcd(x, y)$ is the same as x. This means that the update to the variable x when the system terminates by switching to the mode stop is the desired answer. Thus, the implication

$$(mode \;=\; \mathtt{stop}) \;\to\; (\gcd(m, n) \;=\; x)$$

is an invariant of the transition system $\mathrm{GCD}(m,n)$.

INVARIANT OF TRANSITION SYSTEM

For a transition system T, a *property* φ of T is an *invariant* of T if every reachable state of T satisfies φ.

If we denote the set of reachable states of the transition system T by $Reach(T)$, then a property φ is an invariant exactly when the set-inclusion $Reach(T) \subseteq [\![\varphi]\!]$ holds. The dual of the notion of an invariant property is the concept of a *reachable* property: a property φ of a transition system T is *reachable* if *some* reachable state of T satisfies φ. In other words, a property φ is reachable when the intersection $Reach(T) \cap [\![\varphi]\!]$ is a non-empty set. From the definitions, it follows that:

A property φ of a transition system T is an invariant if and only if the negated property $\neg\varphi$ is not reachable.

As an another example, consider the cruise controller from section 2.4.2 and consider the following property:

$$\varphi_{range} : \; \mathtt{minSpeed} \leq \mathtt{SetSpeed}.s \leq \mathtt{maxSpeed}.$$

It says that the state variable s of the component SetSpeed is guaranteed to be between the threshold values given by minSpeed and maxSpeed. This property is an invariant of the system. This is because whenever the component SetSpeed updates its state variable s, to initialize in response to the input event *cruise*,

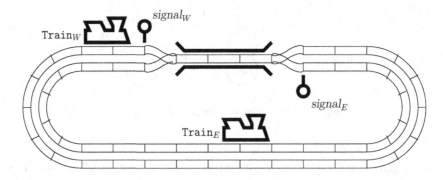

Figure 3.3: Railroad Controller Example

increment in response to the input event *inc*, or decrement in response to the input event *dec*, it checks to ensure that the updated value is in the interval [minSpeed, maxSpeed].

The invariant verification problem is the following: given a transition system T and a property φ, check whether φ is an invariant of the system T. If it is not an invariant, then there must be some state s such that the state s is reachable and violates the property φ. In such a case, for debugging purposes, the analysis technique should produce an execution of T that leads to s. Such an execution is called a *counterexample* to the claim that the property φ is an invariant and, equivalently, a *witness* to the claim that the property $\neg \varphi$ is reachable.

Exercise 3.1: Given two natural numbers m and n, consider the program Mult that multiplies the input numbers using two variables x and y, of type nat, as shown in figure 3.2. Describe the transition system $\text{Mult}(m, n)$ that captures the behavior of this program on input numbers m and n, that is, describe the states, initial states, and transitions. Argue that when the value of the variable x is 0, the value of the variable y must equal the product of the input numbers m and n, that is, the following property is an invariant of this transition system:

$$(mode = \texttt{stop}) \ \rightarrow \ (y = m \cdot n)$$

∎

3.1.2   Role of Requirements in System Design

To illustrate the use of invariants as safety requirements in the design of embedded controllers, let us consider a (toy) system of traffic lights for a railroad.

Specification of the Railroad Controller

Figure 3.3 shows two circular railroad tracks, one for trains that travel clockwise and the other for trains that travel counterclockwise. At one place in the circle,

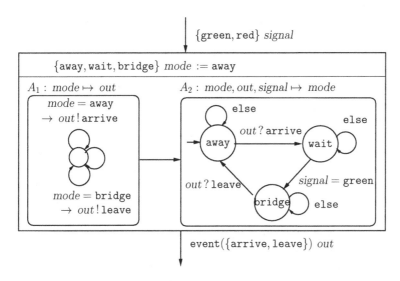

$\{\texttt{green}, \texttt{red}\}$ *signal*

$\{\texttt{away}, \texttt{wait}, \texttt{bridge}\}$ *mode* := $\texttt{away}$

$A_1 : mode \mapsto out$ $A_2 : mode, out, signal \mapsto mode$

mode = $\texttt{away}$
$\rightarrow out!\,\texttt{arrive}$

mode = $\texttt{bridge}$
$\rightarrow out!\,\texttt{leave}$

$\texttt{event}(\{\texttt{arrive}, \texttt{leave}\})$ *out*

Figure 3.4: Modeling the Train as a Nondeterministic Reactive Component

there is a bridge that is not wide enough to accommodate both tracks. The two tracks merge on the bridge, and for controlling the access to the bridge, there is a signal at each entrance. If the signal at the western entrance is green, then a train coming from the west may enter the bridge, and if the signal is red, the train must wait. The signal at the eastern entrance to the bridge controls trains coming from the east in the same fashion.

A train is modeled by the component **Train** in figure 3.4. The state of the train, captured by the enumerated variable *mode*, indicates whether the train is away from the bridge, waiting at the signal, or on the bridge. We use nondeterminism to model the assumption that the train can be away for an unknown period of time: when the train is away, either the state stays unchanged, or the train issues an output event with the value **arrive** and updates the state to waiting. When the train is waiting, it checks the signal. If the signal is red, then the train keeps waiting, and if the signal is green, then the train proceeds onto the bridge. The train can stay on the bridge for an arbitrary number of rounds. When the train exits from the bridge, it issues an output event with the value **leave** and updates the state to **away**.

The reactions of the train component can naturally be described using an extended-state machine with three modes corresponding to *away*, *wait*, and *bridge*. However, specifying the update as a single task would create an await dependency of the output event on the input signal. To avoid this, the component specification of figure 3.4 splits the reaction description into two tasks. The first task A_1 computes the value of the output variable *out*, and this does not depend on the input variable *signal*. The task A_1 is nondeterministic: when

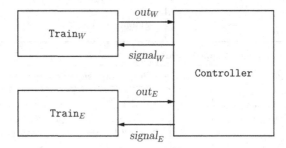

Figure 3.5: Composite System for the Railroad Controller

the mode is *away*, the output can be absent or present with the value `arrive`; when the mode is *wait*, the output is absent; when the mode is *bridge*, the output can be either absent or present with the value `leave`. This description is captured by the single-mode extended-state machine in figure 3.4. Recall that for a mode-switch, absence of a guard condition means that the mode-switch is always enabled (that is, by default, the guard condition is the constant 1 that is satisfied in every state), and absence of an associated update means that state variables do not change and event outputs are absent. The second task A_2 updates the mode based on the output computed by the task A_1 and the value of the input *signal*. In the mode `away`, when the guard-condition *out*? `arrive` holds, the mode is updated to `wait`. The condition `else` on the self-loop is an abbreviation for the negated condition $\neg$ (*out*? `arrive`). In general, the guard-condition `else` on a self-loop on a mode is satisfied exactly when none of the guard-conditions of the mode-switches out of this mode is satisfied. The mode-switches out of the modes `wait` and `bridge` are similar.

Since there are two trains, one traveling clockwise and the other traveling counterclockwise, we create two instances of the train component, Train_W and Train_E.

We are asked to design a deterministic controller that prevents collisions between the two trains by ensuring that at all times, at most one train is on the bridge. More specifically, we want to design a deterministic synchronous reactive component `Controller` with input event variables out_W and out_E and with output variables $signal_W$ and $signal_E$. When composed with the models of the trains, we get the composite system

$$\text{RailRoadSystem} = \text{Controller} \parallel \text{Train}_W \parallel \text{Train}_E$$

shown in figure 3.5. Note that, irrespective of the await dependencies of the controller, there will be no cycles in await dependencies in these three components, and thus the above composition is well defined.

The controller should be designed so that the property

$$\text{TrainSafety}: \ \neg \, (\, mode_W = \text{bridge} \ \wedge \ mode_E = \text{bridge}\,)$$

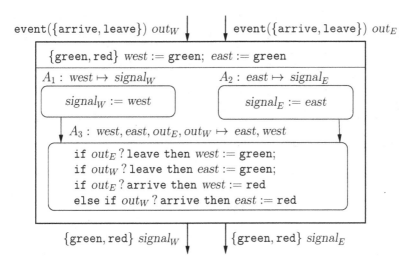

Figure 3.6: A First Attempt at Design of the Railroad Controller

is an invariant of `RailRoadSystem`. Here, the state variables $mode_W$ and $mode_E$ are the state variables of the two instances of the train component.

A First Attempt at the Design of the Railroad Controller

Figure 3.6 shows a first attempt at designing the railroad controller. The controller `Controller1` maintains two state variables *west* and *east* for the states of the two output signals $signal_W$ and $signal_E$, respectively, and in each round, the output variable is set to the value of the corresponding state variable. Initially, both signals are green. A signal is set to red whenever a train approaches the opposite entrance to the bridge, and it is set back to green whenever that train exits from the bridge. If both trains approach the bridge in the same round, then only the west signal turns red, giving priority to the train approaching from the east. The update is split into three tasks: the tasks A_1 and A_2 output the values of the respective signals without waiting for any input, and the task A_3 then updates the state variables based on the input events.

Unfortunately, the resulting railroad system

$$\text{RailRoadSystem1} = \text{Controller1} \parallel \text{Train}_W \parallel \text{Train}_E$$

does not satisfy the desired invariant `TrainSafety`. This is evidenced by the counterexample shown in figure 3.7, which leads to a state with both trains on the bridge. If both trains approach the bridge simultaneously, then the east train is admitted to the bridge with the west signal red and the east signal green. When the east train exits from the bridge, the west signal turns green, allowing the west train to proceed to the bridge. However, the east signal is still

west	east	$mode_W$	$mode_E$	$signal_W$	$signal_E$	out_W	out_E
green	green	away	away				
				green	green	arrive	arrive
red	green	wait	wait				
				red	green	$\perp$	$\perp$
red	green	wait	bridge				
				red	green	$\perp$	leave
green	green	wait	away				
				green	green	$\perp$	arrive
red	green	bridge	wait				
				red	green	$\perp$	$\perp$
red	green	bridge	bridge				

Figure 3.7: An Execution of `RailRoadSystem1` That Violates `TrainSafety`

green. So if the east train returns *before* the west train has left the bridge, the west signal will turn red while admitting the east train onto the bridge, leading to a violation of the safety requirement.

A Second Attempt at the Design of the Railroad Controller

Figure 3.8 shows another attempt at designing the controller. The controller `Controller2`, in addition to the state variables *east* and *west* for the signals, maintains Boolean state variables $near_W$ and $near_E$ to keep track of whether the respective trains need to use the bridge. Initially, $near_W$ is 0. When the west train arrives near the bridge, it is updated to 1, and when the west train leaves the bridge, it is reset to 0. Observe that the state variable $mode_W$ is away precisely when $near_W$ is 0. Analogously, the variable $near_E$ keeps track of the status of the east train.

The controller `Controller2` plays it safe by keeping the two signals red by default. Initially, the state variables *east* and *west* for the signals are red. When a train is away (as indicated by the corresponding *near* variable), the corresponding signal variable is set to red. When the east train is near, the east signal is turned green, provided the west signal is red. Consider the case when both signals are red and both trains issue **arrive**. Then both *near* variables are set to 1. In this case, the east train gets a preference: the variable *east* is changed to green, and this blocks the update of *west*, which is turned green only if the updated value of *east* is red.

For the composite system

$$\text{RailRoadSystem2} \;=\; \text{Controller2} \parallel \text{Train}_W \parallel \text{Train}_E,$$

the property `TrainSafety` is indeed an invariant as desired.

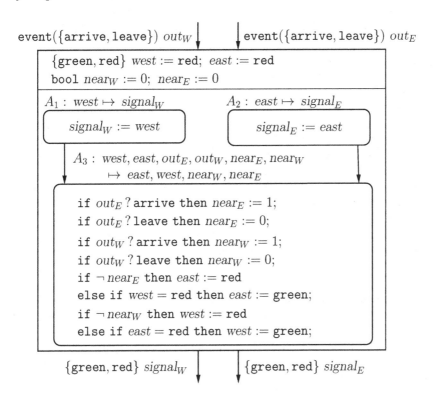

$event(\{arrive, leave\})\ out_W$ $event(\{arrive, leave\})\ out_E$

> $\{green, red\}\ west := red;\ east := red$
> $bool\ near_W := 0;\ near_E := 0$
>
> $A_1 : west \mapsto signal_W$ $A_2 : east \mapsto signal_E$
>
> > $signal_W := west$ $signal_E := east$
>
> $A_3 : west, east, out_E, out_W, near_E, near_W$
> $\qquad \mapsto east, west, near_W, near_E$
>
> > if out_E ? arrive then $near_E := 1$;
> > if out_E ? leave then $near_E := 0$;
> > if out_W ? arrive then $near_W := 1$;
> > if out_W ? leave then $near_W := 0$;
> > if $\neg near_E$ then $east := red$
> > else if $west = red$ then $east := green$;
> > if $\neg near_W$ then $west := red$
> > else if $east = red$ then $west := green$;

$\{green, red\}\ signal_W$ $\{green, red\}\ signal_E$

Figure 3.8: A Safe Controller for the Railroad Problem

Exercise 3.2 : The composed system `RailRoadSystem1` has four state variables, *east* and *west*, each of which can take two values, and *mode_W* and *mode_E*, each of which can take three values. Thus, `RailRoadSystem1` has 36 states. How many of these 36 states are reachable? ■

Exercise 3.3 : The reaction description for the controller `Controller2` consists of three tasks as shown in figure 3.8. Split the task A_3 into four tasks, each of which writes exactly one of the state variables *east*, *west*, *near_W*, and *near_E*. Each task should be described by its read-set, write-set, and update code, along with the necessary precedence constraints. The revised description should have the same set of reactions as the original description. Does this splitting impact output/input await dependencies? If not, what would be the potential benefits and/or drawbacks of the revised description compared to the original description? ■

3.1.3 Safety Monitors

For our railroad crossing example, suppose we have an additional "fairness" requirement that if a train arrives at a bridge and as it waits for its signal to turn green, the other train should not be allowed to enter the bridge repeatedly. More specifically, while a train is waiting with its signal red, the other train should not leave the bridge twice. This is clearly a requirement regarding the sequence of inputs and outputs along an execution of the railroad system, but it cannot be formulated as an invariant directly. It can, however, be stated as an invariant if we add another component, `WestFairMonitor`, shown in figure 3.9.

The monitor is described as an extended-state machine with four possible modes. The mode is initially 0. When the west train arrives, the mode changes to 1. If the east train leaves the bridge, then the mode changes to 2, and if the east train leaves the bridge again, then the mode changes to 3. In modes 1 and 2, if the west signal is turned green, then the mode is reset to 0. If there is an execution in which the monitor's mode gets updated to 3, then it demonstrates a violation of the desired fairness requirement with respect to the west train. The mode 3 of the monitor is marked as an *accepting mode* of the monitor (this is analogous to the final states of automata in theory of formal languages). An execution that reaches this accepting mode corresponds to a counterexample to the desired safety requirement. To check whether there is such a violation, we can determine whether the property `WestFairMonitor.mode` = 3 is reachable in the composite system `RailRoadSystem ∥ WestFairMonitor`. To ensure fairness with respect to the east train, we can compose the system with a symmetric version of the monitor, which can be defined by renaming the input variables of `WestFairMonitor`.

The definition of such monitors is summarized below.

SAFETY MONITOR

A *safety monitor* for a reactive component C with input variables I and output variables O consists of a synchronous reactive component M such that:

- the set of input variables of M is a subset of the variables $I \cup O$,

- the set of output variables of M is disjoint from the variables $I \cup O$,

- and the reaction description of M is given as an extended-state machine, along with a subset F of the modes declared as accepting.

The component C satisfies the monitor specification if the property $M.mode \notin F$ is an invariant of the composed system $C \parallel M$.

The requirement that the input variables of the monitor M are the input/output variables of the component C means that M can *observe* the behavior of C in terms of its interaction with the other components. The requirement that the output variables of M are neither the inputs nor the outputs of C ensures that the behavior of C is not modified by the monitor M and also that it is compatible with M. We design the monitor so that it enters an error mode in the set F

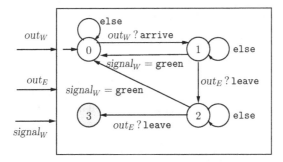

Figure 3.9: Fairness Monitor `WestFairMonitor` for the West Train

when the observed sequence of inputs and corresponding outputs violates the desired safety requirement.

Not all requirements can be expressed as safety monitors. For the railroad crossing example, consider the controller that always keeps both the traffic lights red. Such a controller satisfies the invariant `TrainSafety` as well as the requirement expressed by the safety monitor of figure 3.9. To rule out such solutions that avoid bad situations by not attempting to do anything good, we need to impose additional requirements such as, "If both trains are waiting, then the controller must allow some train to eventually enter the bridge." Such a requirement is called a *response* requirement. In this requirement, we have not asserted any bound on the number of rounds the trains have to wait. As a result, a finite execution in which both trains are waiting in the last, say 10, rounds of the execution cannot be considered a violation of the response requirement. Indeed, hypothetically, there may be a correct implementation of the controller that is slow in its processing of requests and needs 11 rounds to turn the signal to green. In general, no finite execution can demonstrate that the response requirement is truly violated. This is not a safety requirement and, thus, cannot be expressed using monitors and invariants. If we change the requirement to a *bounded response* requirement such as, "If both trains are waiting and both signals are red, then the controller must turn one of the signals to green in the next round," it can be captured by a safety monitor. We will study specification and analysis of response and other forms of liveness requirements in chapter 5.

Exercise 3.4: Consider a component C with an output variable x of type `int`. Design a safety monitor to capture the requirement that the sequence of values output by the component C is strictly increasing (that is, the output in each round should be strictly greater than the output in the preceding round). ■

Exercise 3.5: Does the second attempt to design the railroad controller satisfy the fairness requirement captured by the monitor `WestFairMonitor`? That is, is the property `WestFairMonitor.mode` $\neq 3$ an invariant of the composite system `RailRoadSystem2 ∥ WestFairMonitor`? If not, show a counterexample execution. ■

3.2   Verifying Invariants

In the invariant verification problem, we are given a transition system T and a property φ, and we want to check whether φ is an invariant of T. If not, we should output a counterexample that demonstrates the reachability of a state violating the property. We first describe a general purpose proof methodology for establishing invariants and then consider the challenge of developing automatic tools for solving the invariant verification problem.

3.2.1   Proving Invariants

Inductive Invariants

Consider a transition system T and a property φ. If we can establish that φ holds initially and is preserved during every transition, then by the principle of mathematical induction, it should hold at every state encountered along every execution and thus should be an invariant of T. Showing that the property holds initially amounts to establishing that,

> every initial state satisfies φ.

Showing that the property is preserved by every transition amounts to establishing that,

> if a state s satisfies φ, and (s, t) is a transition of T, then the state t satisfies φ.

Notice the similarity of these two conditions with the classical proofs by induction. To show that a property holds for all natural numbers n, we first show that the property holds for 0 (this is called the *base case*), and then, assuming that the property holds for a number k, we show that the property also holds for the number $k + 1$ (this is called the *inductive case*). In case of properties of reachable states of a transition system, the base case corresponds to proving the property for the initial states, and the inductive case corresponds to, assuming the property holds for an arbitrary state s, proving that the property holds for any state t that has a transition from the state s.

Properties that hold initially and are preserved by the transition relation are called *inductive invariants*.

INDUCTIVE INVARIANT

A property φ of a transition system T is an *inductive invariant* of T if:

1. every initial state s satisfies φ, and

2. if a state s satisfies φ and (s, t) is a transition, then the state t also satisfies φ.

Let us consider the program GCD of figure 3.1. For the transition system $\mathrm{GCD}(m, n)$, consider the property φ_{gcd} given by $\mathrm{gcd}(m, n) = \mathrm{gcd}(\mathrm{x}, \mathrm{y})$. To show that this is an inductive invariant, let us first consider the requirement corresponding to initialization. In this case, there is only one initial state s with $s(x) = m$ and $s(y) = n$, and this initial state clearly satisfies the property. Now let us focus on the inductive case. Consider a state s that satisfies the desired property φ_{gcd}. Let $s(x) = a$ and $s(y) = b$. By assumption, $\gcd(m, n) = \gcd(a, b)$ holds. If $s(mode) = \texttt{stop}$, then the state s has no transition from it. Now suppose $s(mode) = \texttt{loop}$, and consider a transition (s, t) of the system $\texttt{GCD}(m, n)$. To show that the new state t satisfies the property φ_{gcd}, it suffices to show that $\gcd(t(x), t(y))$ equals $\gcd(a, b)$. First, let us consider the case when $a > b > 0$. Then the program decrements x by y, and in this case, $t(x) = a - b$ and $t(y) = b$. So we need to show that $\gcd(a - b, b) = \gcd(a, b)$. This follows from basic properties of arithmetic: when $a > b$, any number that is a divisor of both a and b must also be a divisor of $a - b$, and any number that is a divisor of both a and $a - b$ must also be a divisor of b. In the case when $a \leq b$ and $b > 0$, $t(x) = a$ and $t(y) = b - a$, and by a symmetric argument $\gcd(a, b - a) = \gcd(a, b)$. In the case when either $a = 0$ or $b = 0$, the program switches to the mode $\texttt{stop}$. In such a case, either $a > 0$ and $t(x) = a$ and $t(y) = b$ or $a = 0$ and $t(x) = b$ and $t(y) = b$. In both these cases, verify that $\gcd(t(x), t(y))$ equals $\gcd(a, b)$.

For the cruise-control example from section 2.4.2, the property $\texttt{minSpeed} \leq \texttt{SetSpeed}.s \leq \texttt{maxSpeed}$ is an inductive invariant: it holds in the initial state, and whenever there is a transition (s, t), it holds in state t since the component $\texttt{SetSpeed}$ never updates the value of the state variable s without checking the bounds.

Strengthening Invariants

To illustrate that a property may be an invariant but not an inductive invariant, let us consider the transition system $\texttt{IncDec}(m)$, parameterized by a natural number m, shown in figure 3.10. The system uses two variables x and y, both of type $\texttt{int}$. Initially, x is 0 and y is m, and in each transition, the program increments the variable x and decrements the variable y as long as the value of x does not exceed m.

Consider the property $\varphi_x : 0 \leq x \leq m$, which states that the value of the variable x is always within the range from 0 to m. Let us examine whether the property φ_x is an inductive invariant of the transition system $\texttt{IncDec}(m)$. In the sole initial state of $\texttt{IncDec}(m)$, the value of x is 0, and thus the property φ_x holds initially. A state of $\texttt{IncDec}(m)$ is a valuation of the integer variables x and y and thus is a pair of integers, listing the value of x first. Consider an arbitrary state $s = (a, b)$. We want to show that if the state s satisfies the property φ_x and (s, t) is a transition of $\texttt{IncDec}(m)$, then the state t also satisfies φ_x. Assume that the state s satisfies the property φ_x, that is, $0 \leq a \leq m$. If $a < m$, then in one transition from state s, the program increments x and decrements y, and thus the updated state t is $(a + 1, b - 1)$, and in this case, we can conclude that $0 \leq a + 1 \leq m$, and thus the state t continues to satisfy the property φ_x. If the condition $a < m$ does not hold, then executing the update code does not change

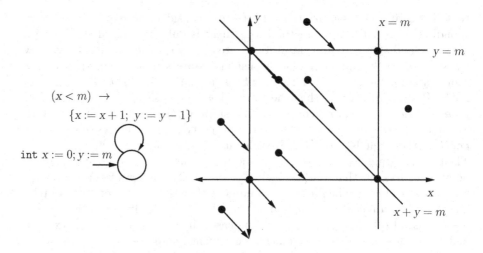

Figure 3.10: Transition System IncDec(m)

the state, and thus the property φ_x continues to hold. Hence, the property φ_x is an inductive invariant of the transition system IncDec(m).

Now let us examine whether the property $\varphi_y : 0 \leq y \leq m$ is an inductive invariant of the transition system IncDec(m). The property φ_y holds in the initial state $(0, m)$. Consider an arbitrary state $s = (a, b)$ of IncDec(m). If the only assumption about the state s is that it satisfies φ_y, that is, $0 \leq b \leq m$, then can we conclude that, after one transition of the program, the value of y will still be in the range $[0, m]$? The answer is negative. In particular, the state $(0, 0)$ satisfies φ_y, and executing one transition from the state $(0, 0)$ gives the state $(1, -1)$, which violates φ_y. In conclusion, the property φ_y is *not* an inductive invariant of the transition system IncDec(m).

Note, however, that the property φ_y is an invariant of IncDec(m): the reachable states of the transition system IncDec(m) are $(0, m), (1, m - 1), \cdots (m, 0)$, all of which satisfy the property φ_y. That is, although the property φ_y is an invariant of the system, *it is not strong enough to be inductive*. The *inductive strengthening* of the property is φ_{xy}:

$$0 \leq y \leq m \ \wedge \ x + y = m.$$

The property φ_{xy} implies φ_y: if a state satisfies the property φ_{xy}, then clearly it also satisfies φ_y. The property φ_{xy} holds in the initial state $(0, m)$. Now consider a state $s = (a, b)$ that satisfies the property φ_{xy}, that is, assume that $a + b = m$ and $0 \leq b \leq m$. If the condition $a < m$ does not hold, then executing the update code does not change the state, and thus the property φ_{xy} continues to hold. Suppose $a < m$. Then there is a transition from the state s to the state $t = (a + 1, b - 1)$. Since $a + b = m$ and $a < m$, we can conclude that $b > 0$. It follows that $0 \leq b - 1 \leq m$ and also $(a + 1) + (b - 1) = m$. Thus, the state t satisfies the property φ_{xy}. Hence, the property φ_{xy} is an inductive invariant of the transition system $\texttt{IncDec}(m)$.

Intuitively, since the program decrements y when the condition $x < m$ holds, it is not possible to show that the property φ_y is preserved by its transitions as this property asserts bounds on the values of y without relating it to the values of x. The stronger property φ_{xy} captures the relevant corelation between the values of the two variables and turns out to be inductive.

Figure 3.10 also shows states and transitions of the system visually and is useful to understand the concepts of reachable states, invariants, and inductive invariants. The reachable states of the system are the states on the line segment joining the points $(0, m)$ and $(m, 0)$. A property is an invariant as long as it includes this line segment. Thus, properties such as $0 \leq x \leq m$, $0 \leq y \leq m$, $x + y = m$, and $x \geq 0$ are all invariants, whereas the property $x < m$ is not an invariant. An inductive invariant is any set of states that contains the reachable line segment and has no transitions crossing its boundary from a state inside to a state outside it. Examples of such inductive invariants are $0 \leq x \leq m$, $x \geq 0$, $x \leq m$, $y \leq m$, $x + y = m$, and $x + y \leq m$. Examples of properties that are *not* inductive invariants include $y \geq 0$ and $0 \leq y \leq m \wedge 0 \leq x \leq m$. Given a property φ, its inductive strengthening is an inductive property that is a subset of φ. Such a strengthening need not be unique: for instance, while φ_{xy} is an inductive strengthening of the property φ_y, so is the property $0 \leq y \leq m \wedge x + y \geq m$.

Proof Rule for Establishing Invariants

The method of inductive strengthening is a general purpose and powerful technique for establishing invariants.

PROOF RULE FOR INVARIANTS

To establish that a property φ is an invariant of the transition system T, find a property ψ such that:

1. ψ is an inductive invariant of T, and

2. the property ψ implies the property φ (that is, a state satisfying ψ is guaranteed to satisfy φ).

If the property ψ is an inductive invariant, then all reachable states of T must satisfy ψ. If ψ implies φ, then every state satisfying ψ must also satisfy φ. It follows that the property φ is an invariant of T if both the assumptions are satisfied. Thus, the above proof rule is *sound*, that is, it is a correct method for establishing invariants.

The proof rule is also *complete* from a theoretical perspective. What this means is that if a property φ is indeed an invariant, then there does exist an inductive property ψ that implies φ, and thus the proof rule can always be used to establish the desired invariant. In particular, let ψ_{reach} be the formula that holds only in those states of T that are reachable. Clearly, if φ is an invariant, then it holds in all reachable states, and thus ψ_{reach} implies φ. The property ψ_{reach} capturing precisely the reachable states is inductive: initial states are reachable, and executing one transition from a reachable state leads to a reachable state. If the assertion language for writing formulas is expressive enough to describe the set of reachable states, then we know that the proof rule gives a complete method.

In practice, to prove invariants of a system in a rigorous manner, the user must identify the inductive strengthening. To establish that ψ is indeed inductive and implies φ, one can either use a "paper-and-pencil" argument or construct a formal proof using an automated theorem prover. Identifying such an inductive strengthening requires expertise, but note that this task can be much easier than understanding the precise set of reachable states of the system. For instance, let us modify the transition system $\texttt{IncDec}(m)$ of figure 3.10 by introducing an additional integer variable z, initialized to 0, and by modifying the self-loop to

$$(x < m) \;\rightarrow\; \{\, x := x + 1;\; y := y - 1;\; z := z + xy \,\}.$$

To prove that the property $0 \leq y \leq m$ is an invariant of the modified system, it still suffices to consider the strengthening $0 \leq y \leq m \land x + y = m$, which turns out to be an inductive invariant for the modified system also. Note that the property characterizing the set of reachable states of the modified system is a lot more complex as it needs to relate the current values of x and z. In other words, to prove that the variable y stays within the bounds of 0 and m using the technique of inductive invariants, it suffices to understand the relationship between x and y, and one can ignore the way the variable z gets updated.

Proof of the Synchronous Leader Election Protocol

To illustrate a more interesting proof, let us consider the central correctness argument for the flooding algorithm for leader election from section 2.4.3. For a set P of nodes and a set E of directed links that induce a strongly connected graph over the nodes, consider the system $\texttt{SyncLE}$ defined as the composition of all the node components $\texttt{SyncLENode}_n$, for each node $n \in P$, each executing the leader election protocol, and the combinational network component $\texttt{Network}_{P,E}$ that transfers the messages in a synchronous manner. The description of the transition system for $\texttt{SyncLE}$ is summarized below.

The set of state variables contains the variables id_n and r_n for each node n. The type of all the state variables is $\texttt{nat}$. Initially, the value of each variable r_n is 1, and the value of each variable id_n equals n.

The transition description of $\texttt{SyncLE}$ uses the variables in_n, out_n, and $status_n$ for each node n, which are the input/output variables of individual components that are used for communication. The update in each round involves the following sequence of steps:

1. The tasks A_1 of all the node components execute: for each node n, if the condition $r_n < N$ holds, then the value id_n is issued on out_n, and r_n is incremented. Otherwise out_n is absent and r_n stays unchanged.

2. The component **Network** executes updating all the variables in_n: for each node n, the value of in_n equals the set containing the values of the variables out_m, such that there is a network edge from node m to node n, and the event out_m is present.

3. The tasks A_2 of all the node components execute: for each node n, the value of id_n is updated to equal the maximum of its current value and the values contained in the set in_n.

Consider the following property φ_{leader} for the transition system **SyncLE**:

> for every node n, after N rounds, the value of the variable id_n equals the highest identifier in P: $r_n = N \ \rightarrow \ id_n = \mathtt{max}\ P$.

Convince yourself that this property is not inductive. To strengthen it, we must assert that, at the beginning of each round j, the value of id_n is the maximum of identifiers of nodes at distance less than j from n. Let $\mathtt{dist}(m, n)$ denote the length of the shortest path from node m to node n according to the links in E. The distance of a node from itself is 0. Consider the following property ψ:

> for every node n, $id_n \ = \ \mathtt{max}\ \{m \mid \mathtt{dist}(m, n) < r_n\}$.

The property ψ does imply the desired invariant φ_{leader}. This is because the distance between any pair of nodes is at most $N - 1$, and hence when r_n equals N, the set $\{m \mid \mathtt{dist}(m, n) < r_n\}$ must equal the set P of all identifiers, and thus the value of id_n must equal the maximum identifier, $\mathtt{max}\ P$.

Let us check whether the property ψ is an inductive invariant of the system **SyncLE**. It holds initially that the only node at distance 0 from a node n is n itself. Unfortunately, it is not strong enough to be preserved in every transition. Consider a state s satisfying ψ and a node n. Suppose the value of the round variable r_n equals j, where $j < N$. Then we know that the value of id_n is $\mathtt{max}\{m \mid \mathtt{dist}(m, n) < j\}$. In one round, the node n receives the current values of id variables of all its neighbors and updates id_n to the maximum of its current value and all the values received. Since the node n increments the variable r_n, we need to show that the updated value of id_n is $\mathtt{max}\{m \mid \mathtt{dist}(m, n) < j + 1\}$. Now any node, say n_1, at distance less than $j + 1$ from n must be at distance less than j from one of the neighbors, say n_2, of n. Do we know that the value of id of node n_2 must be at least j in state s? We know, from the property ψ, that the value of id_{n_2} is the maximum of identifiers of all nodes at distance less than the value of the round variable r_{n_2}. We would be finished with the proof successfully if we could conclude that the value of r_{n_2} in state s is j. But this is not really captured in the property ψ, and the proof fails. The necessary strengthening must also assert that the values of the round variables of all the nodes are equal. The following property ψ_{leader} is indeed an inductive invariant of the system **SyncLE** and implies φ_{leader}:

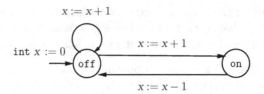

Figure 3.11: Exercise for Inductive Invariants

for every node n, $id_n = \max\{m \mid \mathtt{dist}(m, n) < r_n\}$, and for every
pair of nodes m and n, $r_m = r_n$.

Exercise 3.6: Consider a transition system T with two integer variables x and
y and a Boolean variable z. All the variables are initially 0. The transitions of
the system correspond to executing the conditional statement

$$\mathtt{if}\ (z = 0)\ \mathtt{then}\ \{x := x + 1;\ z := 1\}\ \mathtt{else}\ \{y := y + 1;\ z := 0\}.$$

Consider the property φ given by $(x = y) \vee (x = y + 1)$. Is φ an invariant of
the transition system T? Is φ an inductive invariant of the transition system T?
Find a formula ψ such that ψ is stronger than φ and is an inductive invariant
of the transition system T. Justify your answers. ∎

Exercise 3.7: Recall the transition system $\mathtt{Mult}(m, n)$ from exercise 3.1. First,
show that the invariant property $(mode = \mathtt{stop}) \rightarrow (y = m \cdot n)$ is not an
inductive invariant. Then find a stronger property that is an inductive invariant.
Justify your answers. ∎

Exercise 3.8: Consider the transition system specified by the extended-state
machine of figure 3.11. Consider the property φ given by $x \geq 0$. Show that φ
is *not* an inductive invariant of the system. Find a formula ψ such that ψ is
stronger than φ and is an inductive invariant. Prove your answer. ∎

Exercise 3.9: Consider the system $\mathtt{RailRoadSystem2}$ corresponding to the
controller of figure 3.8. For each of the properties listed below, state whether
the property is an invariant of the system and, if so, whether it is an inductive
invariant. Justify your answers with an explanation.

1. The controller state variable $near_E$ is 0 when the east train is away and 1
   otherwise: $(near_E = 0) \leftrightarrow (mode_E = \mathtt{away})$.

2. When the east train is on the bridge, the controller state variable corresponding to the east signal is green: $mode_E = \mathtt{bridge} \rightarrow east = \mathtt{green}$.

3. The controller variables for the two signals cannot be green simultaneously:
   $\neg(east = \mathtt{green} \wedge west = \mathtt{green})$.

∎

3.2.2   Automated Invariant Verification *

Before we proceed to consider some techniques for checking invariants of transition systems, let us consider invariant verification as a computational problem. Ideally, we would like to *automate* verification. The challenge then is to build a verification tool as shown in figure 3.12: the input to the verifier consists of the description of the transition system T and a property φ, and it decides whether φ is an invariant of T.

Undecidability

In general, the verification problem is *undecidable*. This means that we cannot hope to have a completely algorithmic solution to solve the invariant verification problem, and the ideal verifier of figure 3.12 does not exist. Intuitively, to check whether a property φ is an invariant of a given transition system T, the verification tool needs to check all reachable states of T. Given the description of T, the tool can systematically explore reachable states and check whether each such state satisfies the property φ. However, when the number of states of a system is unbounded, and this is the case for systems described using variables with unbounded types such as **nat**, the verification tool can potentially run forever exploring more and more states without ever being able to conclude that the property holds in all reachable states. Formally, the undecidability of the verification problem follows from the results in computability theory. In fact, if we limit ourselves to transition systems that have only *counter-variables*, then the invariant verification problem remains undecidable, where a counter-variable is a variable of type **nat** that in one transition can only be incremented, decremented, or tested for being equal to zero.

Verification of Finite-State Systems

Now let us restrict our focus on systems whose variables have finite types. For such finite-state systems, the number of all possible states is bounded. In this special case, the invariant verification problem can be automated. The input to the verifier is a description of the transition system in the source modeling language and, thus, is described in a compact manner. In particular, for synchronous reactive components, the set of states is described by listing the names and types of all the variables, and the set of transitions is described by a sequence of assignments and conditional statements. If the transition system T has k state variables, and each variable can take at most m different values, then the bound on the number of states is m^k. The total number of states grows exponentially with the number k of state variables, and as a result, the solution based on examining all states does not lead to a scalable analysis tool. This exponential complexity, however, is intrinsic to the problem.

The precise computational complexity of the invariant verification problem depends on the details of the modeling language used to specify transition systems. As an illustrative case, let us consider transition systems described as sequential

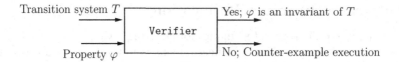

Figure 3.12: Ideal Automated Verification Tool

circuits discussed in section 2.4.1. A sequential circuit is specified using block diagrams connecting Not gates, And gates, and latches. All the variables in such a circuit are of type `bool`. If the circuit has k latches, then the number of possible states is 2^k. Suppose the property φ is specified as a Boolean expression over the state variables corresponding to the latches. In this case, the invariant verification problem is a canonical problem for the complexity class PSPACE of computational problems. PSPACE stands for problems solvable using space polynomial in the size of the input description. The upper bound of PSPACE means that there is an algorithm to solve the invariant verification problem for sequential circuits that requires memory polynomial in the number of latches and gates used in the circuit description. Furthermore, the invariant verification problem is computationally hard for the class PSPACE, implying that every other problem in this class can be reformulated as an invariant verification for sequential circuits.

Theorem 3.1 [Complexity of Finite-state Invariant Verification] *Given a sequential circuit C represented as a block diagram over Not gates, And gates, and latches, and a property φ specified as a Boolean formula over the state variables corresponding to the latches, the computational problem of checking whether the property φ is an invariant of the transition system corresponding to C is* PSPACE-*complete.* ■

Understanding the PSPACE complexity bounds in a rigorous manner would require a detour into computational complexity theory and is beyond the scope of this textbook. We will just note some facts useful in our context. First, the characterization of the complexity of the invariant verification problem is robust: the invariant verification problem is PSPACE-complete for finite-state systems for all typical choices of modeling languages used in practice. The complexity stays the same whether systems are deterministic or nondeterministic. Second, the class PSPACE of problems is a superset of the class NP of problems, which stands for problems that can be solved by *nondeterministic* algorithms in time polynomial in the size of the input. The canonical problem for the class NP is analysis of *combinational* circuits. Consider a combinational circuit C represented as a block diagram over Not and And gates, such that C has m Boolean input variables $x_1, \ldots x_m$ and a Boolean output variable y. Suppose we want to check whether there is an input, that is, an assignment of 0/1 values to the m input variables, so that the output of C is 1. Given a specific input, computing the corresponding output is easy, and we can develop an efficient algorithm for this purpose that would require time linear in the size of the circuit (that is, the number of gates). However, to check whether there exists *some* input for which the corresponding output is 1, the most obvious strategy to try all inputs would lead to an exponential-time algorithm since there are 2^m possible inputs. This structure makes this analysis problem for combinational circuits belong to the class NP. Furthermore, it is a canonical representative of all NP problems: if a combinational circuit analysis problem can be solved efficiently, then so can every problem in the class NP. Thus, the combinational circuit analysis problem is an NP-complete problem. Whether such a problem can be solved by a polynomial-time algorithm remains a long-standing open problem in computer science. For all practical purposes, we can assume that efficient algorithms do not exist for NP-complete problems and, thus, also not for PSPACE-complete problems, such as the invariant verification problem for finite-state systems. Finally, note that the combinational circuit analysis problem is representative of computing reactions from a given state. For a finite-state system corresponding to a synchronous reactive component, given two states s and t, determining whether there is a transition from state s to state t is typically NP-complete as it would require examining all possible input combinations, but for a specific input i, determining whether state s on input i gets updated to state t can be determined efficiently by executing the update code.

In summary, for finite-state reactive components, analyzing what happens within a round is computationally hard (NP-complete) due to the multitude of combinations of values of input variables, and analyzing what happens in an execution, across multiple rounds, has additional computational difficulty (PSPACE-complete) due to the multitude of combinations of values of state variables.

3.2.3   Simulation-Based Analysis

The most commonly used industrial technique for analyzing systems is exploration using simulation. Given a user-specified value k for the number of steps of the simulation, the algorithm generates an execution of the transition system containing k transitions and checks whether the invariant holds for every state visited during this execution. In general, the transition system has many executions of a given length. For instance, in transition systems corresponding to synchronous reactive components, each state can have multiple successors due to inputs as well as nondeterministic choices within the reaction description. In such a case, the simulator must resolve the choice in some way, for instance, by using randomization.

The simulation-based algorithm can process transition systems represented in many different possible ways ranging from source code to internal representation. What is needed is a way of generating an initial state from the representation and a way of generating a successor state of a given state. More specifically, the simulation-based algorithm relies on the implementation of the following data structures and operations:

Input: A transition system T, a property φ, and an integer $k > 0$;
Output: If a state violating φ is encountered, return a counterexample;

```
array [state] exec;
nat j := 0;
state s := ChooseInitState(T);

if s = null then return;
exec[j] := s;
if Satisfies(s, φ) = 0 then return exec;
for j = 1 to k do {
    s := ChooseSuccState(s, T);
    if s = null then return;
    exec[j] := s;
    if Satisfies(s, φ) = 0 then return exec;
    }.
```

Figure 3.13: Simulation-based Invariant Falsification

- States of the transition system are of the type state. The constant null specifies a dummy state.

- Given a transition system T, *ChooseInitState*(T) returns *some* initial state and the dummy state null if T has no initial states.

- Given a transition system T and a state s, *ChooseSuccState*(s, T) returns *some* successor state of s, that is, some state t such that (s, t) is a transition of T, and the dummy state null if s has no outgoing transitions.

- Given a property φ and a state s, *Satisfies*(s, φ) returns 1 if the state s satisfies the property φ, and 0 otherwise.

The simulation algorithm is presented in figure 3.13. The variable *exec* is an array of states, and the algorithm fills its entries one by one for the specified number of steps. At each step, the function *Satisfies* is used to check whether the current state violates the desired invariant. Note that if no violation is found, the algorithm cannot make any conclusions about whether the invariant holds. In such a case, we can run the algorithm repeatedly to gain more confidence in the correctness of the system. However, such analysis cannot prove that the system indeed satisfies the invariant. This form of analysis, which can conclusively only disprove the claim that the system satisfies the specified correctness requirement, is called *falsification*.

Representing Transition Systems

The actual running time of the simulation-based algorithm depends on how long it takes to execute functions such as *ChooseSuccState* that depend on the

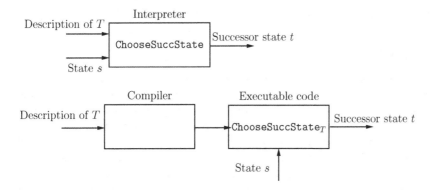

Figure 3.14: Interpretation vs. Compilation for Simulation-based Analysis

representation of the transition system. One possible representation for the transition system, then, is the original description of the system in the source modeling language. For transition systems corresponding to synchronous reactive components, a system description is typically a parallel composition of compatible components, where each component consists of one or more tasks, each with an update description given as straight-line code, along with precedence constraints among tasks. With this choice, the implementation of the function *ChooseSuccState* can be as follows. Given a state, it first picks arbitrary values for the input variables. It then orders all the tasks of all the components in a manner consistent with all the precedence constraints. Then the update specifications of the tasks are executed in the chosen order. If there is a nondeterministic assignment (using the **choose** command) in any update specification, an arbitrary choice is made. This style of implementation is called *model interpretation*: the different constructs in the model description are interpreted as needed during the execution of *ChooseSuccState*.

An alternative strategy is *model compilation*: the original description of the system in the source modeling language is compiled to executable functions for operations such as *ChooseSuccState* that are specialized for this particular transition system. That is, given the description of the transition system T, the model compiler generates an executable function for $ChooseSuccState_T$. The input to $ChooseSuccState_T$ is a state of T, and its output will be a successor-state of the input state. The two approaches of interpretation and compilation are illustrated in figure 3.14. In the context of synchronous reactive components, the compiled version of $ChooseSuccState_T$ contains a fixed order of execution of all the tasks: this order needs to be consistent with all the precedence constraints but is chosen just once when the code for $ChooseSuccState_T$ is generated. Nondeterministic constructs such as **choose** in the update specification of individual tasks are replaced with suitably chosen calls to random-number generator routines in the target language. The main benefit of the model-compilation approach is performance: individual calls to $ChooseSuccState_T$ do not need to process the internal representation of the source model and thus execute faster.

State Compaction

The memory used by an analysis algorithm that stores states is obviously affected by how an individual state is represented and stored. For the simulation algorithm of figure 3.13, the states are stored in the array *exec*. The most natural data structure for representing a state is a record type with a field for each of the state variables. However, this is too wasteful: if a system has, say, 50 state variables, allocating a word to each field on a 32-bit machine would mean 400 bytes per state, and storing thousands of states would be impractical. As a result, a state is represented using low-level bit encoding. For each state variable, the analysis tool first computes an upper bound on the number of bits that are sufficient to encode all possible values. A Boolean variable needs just one bit, a variable ranging over an enumerated type with three values needs two bits, and for a variable storing speed of a car, in miles per hour up to one decimal point, 10 bits should suffice. The state then is encoded as a sequence of bits.

3.3   Enumerative Search *

Given a transition system T and a property, to check whether the property is an invariant of the transition system, we can check whether the negation of the property is reachable. This problem can be viewed as finding a path in a graph whose vertices correspond to states of T and whose edges correspond to transitions of T. In classical graph search algorithms, the input graph is represented by listing all its vertices and edges. In invariant verification, the graph is given implicitly by listing the state variables, their initialization, and the update code for the transitions and may not be finite. Hence, we do not want to build the graph explicitly and explore the graph only as much as needed.

Reachable Subgraph

For a transition system T, the graph consisting of the reachable states of T and transitions out of these states constitutes the reachable subgraph of the system. To check whether a property is an invariant, it suffices to examine only the reachable subgraph.

Let us revisit the controller `Controller2` for the railroad crossing example of figure 3.8. For the system `RailRoadSystem2`, there are six state variables: $mode_W$ and $mode_E$, each of which ranges over $\{\texttt{away}, \texttt{wait}, \texttt{bridge}\}$; *west* and *east*, each of which ranges over $\{\texttt{green}, \texttt{red}\}$; and Boolean variables $near_W$ and $near_E$. As a result, the system has 144 possible states. It turns out that few of these states are reachable. Figure 3.15 shows the *reachable* portion of the graph. Each state is denoted by listing the values of the variables $mode_W$, $mode_E$, $near_W$, $near_E$, *west*, and *east*, in that order. We use a, w, b, g, and r as abbreviations for the values `away`, `wait`, `bridge`, `green`, and `red`, respectively. The initial state then

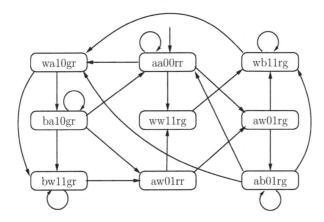

Figure 3.15: Reachable Subgraph of `RailRoadSystem2`

is *aa00rr* denoting that both trains are away, both *near* variables are 0, and both signals are **red**. The four transitions out of this state correspond to the possibilities of one, both, or neither of the two trains arriving. One of the transitions is a self-loop to *aa00rr*, and the other three lead to states *wa10gr*, *aw01rg*, and *ww11rg*. We then systematically consider all the possible transitions from these three new states and continue until no new states are discovered. Only 9 out of 144 states are found to be reachable.

As another example, consider the synchronous leader election of section 2.4.3. Suppose we choose values for the set P of node identifiers and the set E of network links. The component `SyncLE` still has infinitely many states since the type of the variables such as r and *id* of each of the node components is **nat**. However, during an execution of the system, the value of the round variable r ranges over $\{1, 2, \ldots N\}$, where N is the upper bound on the number of nodes in the network, and the value of the identifier variable *id* ranges over P. Thus, for a given network, the number of reachable states of the system is finite.

These examples indicate that the analysis algorithm for the invariant verification problem should explore only the reachable states of the system starting with the initial states. Such a search procedure is called *on-the-fly* since it examines states and transitions in an incremental manner. It is *enumerative* as it processes states individually.

On-the-Fly Depth-First Search

As in the case of simulation-based exploration, our presentation of the on-the-fly algorithm does not rely on a specific representation of transition systems. In the case of enumerative search, what we need is that there is a way to systematically enumerate all the initial states of a transition system, and given a state s, there is a way to systematically enumerate all the successors of s, that is, states t such that (s, t) is a transition. As in the case of simulation-based exploration, states of the transition system are of type `state`. We will use the following functions that access the representation of transition systems:

- Given a transition system T, *FirstInitState(T)* returns the first initial state of T, according to the chosen enumeration of the initial states, and the dummy state `null` if T has no initial states.

- Given a transition system T and an initial state s, *NextInitState(s, T)* returns the initial state following s in the chosen enumeration of the initial states and `null` if no such state exists.

- Given a transition system T and a state s of T, *FirstSuccState(s, T)* returns the first successor state of s, according to the chosen enumeration of the set of states t such that (s, t) is a transition of T and the dummy state `null` if s has no outgoing transitions.

- Given a transition system T and states s and t, *NextSuccState(s, t, T)* returns the state following t in the chosen enumeration of the set of successor states of s and the dummy state `null` if no such state exists.

For instance, if T has a state variable x of type `nat` and the initialization specifies that $x \geq 10$, then T has infinitely many initial states that can be enumerated as $10, 11, 12, \ldots$. In such a case, *FirstInitState* can return 10, and for $n \geq 10$, *NextInitState* on input n can return $n + 1$. The ability to systematically list all the initial states and the successors of a state is necessary for enumerative search. A transition system T for which the set of initial states, and the set of successor states of each state, can be effectively enumerated is called *countably branching*: whenever there is a choice in extending an execution, the number of choices is countable. If T has a state variable x of type `real` and the initialization specifies that $0 \leq x \leq 1$, then there are uncountably many initial states, one for every real number in the interval $[0, 1]$. This transition system is not countably branching. For such a system, it would not be possible to apply a systematic search exploring states one by one. Note that the simulation-based exploration is still possible in such a case, as one can implement *ChooseInitState* to return a randomly chosen real number up to a given precision of the floating-point representation of numbers.

The classical depth-first-search algorithm for on-the-fly exploration of (countably branching) transition systems is depicted in figure 3.16. It relies on the following data structures:

- The variable *Reach*, of type `set(state)`, stores the set of reachable states. The operations used on this set data structure are: (1) an initialization constant `EmptySet` that corresponds to the empty set; (2) a membership test `Contains` that takes a set and a state and returns 1 if the input state belongs to the input set and 0 otherwise; and (3) an insertion procedure `Insert` that takes a state and a set and updates the input set by adding the input state to it.

Input: A transition system T and property φ;
Output: If φ is reachable in T, return a witness, else return 0;

```
set(state) Reach := EmptySet;
stack(state) Pending := EmptyStack;
state s := FirstInitState(T);

while s ≠ null do {
  if Contains(Reach, s) = 0 then
    if DFS(s) = 1 then return Reverse(Pending);
  s := NextInitState(s, T);
  };
return 0.

bool function DFS(state s)
  Insert(s, Reach);
  Push(s, Pending);
  if Satisfies(s, φ) = 1 then return 1;
  state t := FirstSuccState(s, T);
  while t ≠ null do {
    if Contains(Reach, t) = 0 then
      if DFS(t) = 1 then return 1;
    t := NextSuccState(s, t, T);
    };
  Pop(Pending);
  return 0.
```

Figure 3.16: On-the-fly Depth-first Search Algorithm for Reachability

- The variable *Pending* stores the sequence of states from which exploration is in progress, and is of type `stack(state)`. The operations used on this stack data structure are: (1) an initialization constant `EmptyStack` that corresponds to the empty stack; (2) a procedure `Push` that takes a state and a stack and updates the input stack by adding the input state at its top; (3) a procedure `Pop` that takes a stack and updates it by removing the top element, if any; and (4) the function `Reverse` that takes a stack and returns the sequence of states it contains from bottom to top.

The algorithm maintains the set *Reach* of states it has encountered so far. All the states in this set are guaranteed to be reachable states of the transition system T. The initial states of the transition system are supplied to the algorithm one by one by the functions *FirstInitState* and *NextInitState*. The algorithm initiates search from every initial state that is not already visited by calling the recursive function *DFS*. When *DFS* is called with an input state s, it adds it to the set *Reach* and pushes it onto the top of the stack *Pending*. At any time, the stack contains a sequence of states such that the state at the bottom is an initial state, and every state has a transition from the state immediately below it. If the input state s satisfies the property φ, then the algorithm has discovered φ to be reachable, and all pending invocations of *DFS* terminate. The stack contains an execution of the transition system that demonstrates reachability of a state satisfying the property φ, and thus the reversed stack can be output as a witness execution. If the input state s does not satisfy the property φ, then the algorithm examines all the outgoing transitions from s. The successor states of s are supplied to the algorithm one by one by the functions *FirstSuccState* and *NextSuccState*. The algorithm calls *DFS* recursively from those successor states that are not already visited. If all the *DFS* calls terminate, then the algorithm has visited all the reachable states without encountering a state satisfying the property φ, and it returns 0.

Illustrative Example

Figure 3.17 shows a possible execution of the depth-first search algorithm on the transition system corresponding to the component `RailRoadSystem2` (see figure 3.15 for the reachable graph).

Initially, the algorithm invokes *DFS* with input state $aa00rr$ with *Reach* equal to the empty set and *Pending* equal to the empty stack. In figure 3.17, each state is assigned a unique number that indicates the order in which states are discovered and added to *Reach*. Thus, the state $aa00rr$ has number 1. The ordering of the successor states from a given state is depicted left to right in figure 3.17. That is, the first successor of $aa00rr$ is $aa00rr$. Since this state is already visited, no new call to *DFS* is initiated. Then the next successor of $aa00rr$, the state $wa10gr$, is examined. At this point, *Reach* contains only the state $aa00rr$, the stack also contains just this state, and the function *DFS* is invoked again with input state $wa10gr$, whose DFS discovery number is 2. Figure 3.17 shows the values of the set *Reach* of visited states and the stack *Pending* when a new call to *DFS* is made (for brevity, states are represented by their DFS discovery numbers). Note that when the call $DFS(wa10gr)$ returns, the set *Reach* contains all the states; as a result, examining the remaining two successors of $aa00rr$, states $aw01rg$ and $ww11rg$ in that order, do not initiate new calls to *DFS*.

When $DFS(aw01rr)$ first examines the successor $aw01rg$, it calls $DFS(aw01rg)$. This call explores all states that can be reached from $aw01rg$ and then returns. The value of *Reach* contains all these states. Since the next successor of $aw01rr$ is $ww11rg$ and not yet visited, $DFS(ww11rg)$ gets called. For two calls such as $DFS(aw01rg)$ and $DFS(ww11rg)$, which share the immediate parent calling context ($DFS(aw01rr)$ in this case), at the time the call, the value of the stack *Pending* is the same (and captures an execution leading to the input state for the call), but the value of the variable *Reach* has changed (the one on the right is guaranteed to be a superset of the one on the left).

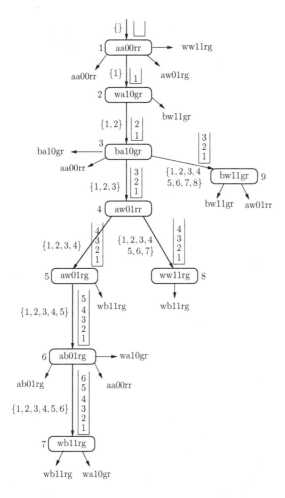

Figure 3.17: Sample Execution of Depth-first Search on `RailRoadSystem2`

Analysis of the Depth-First Search

For a reachable state s, the algorithm calls *DFS* with input s at most once, and thus it processes every transition out of a reachable state at most once. If the number of reachable states of a transition system is finite, then the algorithm is guaranteed to terminate, and its running time will be linearly proportional to the number of reachable states and transitions. Even when the number of reachable states is not finite, the algorithm may encounter a state satisfying φ and terminate with a witness execution. However, if the property φ is not reachable and the number of reachable states is not finite, then the algorithm will just keep on examining more and more states until it runs out of available memory. These properties are summarized in the following theorem:

Theorem 3.2 [On-the-fly Depth-First Search for Invariant Verification] *Given a countably branching transition system T and a property φ, the depth-first search algorithm of figure 3.16 has the following guarantees:*

1. *If the algorithm returns 0, then the property φ is not reachable in T.*

2. *If the algorithm returns a sequence of states, then its output is a witness execution demonstrating the reachability of the property φ.*

3. *If the number of reachable states of T is finite, then the algorithm terminates, and the number of calls to DFS is bounded by the number of reachable states.*

■

We conclude the discussion of the enumerative search by discussing some commonly used implementation techniques employed by enumerative model checkers for improving the efficiency of the depth-first search algorithm. As in the case of simulation-based exploration, the representation of the transition system should allow fast computation of initial states and successor states of a given state, and memory requirements of the algorithm can be reduced by compact encoding of states.

Bit-State Hashing

The most commonly used data structure for storing the set *Reach* of states already visited is a *hash-table*. A hash-table consists of a hashing function that maps each state to an integer between 0 and N, for a suitably chosen positive integer N, and an array of length N, whose each entry is a list of states. Initially, all lists are empty. To insert a state s in the hash-table, first the hashing function is used to map s to an index j, and then the state s is added to the list at the jth entry of the hash-table. To check whether a state s is already in the hash-table, the hashing function is used to map s to an index j, and the list at the jth entry is scanned to check whether it contains s. If the hashing function is chosen appropriately, then the number of states mapped to the same index is small; as a result, both `Insert` and `Contains` take near-constant time.

While hashing is an effective technique to store the set of explored states, often the number of reachable states is too large to be stored in memory. In such cases, an approximate strategy, known as *bit-state hashing*, can be used. This approach uses a hash-table of size N whose jth entry is a single bit. Initially all bits are 0. The insertion of a state s, which is mapped to an integer j by the hash function, is implemented by setting the jth bit of the hash-table to 1. The membership test for a state s returns the value of the bit stored at the index corresponding to s. This scheme does not handle hash collisions correctly. Suppose that two states s and t are mapped to the same index j, and state s is inserted in the hash-table first. When the state t is encountered, as the jth bit of the hash-table is already set to 1, the membership test `Contains` incorrectly returns a positive answer for the state t. Consequently, the depth-first search algorithm does not explore the successors of t. Hence, only a fraction of the set of reachable states is explored. Thus, when the algorithm returns 0, we cannot be sure that the property φ is not reachable, but when it returns a sequence of states, it is guaranteed to be a witness to the reachability of φ. What fraction of the reachable states is explored by bit-state hashing depends on the choice of the table size and the hash function. The performance of bit-state hashing can be improved dramatically by using two bit-state hash-tables that employ independent hash functions: to insert a state, the corresponding bits in both the hash-tables are set to 1, and the membership test checks whether the bits corresponding to the input state in both the hash-tables are 1.

Exercise 3.10: Consider the reactive component `Switch` of figure 2.2. How many reachable states does it have? Draw the reachable subgraph for the corresponding transition system. ■

Exercise 3.11: To save memory, sometimes the on-the-fly depth-first search algorithm is modified so that it does not store any states. That is, we modify the algorithm of figure 3.16 by removing the set *Reach*, and when a state is encountered, there is no test to check whether it was visited before. This form of search is called *stateless search*. Show that the partial correctness of the algorithm is still preserved: the claims (1) and (2) of theorem 3.2 continue to hold. How is the termination (claim (3) of theorem 3.2) affected? ■

Exercise 3.12 *: A breadth-first search algorithm first examines all the initial states, then all the successors of the initial states, and so on. To implement such an algorithm, we need two functions on the representation of transition systems: *InitStates* returns a list of initial states of a given transition system, and *SuccStates* returns a list of the successors of a given state of a transition system. Such a representation is feasible for *finitely branching* transition systems, namely, transition systems for which the number of initial states is finite, and each state has only finitely many successor states. Given a finitely branching transition system T and a property φ, develop a breadth-first search algorithm to check whether φ is reachable in T, and state precisely its correctness and termination guarantees. ■

3.4 Symbolic Search

As the invariant-verification problem is computationally hard, we cannot hope to find an efficient scalable solution. There are, however, heuristics that perform well on many instances of the invariant-verification problem that occur in practice. One such heuristic is based on a symbolic reachability analysis of the transition system. Instead of explicitly processing states one at a time, a symbolic search algorithm processes sets of states represented by constraints. For example, for an integer variable x, the constraint $20 \leq x \leq 99$ concisely represents the set $\{20, 21, \dots, 99\}$ of 80 states. Such a symbolic representation can be succinct, and with suitable operations that can manipulate the symbolic representation, we can develop a symbolic search algorithm that can solve the invariant verification problem.

3.4.1 Symbolic Transition Systems

For symbolic analysis, the initial states and transitions can be described by formulas, that is, Boolean-valued expressions, over state variables that capture constraints on initialization and update.

Initialization and Transition Formulas

Consider a transition system T with state variables S. The initialization can be expressed by a Boolean expression φ_I over the variables S. The set of initial states then contains the states satisfying this expression. Let us revisit the transition system $\text{GCD}(m, n)$ (see figure 3.1) corresponding to the program that computes the gcd function. It has three state variables x, y, and *mode*. The initialization formula is $x = m \wedge y = n \wedge mode = \texttt{loop}$.

To express the transition description *Trans* as a formula, we need to relate the values of state variables *before* the transition to the values of the state variables *after* the transition. For a variable v, we use v' to denote a *primed* version of v with the same type as v, and this primed copy is used to capture the value after the transition. With this convention, the transition description is expressed by a Boolean expression φ_T over the state variables S and the corresponding set S' of primed state variables.

Let us develop the transition formula for the GCD program step by step. The extended-state machine of figure 3.1 has two mode-switches, and a transition corresponds to executing one of these switches. As a result, the transition formula is a disjunction of two formulas, with one disjunct each contributed by each mode-switch. Let us first focus on the self-loop.

Consider the assignment $x := x - y$. This translates to the logical formula $x' = x - y$ that relates the updated value of x to the old values of x and y. The assumption that "a state variable that is not updated explicitly stays unchanged" needs to be captured by explicit constraints in the logical formula. Thus, the formula corresponding to the decrement of x by y, with y implicitly unchanged, is $(x' = x - y) \wedge (y' = y)$. The conditional statement "$\texttt{if }(x > y) \texttt{ then } x := x - y$" is captured by the formula

$$(x > y) \wedge (x' = x - y) \wedge (y' = y).$$

Such a formula is evaluated over a pair of states of the system, where the first state is used to obtain values for the variables x and y, and the second state is used to obtain values for the variables x' and y'. A pair of states (s, t) satisfies this transition formula precisely when $s(x) > s(y)$ and $t(x) = s(x) - s(y)$ and $t(y) = s(y)$. Notice that if we simply omit the conjunct $y' = y$, the formula would have the unintended meaning that the variable y can be updated to any arbitrary value when the condition $x > y$ holds.

The if-then-else statement "`if` $(x > y)$ `then` $x := x - y$ `else` $y := y - x$" corresponds to the formula ψ that is the disjunction of the two cases:

$$[(x > y) \wedge (x' = x - y) \wedge (y' = y)] \vee [\neg(x > y) \wedge (y' = y - x) \wedge (x' = x)].$$

The above pattern for mapping conditional statements to transition formulas is typical: the formula is a disjunction of cases, where each case is a conjunction of the condition and the update corresponding to this condition.

For the self-loop of the GCD program, the guard condition is $(x > 0 \wedge y > 0)$, and this mode-switch is from the mode `loop` to itself. Thus, the contribution of this self-loop to the transition formula is the formula ψ_1:

$$[(x > 0) \wedge (y > 0) \wedge (mode = \texttt{loop}) \wedge \psi \wedge (mode' = \texttt{loop})].$$

The contribution of the mode-switch from `loop` to `stop` can be computed in a similar manner. In particular, the update code "`if` $(x = 0)$ `then` $x := y$" translates to the formula ψ':

$$[(x = 0) \wedge (x' = y) \wedge (y' = y)] \vee [\neg(x = 0) \wedge (x' = x) \wedge (y' = y)],$$

and the contribution of this mode-switch to the transition formula is the formula ψ_2:

$$[\neg (x > 0 \wedge y > 0) \wedge (mode = \texttt{loop}) \wedge \psi' \wedge (mode' = \texttt{stop})].$$

The desired transition formula φ_T of the transition system $\texttt{GCD}(m, n)$ is: $\psi_1 \vee \psi_2$.

This style of specification of the transition relation is called *declarative*. Unlike the more familiar *operational* style that prescribes the sequence of statements to be executed, the declarative specifications only capture the constraints on the relationship between the old and new values of variables. Unlike an assignment, which is an operational and executable description, equality, which is a declarative and logical specification, does not have a meaningful distinction between the left-hand side and right-hand side. In particular, the expressions $x' = x - y$ and $x - y = x'$ are logically equivalent and express exactly the same constraint. Similarly, since logical conjunction is commutative, the formula $(x' = x - y) \wedge (y' = y)$ expresses exactly the same constraint as $(y' = y) \wedge (x' = x - y)$.

Reaction Formulas

For transition systems described by synchronous reactive components, the symbolic description of the transition system can be obtained by the analogous symbolic description of the corresponding reactive component. For a synchronous reactive component, the initialization can be captured by a formula φ_I over the state variables, and its reactions can be specified by a formula φ_R over (un-primed) state variables (denoting the state at the beginning of a round), input and output variables, and primed state variables (denoting the state at the end of a round).

Let us revisit our first reactive component, Delay, of figure 2.1. The initialization formula φ_I for the component Delay is $x = 0$. Note that if we want to specify that the initial value of x may be either 0 or 1, the corresponding initial assignment $x := $ choose $\{0, 1\}$ is captured by the formula 1 (every state satisfies the constant 1). Indeed, the constraint-based or declarative style can be more convenient to specify nondeterminism.

For the Delay component, the reaction formula φ_R is

$$(out = x) \;\wedge\; (x' = in)$$

that captures the relationship among old state, input, output, and updated state. In general, the reaction formula φ_R for a component C with state variables S, input variables I, and output variables O is a formula over the variables $S \cup I \cup O \cup S'$. For states s, t, input i, and output o of C, $s \xrightarrow{i/o} t$ is a reaction of C precisely when the formula φ_R is satisfied when we use state s to assign values to variables in S, input i to assign values to variables in I, output o to assign values to variables in O, and state t to assign values to the corresponding primed variables in S'.

Given a symbolic description of a synchronous reactive component C, the symbolic description of the corresponding transition system T can be obtained easily. The initialization formula for T is the same as the initialization formula for C. Recall that, for states s and t, (s, t) is a transition of T precisely when there *exists some* input i and output o such that $s \xrightarrow{i/o} t$ is a reaction of C. This relationship between transitions of T and reactions of C can be naturally expressed using the operation of *existential quantification* for logical formulas.

If f is a Boolean formula over a set V of variables and x is a variable in V, then $\exists x. f$ is a Boolean formula over $V \setminus \{x\}$. A valuation q over $V \setminus \{x\}$ satisfies the quantified formula $\exists x. f$ if q can be extended by assigning some value to x to satisfy f, that is, there exists a valuation s over V such that $s(f) = 1$ and $s(y) = q(y)$ for each variable y in $V \setminus \{x\}$. For example, if x and y are Boolean variables, then the formula $\exists x. (x \wedge y)$ expresses a constraint only over the variable y and is equivalent to the formula y (that is, $\exists x. (x \wedge y)$ evaluates to 1 exactly when y is assigned 1). Similarly, the formula $\exists x. (x \vee y)$ is equivalent to 1 (that is, this formula evaluates to 1 independent of the value of y).

The transition formula for the transition system corresponding to Delay is

$$\exists in. \exists out. \left[(out = x) \;\wedge\; (x' = in) \right].$$

This formula simplifies to the logical constant 1 since the formula is satisfied no matter what the values of x and x' are. This corresponds to the fact that for this transition system, there is a transition between every pair of states.

More generally, if φ_R is the reaction formula for the component C, then the transition formula φ_T for the corresponding transition system is obtained by existentially quantifying all the input and output variables: $\exists I. \exists O. \varphi_R$.

As another example, consider the component **TriggeredCopy** of figure 2.5. The only state variable is x of type **nat**, the initialization formula is $x = 0$, and the reaction formula is

$$(in? \wedge out = in \wedge x' = x + 1) \vee (\neg in? \wedge out = \bot \wedge x' = x).$$

The assumption that when the input event is absent, the component leaves the state unchanged and the output is absent, is captured explicitly in the reaction formula as a separate case. The corresponding transition formula is obtained by existentially quantifying the input and output variable:

$$\exists in, out. \, [(in? \wedge out = in \wedge x' = x + 1) \vee (\neg in? \wedge out = \bot \wedge x' = x)].$$

This transition formula can be simplified to a logically equivalent formula

$$(x' = x + 1) \vee (x' = x).$$

Composing Symbolic Representations

In section 2.3, we studied how complex components can be combined using the operations of input/output variable renaming, parallel composition, and output hiding. The symbolic description of the resulting component can be obtained naturally from the symbolic descriptions of the original components. We will use the block diagram for **DoubleDelay** from figure 2.15 to illustrate this.

Renaming of variables is useful to create instances of components so that there are no name conflicts among state variables of different components, and common names for input/output variables indicate input/output connections. Given the initialization and reaction formulas for the original component, the corresponding formulas for the instantiated component can be obtained by textual substitution. For instance, the component **Delay1** is obtained from the component **Delay** by renaming the state variable x to x_1 and the output *out* to *temp*. The initialization formula for **Delay1** then is $x_1 = 0$, and the reaction formula is $(temp = x_1) \wedge (x_1' = in)$. Similarly, the initialization formula for **Delay2** is $x_2 = 0$, and the reaction formula is $(out = x_2) \wedge (x_2' = temp)$.

Consider two compatible components C_1 and C_2. If φ_I^1 and φ_I^2 are the respective initialization formulas for C_1 and C_2, then the initialization formula for the product $C_1 \| C_2$ is simply the conjunction $\varphi_I^1 \wedge \varphi_I^2$. This captures the fact that the formula φ_I^1 constrains the initial values of the state variables of C_1, and the formula φ_I^2 constrains the initial values of the state variables of C_2. Similarly, if φ_R^1 and φ_R^2 are the respective reaction formulas for C_1 and C_2, then the reaction formula for the product $C_1 \| C_2$ is the conjunction $\varphi_R^1 \wedge \varphi_R^2$. This again captures the intuition that in synchronous composition, state s of the composite, on input i, can react with output o updating the state to t if the values assigned by s, i, o, and t are consistent with the reaction descriptions of the two original components.

In our running example, the composition of Delay1 and Delay2 gives the component with state variables $\{x_1, x_2\}$, output variables $\{temp, out\}$, input variables $\{in\}$, initialization formula $x_1 = 0 \wedge x_2 = 0$, and reaction formula

$$temp = x_1 \ \wedge \ x_1' = in \ \wedge \ out = x_2 \ \wedge \ x_2' = temp.$$

If y is an output variable of a component C, then we can use hiding to ensure that y is no longer an output that is observable outside. The initialization formula of the resulting component $C \setminus y$ is the same as the initialization formula for C, and if φ_R is the reaction formula for C, then the reaction formula for the result is $\exists y. \varphi_R$. In our example, if we hide the intermediate output $temp$ of Delay1 $\|$ Delay2, then we get the desired composite component DoubleDelay. Its initialization formula is

$$x_1 = 0 \ \wedge \ x_2 = 0,$$

and the reaction formula is

$$\exists\, temp. \ (\ temp = x_1 \ \wedge \ x_1' = in \ \wedge \ out = x_2 \ \wedge \ x_2' = temp \ ),$$

which can be simplified to an equivalent formula

$$x_1' = in \ \wedge \ out = x_2 \ \wedge \ x_2' = x_1.$$

To obtain the transition formula for the transition system corresponding to the component DoubleDelay, we can existentially quantify the variables in and out from the above formula. The resulting transition formula is equivalent to $x_2' = x_1$.

Exercise 3.13: Consider the transition system $\mathtt{Mult}(m, n)$ described in exercise 3.1. Describe this transition system symbolically using initialization and transition formulas. ■

Exercise 3.14: Consider the description of the component Switch given as an extended-state machine in figure 2.2. Give the initialization and reaction formulas corresponding to Switch. Obtain the transition formula for the corresponding transition system in as simplified form as possible. ■

Exercise 3.15*: Let C_1 and C_2 be two compatible reactive components with reaction formulas are φ_R^1 and φ_R^2, respectively. We have argued that the reaction formula for the product $C_1 \| C_2$ is $\varphi_R^1 \wedge \varphi_R^2$. Let φ_T^1 and φ_T^2 be the transition formulas for the transition systems corresponding to C_1 and C_2, respectively. Can we conclude that the transition formula for the transition system corresponding to the product $C_1 \| C_2$ is $\varphi_T^1 \wedge \varphi_T^2$? Justify your answer. ■

3.4.2 Symbolic Breadth-First Search

Before developing the symbolic algorithm for invariant verification, let us identify the operations that we need for symbolic search.

Operations on Regions

We call a symbolically represented set of states a *region*. Given a set V of typed variables, a set of states over V is represented as a region of type `reg`. In the symbolic representation of a transition system with state variables S, the initial states are represented by a region over S, and the transitions are represented by a region over $S \cup S'$. We have already considered a specific instance of such a representation, namely, Boolean formulas over the variables. While representation using formulas is useful for understanding the symbolic algorithm, the algorithm works for any choice as long as the primitives discussed below are implemented.

The data type `reg` for regions supports the following operations:

- Given regions A and B, $\texttt{Disj}(A, B)$ returns the region that contains those states that are in either A or B.

- Given regions A and B, $\texttt{Conj}(A, B)$ returns the region that contains those states that are in both regions A and B.

- Given regions A and B, $\texttt{Diff}(A, B)$ returns the region that contains those states that are in A but not in B.

- Given a region A, $\texttt{IsEmpty}(A)$ returns 1 if the region A contains no states and 0 otherwise.

- Given a region A over V and a set $X \subseteq V$ of variables, $\texttt{Exists}(A, X)$ returns the region A projected onto over the variables $V \setminus X$. The result contains a valuation s over $V \setminus X$ precisely when there exists some valuation t over X such that the valuation over V obtained by combining s and t is in A.

- Given a region A over variables V, a list of variables $X = \{x_1, \ldots x_n\}$ in V and a list of variables $Y = \{y_1, \ldots y_n\}$ not in V, such that each variable y_j is of the same type as the corresponding variable x_j, $\texttt{Rename}(A, X, Y)$ returns the region obtained by renaming each variable x_j to y_j. Thus, the result contains a valuation t over the variables $(V \cup Y) \setminus X$ exactly when there exists some valuation s in A such that $t(y_j) = s(x_j)$, for $j = 1, \ldots n$, and $t(z) = s(z)$ for all variables z in $V \setminus X$.

Image Computation

The core of symbolic search is *image computation*: given a region A over state variables, we want to compute the region that contains all the states that can be reached from the states in A using one transition. The desired operation Post can be implemented using the operations intersection, renaming, and existential quantification as follows. Given a region A, we first conjoin it with *Trans*, a region over unprimed and primed state variables containing all the transitions. The intersection $\texttt{Conj}(A, \textit{Trans})$ is a region over $S \cup S'$ and contains all the transitions that originate in states in A. Then we project the result onto the set S' of primed state variables by existentially quantifying the variables in S. The result is the region containing states that can be reached from states in A using one transition. However, it is a region over the primed variables. Renaming each primed variable x' to x gives us the desired region $\texttt{Post}(A)$.

SYMBOLIC IMAGE COMPUTATION

Consider a transition system with state variables S and transition specification given by the region *Trans* over $S \cup S'$. Given a region A over S, the *post-image* of A, defined by

$$\texttt{Post}(A, \textit{Trans}) = \texttt{Rename}(\texttt{Exists}(\texttt{Conj}(A, \textit{Trans}), S), S', S)$$

is a region over S that contains precisely those states t for which there is a transition (s, t) for some state s in A.

Examples of Image Computation

As an example, suppose the system has a single variable x of type real, and the transition region is given by the formula $x' = 2x + 1$ (this corresponds to the assignment $x := 2x+1$). Consider the region A given by the formula $0 \leq x \leq 10$. In the first step of the image computation, we conjoin A with *Trans*, and this gives the region $0 \leq x \leq 10 \wedge x' = 2x+1$. Applying existential quantification of x and simplifying the result, we get $1 \leq x' \leq 21$. The final step renames x' to x, and we get the region $1 \leq x \leq 21$. Verify that the constraint $1 \leq x \leq 21$ indeed describes all possible values of x after executing the assignment $x := 2x + 1$, given that $0 \leq x \leq 10$ describes all possible values of x before executing the assignment.

As a second example, consider a transition system with two integer variables x and y. Suppose the transitions of the system correspond to executing the statement:

$$\texttt{if } (y > 0) \texttt{ then } x := x + 1 \texttt{ else } y := y - 1.$$

This statement translates to the following transition formula:

$$[(y > 0) \wedge (x' = x + 1) \wedge (y' = y)] \vee [(y \leq 0) \wedge (x' = x) \wedge (y' = y - 1)].$$

Consider a region A of this transition system described by the formula $2 \leq x - y \leq 5$. To compute the post-image of this region, we take its conjunction with the transition formula, and this gives:

$$[(y > 0) \wedge (x' = x + 1) \wedge (y' = y) \wedge (2 \leq x - y \leq 5)]$$
$$\vee \ [(y \leq 0) \wedge (x' = x) \wedge (y' = y - 1) \wedge (2 \leq x - y \leq 5)].$$

If we existentially quantify the variable x from this formula, then we get

$$[(y > 0) \wedge (y' = y) \wedge (2 \le x' - 1 - y \le 5)]$$
$$\vee \ [(y \le 0) \wedge (y' = y - 1) \wedge (2 \le x' - y \le 5)].$$

Now let us existentially quantify the variable y from this formula, and we get

$$[(y' > 0) \wedge (2 \le x' - 1 - y' \le 5)] \vee [(y' + 1 \le 0) \wedge (2 \le x' - y' - 1 \le 5)].$$

By renaming x' and y' to x and y, respectively, and some simplification, we obtain

$$(3 \le x - y \le 6) \wedge (y \ne 0)$$

as the post-image of the region A.

Consider the synchronous reactive component `3BitCounter` of section 2.4.1 (see figure 2.27). The component has two input variables, *inc* and *start*, and three output variables, out_0, out_1, and out_2. Let the state variables of the three latches corresponding to the three bits be x_0, x_1, and x_2. The reaction formula φ_R for the component is equivalent to:

$$out_0 = x_0 \ \wedge \ out_1 = x_1 \ \wedge \ out_2 = x_2 \ \wedge$$

$$\left[\begin{array}{l} (start = 1 \ \wedge \ x'_0 = 0 \ \wedge \ x'_1 = 0 \ \wedge \ x'_2 = 0) \ \vee \\ (start = 0 \ \wedge \ inc = 0 \ \wedge \ x'_0 = x_0 \ \wedge \ x'_1 = x_1 \ \wedge \ x'_2 = x_2) \ \vee \\ (start = 0 \ \wedge \ inc = 1 \ \wedge \ x'_0 = \neg x_0 \ \wedge \ x'_1 = x_0 \oplus x_1 \ \wedge \ x'_2 = (x_0 \wedge x_1) \oplus x_2) \end{array} \right]$$

This formula expresses the constraint that each output bit out_j is the same as the (old) state x_j, and either (1) the input *start* is high, and all the updated state bits are 0, or (2) both the inputs are low, and the state stays unchanged, or (3) the input *start* is low and *inc* is high, and counter increments. The increment condition for the counter means that the low-order bit out_0 flips, the new value of middle bit out_1 is the exclusive-or (denoted by the operator $\oplus$) of the two low-order bits, and the new value of the high-order bit out_2 is the exclusive-or of the old value of out_2 and the conjunction of the other two bits. For the corresponding transition system, the transition formula φ_T is

$$\exists \, inc, start, out_0, out_1, out_2 \,.\, \varphi_R$$

which simplifies to

$$(x'_0 = 0 \ \wedge \ x'_1 = 0 \ \wedge \ x'_2 = 0) \ \vee$$
$$(x'_0 = x_0 \ \wedge \ x'_1 = x_1 \ \wedge \ x'_2 = x_2) \ \vee$$
$$(x'_0 = \neg x_0 \ \wedge \ x'_1 = x_0 \oplus x_1 \ \wedge \ x'_2 = (x_0 \wedge x_1) \oplus x_2)$$

Now consider the region A given by the formula $x_0 = 0 \wedge x_1 = 0$; that is, the region A consists of states in which the two lower order bits are 0, but the higher order bit may be 0 or 1. To compute the image of this region, we first conjoin it with the transition formula φ_T, and this simplifies to

$$(x'_0 = 0 \ \wedge \ x'_1 = 0 \ \wedge \ x'_2 = 0) \ \vee \ (x'_1 = 0 \ \wedge \ x'_2 = x_2)$$

Input: A transition system T given by regions $Init$ for initial states
and $Trans$ for transitions, and a region φ for the property.
Output: If φ is reachable in T, return 1, else return 0.

```
reg Reach := Init;
reg New := Init;
while IsEmpty(New) = 0 do {
   if IsEmpty(Conj(New, φ)) = 0 then return 1;
   New := Diff(Post(New, Trans), Reach);
   Reach := Disj(Reach, New);
   };
return 0.
```

Figure 3.18: Symbolic Breadth-first Search Algorithm for Reachability

Existentially quantifying the old state variable x_2 and simplifying the result
give us $x'_1 = 0$. Finally, renaming the primed variables to unprimed ones
gives the region $x_1 = 0$. This correctly captures the effect of executing one
round of the three-bit counter: the set of successors of the region $\{000, 100\}$ is
$\{000, 001, 100, 101\}$.

Iterative Image Computation

Now we are ready to describe the symbolic breadth-first search algorithm. The
algorithm of figure 3.18 computes successive approximations of the set of reachable states by repeatedly applying the image computation, starting with the
initial region. The region $Reach$ stores the set of states found reachable so far,
and the region New represents the states newly found reachable. The successive
values of New capture the minimum number of transitions needed to reach a
state: in the j-th iteration of the loop, New contains precisely those states s
for which the shortest execution from an initial state leading to s involves j
transitions. Initially, both the regions $Reach$ and New are set to $Init$. If any
state in New satisfies the property φ, that is, the intersection of the regions
New and φ is non-empty, then the algorithm stops. If the region New is empty,
then the algorithm can terminate reporting that no reachable state satisfies the
property φ (that is, the negated property $\neg \varphi$ is an invariant). Otherwise, to
find the states that are reachable in one more step, the algorithm applies the
Post operator to the region New and removes those states that were already
known to be reachable using the set-difference operation on regions. The values
of the region $Reach$ in the successive iterations of the algorithm are depicted in
figure 3.19.

To illustrate how the symbolic breadth-first search algorithm works, let us revisit
the transition system corresponding to the component `RailRoadSystem2` (see
figure 3.15). Initially, $Reach = New = Init = \{aa00rr\}$. After one iteration of

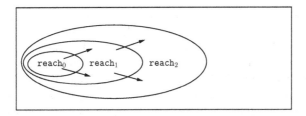

Figure 3.19: Symbolic Breadth-First Computation of Reachable States

the loop,

$$Reach \;=\; \{\, aa00rr,\, wa10gr,\, ww11rg,\, aw01rg \,\};$$
$$New \;=\; \{\, wa10gr,\, ww11rg,\, aw01rg \,\}.$$

After the second iteration of the loop,

$$New = \{\, ba10gr,\, bw11gr,\, ab01rg,\, wb11rg \,\}.$$

After the third iteration of the loop, *Reach* contains all the nine reachable states, and *New* equals $\{\, aw01rr \,\}$. After the fourth iteration, *New* becomes empty, and *Reach* stays unchanged. As a result, the algorithm stops.

If the symbolic breadth-first search algorithm terminates, then its answer is correct. If the property φ is reachable and the shortest witness contains j transitions, then after j iterations of the while-loop, the algorithm terminates. If the property is not reachable, then the algorithm can terminate only if it discovers all the reachable states after a finite number of iterations. This is guaranteed if the number of reachable states of T is finite.

Theorem 3.3 [Symbolic Breadth-first Search for Reachability] *Given a symbolic representation of a transition system T and a property φ, the symbolic breadth-first search algorithm of figure 3.18 has the following guarantees:*

1. *If the algorithm terminates, then the returned value correctly indicates whether the property φ is reachable in T.*

2. *If the property φ is reachable in the transition system T, then the algorithm terminates after j iterations of the while-loop, where j is the length of the shortest witness to the reachability of φ.*

3. *If there exists a natural number j, such that every reachable state of T is reachable by an execution with at most j transitions, then the algorithm terminates after at most j iterations of the while-loop.*

A natural choice for a symbolic representation of regions is formulas. In particular, for transition systems with only Boolean variables, we can use formulas over Boolean variables as a data type for regions. The operations such as `Disj` and `Conj` correspond to the logical connectives such as disjunction and conjunction. To take set-difference of regions A and B, we can take conjunction of A with the negation of B. Renaming corresponds to textual substitution. Finally, existential-quantifier elimination `Exists`(A, x), where x is a Boolean variable, corresponds to $A_0 \lor A_1$, where the formula A_0 is obtained from A by replacing each occurrence of the variable x with the constant 0, and the formula A_1 is obtained from A by replacing each occurrence of x with 1. All these operations individually can be implemented efficiently over formulas. However, the check `IsEmpty`(A) corresponds to checking satisfiability of the formula A, that is, whether there *exists* a 0/1 assignment to Boolean variables so that the formula A evaluates to 1. This test is computationally expensive. When A has Boolean variables and logical connectives of negation, conjunction, and disjunction but no existential quantification, this is exactly the canonical NP-complete problem known as *propositional satisfiability* or `SAT`. Furthermore, in the context of the iterative breadth-first search of our symbolic algorithm of figure 3.18, the main drawback of representing the regions *Reach* and *New* as formulas is that there is no guaranteed way to simplify the formula representing the reachable states at each step. The formulas will get more and more complex due to the operations applied in each iteration of the while-loop, and their size grows exponentially with the number of iterations. The data structure of ordered binary decision diagrams, to be discussed in the next section, offers a possible remedy.

Exercise 3.16: Consider the symbolic image computation for a transition system with two real-valued variables x and y and transition description given by the formula $x' = x + 1 \land y' = x$. Suppose the region A is described by the formula $0 \leq x \leq 4 \land y \leq 7$. Compute the formula describing the post-image of A. ∎

Exercise 3.17: Consider a transition system T with two integer variables x and y. The transitions of the system correspond to executing the statement:

$$\text{if } (x < y) \text{ then } x := x + y \text{ else } y := y + 1.$$

Write the transition formula over the variables x, y, x', and y' that captures the transition relation of the system. Consider a region A of the above transition system described by the formula $0 \leq x \leq 5$. Compute the formula describing the post-image of A. ∎

Exercise 3.18*: The symbolic breadth-first search algorithm of figure 3.18 is a *forward search* algorithm that computes the set of states reachable from the initial states by repeatedly applying the image computation operator `Post`. Define a *pre-image* computation `Pre` using the symbolic operations such as `Conj`, `Rename`, and `Exists` so that given a region A, `Pre`$(A, \textit{Trans})$ is the region over S that contains precisely those states s for which there is a transition (s, t)

for some state t in A. Develop a *backward search* algorithm for the invariant verification problem that starts with the states that violate the desired invariant and computes the set of states that can reach the violating states by repeatedly applying the pre-image computation operator `Pre`. ∎

Exercise 3.19 *: Suppose we want to modify the symbolic breadth-first search algorithm of figure 3.18 so that when it finds the property φ to be reachable, it outputs a witness execution. Which additional operations on regions will be needed for this purpose? Using these operations, modify the algorithm so that it outputs a witness execution. ∎

3.4.3 Reduced Ordered Binary Decision Diagrams *

Reduced ordered binary decision diagrams (ROBDDs) provide a compact and canonical representation for formulas over Boolean variables (or, equivalently, for Boolean functions).

Ordered Binary Decision Diagrams

Let V be a set containing k Boolean variables. A Boolean formula f over V represents a function from bool^k to bool. For a variable x in V, the following equivalence, called the *Shannon expansion* of f around the variable x, holds:

$$f \equiv (\neg x \,\wedge\, f[x \mapsto 0]) \,\vee\, (x \,\wedge\, f[x \mapsto 1]).$$

Here, the formulas $f[x \mapsto 0]$ and $f[x \mapsto 1]$ are obtained from f by substituting the variable x by the constants 0 and 1, respectively. The formulas $f[x \mapsto 0]$ and $f[x \mapsto 1]$ do not refer to the variable x and are thus Boolean functions with domain bool^{k-1}. As a result, the Shannon expansion can be used to recursively simplify a Boolean function. This suggests representing Boolean functions as decision diagrams.

A (binary) decision diagram is a directed acyclic graph with two types of vertices: *terminal* and *internal*. The terminal vertices have no outgoing edges, and are labeled with one of the Boolean constants, 0 or 1. Each internal vertex is labeled with a variable in V and has two outgoing edges: a *left* edge and a *right* edge. Every path from an internal vertex to a terminal vertex contains, for each variable x, at most one vertex labeled with x. Each vertex u represents a Boolean function $f(u)$. Given a valuation q for all the variables in V, the value of the Boolean function $f(u)$ is obtained by traversing a path starting from u as follows. Consider an internal vertex v labeled with x. If $q(x)$ is 0, then we choose the left successor of v; if $q(x)$ is 1, then we choose the right successor of v. If the path terminates in a terminal vertex labeled with 0, then the function $f(u)$ evaluates to 0 according to the assignment q; if the path terminates in a terminal vertex labeled with 1, then the function $f(u)$ evaluates to 1 according to q.

Ordered (binary) decision diagrams (OBDDs) are decision diagrams in which we choose a linear order $<$ over the variables V and require that the labels of internal vertices appear in an order that is consistent with $<$. Note that there is no requirement that every variable should appear as a vertex label along a path from the root to a terminal vertex, but simply that the sequence of vertex labels along a path from the root to a terminal vertex is monotonically increasing according to $<$. The semantics of OBDDs is defined by associating Boolean formulas with each of the vertices. The definition is formalized below.

ORDERED BINARY DECISION DIAGRAM

Let V be a finite set of Boolean variables and $<$ be a total order over V. An *ordered binary decision diagram* B over $(V, <)$ consists of:

1. *Vertices:* a finite set U of vertices that is partitioned into two sets: *internal* vertices U^I and *terminal* vertices U^T;

2. *Root:* a root vertex u^0 in U;

3. *Labeling:* a labeling function *label* that labels each internal vertex with a variable in V and each terminal vertex with a constant in $\{0, 1\}$;

4. *Left edges:* a left-child function *left* that maps each internal vertex u to a vertex $left(u)$, such that if $left(u)$ is an internal vertex, then $label(u) < label(left(u))$; and

5. *Right edges:* a right-child function *right* that maps each internal vertex u to a vertex $right(u)$, such that if $right(u)$ is an internal vertex, then $label(u) < label(right(u))$.

For such an OBDD, each vertex u has an associated Boolean formula over V: $f(u)$ equals $label(u)$ if u is a terminal vertex, and equals

$$[\neg\, label(u) \;\wedge\; f(left(u))] \;\vee\; [label(u) \;\wedge\; f(right(u))]$$

otherwise. The Boolean formula $f(B)$ associated with the OBDD B is the formula $f(u^0)$ associated with the root u^0.

A Boolean constant is represented by an OBDD that contains a single terminal vertex labeled with that constant. Figure 3.20 shows one possible OBDD for the formula $(x \wedge y) \vee (x' \wedge y')$ with the ordering $x < y < x' < y'$. A left edge is shown as an arrow ending with an empty circle, and a right edge is shown as an arrow ending with a filled circle. The OBDD of figure 3.20 is, in fact, a tree. Figure 3.21 shows a more compact OBDD for the same formula with the same ordering of variables.

Isomorphism and Equivalence

Two OBDDs B and C are *isomorphic* if the corresponding labeled graphs are isomorphic. Two OBDDs B and C are *equivalent* if the Boolean formulas $f(B)$

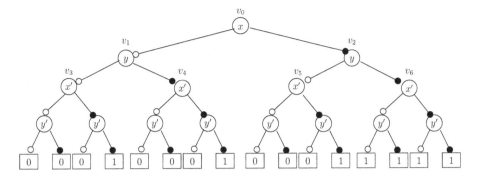

Figure 3.20: Ordered Binary Tree for $(x \wedge y) \vee (x' \wedge y')$

and $f(C)$ are equivalent. Thus, isomorphism means that the two OBDDs are structurally the same, and equivalence means that the two OBDDs are semantically the same. Clearly, isomorphic OBDDs represent the same formulas and are, thus, equivalent. The converse need not hold: the ordered binary decision diagrams of figures 3.20 and 3.21 are not isomorphic but are equivalent.

The notions of equivalence and isomorphism also extend to individual vertices. If B is an OBDD over $(V, <)$ and u is a vertex of B, then the subgraph rooted at u consisting of the vertices and edges that are reachable from u is also an OBDD over $(V, <)$. In figure 3.20, the subgraph rooted at vertex v_3 is an OBDD that represents the Boolean formula $x' \wedge y'$. Two vertices u and v of the OBDD B are isomorphic if the subgraphs rooted at u and v are isomorphic. Similarly, two vertices u and v are equivalent if the subgraphs rooted at u and v are equivalent. Since OBDDs are acyclic graphs, the notion of isomorphism can be defined by the following rules: (1) two terminal vertices u and v are isomorphic if $label(u) = label(v)$, and (2) two internal vertices u and v are isomorphic if $label(u) = label(v)$ and the left successors $left(u)$ and $left(v)$ are isomorphic and the right successors $right(u)$ and $right(v)$ are isomorphic. In figure 3.20, all terminal vertices with label 0 are isomorphic to one another. The subgraphs rooted at vertices v_3, v_4, and v_5 are isomorphic. In contrast, the vertices v_5 and v_6 are not isomorphic to each other.

Reduced Ordered Binary Decision Diagrams

A *reduced OBDD* (ROBDD) is obtained from an OBDD by repeatedly applying the following two steps:

1. Merge isomorphic vertices into one.

2. Eliminate internal vertices with identical left and right children.

Each step reduces the number of vertices while preserving equivalence. For instance, consider the OBDD of figure 3.20. Since vertices v_3 and v_4 are iso-

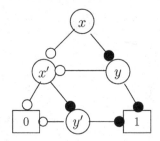

Figure 3.21: Reduced Ordered Binary Decision Diagram for $(x \wedge y) \vee (x' \wedge y')$

morphic, we can delete one of them, say v_4, along with the subtree rooted at v_4 and redirect the right edge of the vertex v_1 to v_3. Now, since both edges of the vertex v_1 point to v_3, we can delete the vertex v_1, redirecting the left edge of the root v_0 to v_3. Continuing in this manner, we obtain the ROBDD of figure 3.21. It turns out that the above transformations are sufficient to obtain a *canonical* form: the final ROBDD is the same irrespective of the specific order in which the reductions are applied, and if we start with two non-isomorphic but equivalent OBDDs, we get the same final ROBDD.

REDUCED ORDERED BINARY DECISION DIAGRAM

A ROBDD over a totally ordered set $(V, <)$ of Boolean variables is an ordered binary decision diagram $B = (U, u^0, label, left, right)$ over $(V, <)$ such that:

1. there are no two distinct vertices u and v in U with $label(u) = label(v)$ and $left(u) = left(v)$ and $right(u) = right(v)$, and

2. for every internal vertex u, the two children $left(u)$ and $right(u)$ are distinct vertices.

Direct Construction of ROBDDs: An Example

Let us consider the Boolean formula

$$\varphi_0 : \ (x \vee \neg y) \ \wedge \ (y \vee z)$$

and suppose the variable ordering is $x < y < z$. Instead of first building an OBDD and then reducing it using the two reduction rules, let us try to build a ROBDD directly. The first step is illustrated in figure 3.22(a), which shows the root vertex v_0 labeled with the variable x. Note that the root vertex must be labeled with x because x is the first variable in the chosen ordering, and the value of the formula φ_0 depends on the value of x. The left successor of the vertex v_0 should be a vertex, say v_1, that represents the formula $\varphi_1 = \varphi_0[x \mapsto 0]$, which simplifies to the formula $\neg y \wedge z$, and the right successor of the vertex v_0

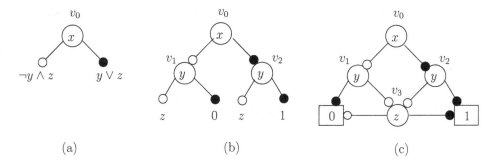

Figure 3.22: Building ROBDD for $(x \lor \neg y) \land (y \lor z)$

should be a vertex, say v_2, that represents the formula $\varphi_2 = \varphi_0[x \mapsto 1]$, which simplifies to the formula $y \lor z$.

The vertex v_1 corresponds to the formula $\varphi_1 : \neg y \land z$, and since the variable y is the next variable in the chosen ordering, the vertex v_1 is labeled with y. Its left successor should be a vertex, say v_3, that represents the formula $\varphi_1[y \mapsto 0]$, which is equivalent to the formula z, and the right successor of the vertex v_1 should be a vertex that represents the formula $\varphi_1[y \mapsto 1]$, which is equivalent to the constant 0, and hence must be a terminal vertex.

The ROBDD corresponding to the vertex v_2 that represents the formula $\varphi_2 : y \lor z$ is developed using a similar logic (see figure 3.22(b)). The vertex v_2 is labeled with the variable y. Its left successor should be a vertex that represents the formula $\varphi_2[y \mapsto 0]$, which is equivalent to the formula z. Since the vertex v_3 already represents the formula z and the reduction rules require as much sharing as possible, the left child of the vertex v_2 must be v_3. The right successor of the vertex v_2 should be a vertex that represents the formula $\varphi_2[y \mapsto 1]$, which is equivalent to the constant 1 and, hence, must be a terminal vertex.

Finally, the vertex v_3 corresponding to the formula z has label z, the terminal vertex 0 as its left successor, and the terminal vertex 1 as its right successor. This completes the desired ROBDD for the formula φ_0 as shown in figure 3.22(c).

Properties of ROBDDs

The next theorem asserts the basic facts about representing Boolean functions using ROBDDs. Every Boolean function has a unique, up to isomorphism, representation as a ROBDD. Furthermore, the ROBDD of a Boolean function has the least number of vertices among all OBDDs for the same function once we fix the variable ordering.

Theorem 3.4 [Existence, Uniqueness, and Minimality of ROBDDs] *Let V be a set of variables and $<$ be a total order over V.*

1. If f is a Boolean formula over V, then there exists a ROBDD B over $(V, <)$ such that $f(B)$ and f are equivalent.

2. If B and C be two ROBDDs over $(V, <)$, then they are equivalent if and only if they are isomorphic.

3. If B is a ROBDD over $(V, <)$ and C is an OBDD over $(V, <)$, such that the two are equivalent, then C contains at least as many vertices as B.

∎

Checking equivalence of two ROBDDs with the same variable ordering corresponds to checking isomorphism and, hence, can be performed in time linear in the number of vertices. The Boolean constant 0 is represented by a ROBDD with a single terminal vertex labeled with 0, and the Boolean constant 1 is represented by a ROBDD with a single terminal vertex labeled with 1. A Boolean formula represented by a ROBDD B is satisfiable if and only if the root of B is not a terminal vertex labeled with 0. A Boolean formula represented by a ROBDD B is valid if and only if the root of B is a terminal vertex labeled with 1. Thus, checking satisfiability or validity of Boolean formulas is particularly easy if we use ROBDD representation.

The size of the ROBDD representation of a Boolean formula may be exponential in the number of variables. The size of the ROBDD representing a given formula depends on the choice of the ordering of variables. Consider the formula $(x = y) \land (x' = y')$. Figure 3.23 shows two ROBDDs for two different orderings of the variables. This example illustrates that the ordering can influence the size dramatically: one ordering may result in a ROBDD whose size is linear in the number of variables, whereas another ordering may result in a ROBDD whose size is exponential in the number of variables. While choosing an optimal ordering of variables can lead to exponential saving, computing the optimal ordering is a computationally hard problem.

There are Boolean functions whose ROBDD representation does not depend on the chosen ordering, and the ROBDD representation of some functions is exponential in the number of variables, irrespective of the ordering. An example of the former variety is the parity function, whereas that of the latter variety is the multiplication function:

- **Parity.** Given a valuation s to a set V of Boolean variables, the parity function returns 1 precisely when the number of variables x with $s(x) = 1$ is even. If V contains k variables, then irrespective of the chosen ordering, the ROBDD for the parity function contains $2k + 1$ vertices.

- **Multiplication.** Consider the set of variables $\{x_0, \ldots, x_{k-1}, y_0, \ldots, y_{k-1}\}$, and for $0 \leq j < 2k$, let $Mult_j$ denote the Boolean function that denotes the jth bit of the product of the two k-bit inputs, one encoded by the x-bits and another encoded by the y-bits. For every ordering $<$ of the variables, the total number of vertices in the ROBDDs representing all the functions

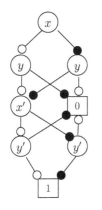
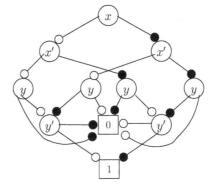

Ordering: $x < y < x' < y'$ *Ordering:* $x < x' < y < y'$

Figure 3.23: Two ROBDDs for $(x = y) \wedge (x' = y')$

$Mult_j$ is guaranteed to grow exponentially with k. More precisely, the following *lower bound* result has been established: there exists an index $0 \leq j < 2k$ such that the ROBDD for $Mult_j$ has at least $2^{k/8}$ vertices.

Shared Data Structure for ROBDDs

Let us turn our attention to implementing regions as ROBDDs. Every vertex of a ROBDD is a ROBDD rooted at that vertex. This suggests that a ROBDD can be represented by an index to a global data structure that stores vertices of all the ROBDDs such that no two vertices are isomorphic. There are two significant advantages to this scheme, as opposed to maintaining each ROBDD as an individual data structure. First, checking isomorphism, and hence equivalence, corresponds to comparing indices and does not require traversal of the ROBDDs. Second, two non-isomorphic ROBDDS may have isomorphic subgraphs and, hence, can share vertices.

Let V be an ordered set of k Boolean variables. The type of ROBDDs is **bdd**, which is either a Boolean constant (denoting a terminal vertex) or a pointer to an entry in the global data structure *BDDPool*. The type of *BDDPool* is **set(bddnode)**, and it stores the (internal) vertices of ROBDDs. An internal vertex records a variable label, which ranges over the type **nat**$[1, k]$ containing the numbers $\{1, \ldots k\}$, and a left and a right pointer, each of which has type **bdd**. Thus, the vertices of ROBDDs have type **bddnode**, which equals **nat**$[1, k] \times$ **bdd** $\times$ **bdd**. The type **bddnode** supports the following operations:

- The operation **Label**(u), for an internal vertex u, returns the first component of u, which is the number of the variable labeling u.

- The operation **Left**(u), for an internal vertex u, returns the second component of u, which is either a Boolean constant or a pointer to the left

```
function AddVertex
```
Input: Variable label j in $\mathbf{nat}[1,k]$, ROBDDs B_0, B_1 of type $\mathbf{bdd}$.
Output: ROBDD B such that $f(B)$ is equivalent to $(\neg x_j \wedge f(B_0)) \vee$
$\quad (x_j \wedge f(B_1))$.

```
if B₀ = B₁ then return B₀;
if Contains((j, B₀, B₁), BDDPool) = 0 then
   Insert((j, B₀, B₁), BDDPool);
return Index((j, B₀, B₁)).
```

Figure 3.24: Creating ROBDD Vertices

successor of u.

- The operation $\mathtt{Right}(u)$, for an internal vertex u, returns the third component of u, which is either a Boolean constant or a pointer to the right successor of u.

The type $\mathbf{set}(\mathbf{bddnode})$, besides usual operations such as $\mathtt{Insert}$ and $\mathtt{Contains}$, also supports

- For an internal vertex u in $BDDPool$, $\mathtt{Index}(u)$ returns a pointer to u.

- For a pointer B, the operation $BDDPool[B]$ returns the vertex that B points to.

For such a representation, given a pointer B of type $\mathbf{bdd}$, we write $f(B)$ to denote the Boolean function associated with the ROBDD that B points to. To avoid duplication of isomorphic nodes while manipulating ROBDDs, it is necessary that new vertices are created using the function $AddVertex$ of figure 3.24. If no two vertices in the global set $BDDPool$ are isomorphic before an invocation of the function $AddVertex$, then even after the invocation, no two vertices in $BDDPool$ are isomorphic.

As an illustrative example, let us examine the snapshot of the global data structure $BDDPool$ shown in figure 3.25. Each row shows an internal vertex stored in this data structure. For example, the ROBDD B_0 points to a vertex labeled with variable x_4 whose left successor is the terminal vertex 0 and right successor is the terminal vertex 1, and the ROBDD B_4 points to a vertex labeled with variable x_1 whose left successor is the terminal vertex 0 and right successor is the ROBDD B_2. The Boolean functions corresponding to each of the vertices are listed below:

$$
\begin{aligned}
f(B_0) &= x_4, \\
f(B_1) &= x_2 \vee x_4, \\
f(B_2) &= \neg x_3 \vee x_4, \\
f(B_3) &= x_1 \wedge (x_2 \vee x_4),
\end{aligned}
$$

Index	Label	Left	Right
B_0	4	0	1
B_1	2	B_0	1
B_2	3	1	B_0
B_3	1	0	B_1
B_4	1	0	B_2

Figure 3.25: Illustrative Snapshot of the Data Structure *BDDPool*

$$f(B_4) \;\; = \;\; x_1 \,\wedge\, (\neg x_3 \vee x_4).$$

Observe that the ROBDDs B_3 and B_4 share the ROBDD B_0.

Operations on ROBDDs

To construct a ROBDD representation of a given Boolean formula and implement the primitives of the symbolic reachability algorithm, we need a way to compute conjunctions and disjunctions of ROBDDs. We give a recursive algorithm for obtaining conjunction of ROBDDs. The algorithm is shown in figure 3.26.

Consider two ROBDDs, B and B', and suppose we wish to compute the conjunction $f(B) \wedge f(B')$. If one of them is a Boolean constant, then the result can be determined immediately. For instance, if B is the terminal constant 0, then the conjunction is also the terminal constant 0. If B is the terminal constant 1, then the conjunction is equivalent to $f(B')$, and thus the algorithm can return B'. Also, note that when both the ROBDDs are the same, we can use the fact $f \wedge f$ always equals f, and thus the result coincides with the input argument.

The interesting case is when both ROBDDs are pointers to distinct internal vertices, say u and u', respectively. Let j be the minimum of the indices labeling u and u'. Then x_j is the least variable that the function $f(u) \wedge f(u')$ can depend on. The label of the root of the conjunction is j, the left successor is the ROBDD for $(f(u) \wedge f(u'))[x_j \mapsto 0]$, and the right successor is the ROBDD for $(f(u) \wedge f(u'))[x_j \mapsto 1]$. Let us consider the left successor. Observe the equivalence

$$(f(u) \wedge f(u'))[x_j \mapsto 0] \;\equiv\; f(u)[x_j \mapsto 0] \,\wedge\, f(u')[x_j \mapsto 0].$$

If u is labeled with j, the ROBDD for $f(u)[x_j \mapsto 0]$ is the left successor of u. If the label of u exceeds j, then the function $f(u)$ does not depend on x_j, and the ROBDD for $f(u)[x_j \mapsto 0]$ is u itself. The ROBDD for $f(u')[x_j \mapsto 0]$ is computed similarly, and then the function *Conj* is applied recursively to compute the conjunction according to the expression above.

Let us apply this scheme to compute the conjunction of the ROBDDs B_3 and B_4 from figure 3.25. The vertex corresponding to B_3 has label x_1, left successor 0, and right successor B_1, while the vertex corresponding to B_4 has label x_1, left successor 0, and right successor B_2. As a result,

$$Conj(B_3, B_4) \ = \ AddVertex(1, Conj(0,0), Conj(B_1, B_2)).$$

The call $Conj(0,0)$ returns 0 using the rule for the constant ROBDDs. To compute the conjunction of B_1 and B_2, the algorithm examines the corresponding vertices: the vertex corresponding to B_1 has label x_2, left successor B_0, and right successor 1, while the vertex corresponding to B_2 has a higher label x_3. This leads to:

$$Conj(B_1, B_2) \ = \ AddVertex(2, Conj(B_0, B_2), Conj(1, B_2)).$$

This generates two recursive calls to $Conj$ again: the second call $Conj(1, B_2)$ returns immediately with the answer B_2 using rules for simplification when one of the arguments is a constant. The first call $Conj(B_0, B_2)$ requires examination of the corresponding vertices: the vertex corresponding to B_0 has label x_4, while the vertex corresponding to B_2 has label x_3, left successor 1, and right successor B_0. This leads to:

$$Conj(B_0, B_2) \ = \ AddVertex(3, Conj(B_0, 1), Conj(B_0, B_0)).$$

In this case, both the recursive calls to `Conj` return immediately: $Conj(B_0, 1)$ returns B_0 and $Conj(B_0, B_0)$ also returns B_0 using the rule for the conjunction of identical ROBDDs. As a result, a call is made to $AddVertex(3, B_0, B_0)$. This does not create a new vertex since the reduction rules do not allow both left and right successors to be the same. The call $AddVertex(3, B_0, B_0)$ simply returns B_0:

$$Conj(B_0, B_2) \ = \ AddVertex(3, B_0, B_0) \ = \ B_0.$$

Now, $Conj(B_1, B_2)$ calls $AddVertex(2, B_0, B_2)$. The data structure $BDDPool$ does not contain a vertex with label 2, left successor B_0, and right successor B_2. As a result, $AddVertex$ will create a new entry $(2, B_0, B_2)$, with the index B_5:

$$Conj(B_1, B_2) \ = \ AddVertex(2, B_0, B_2) \ = \ B_5.$$

Finally, $Conj(B_3, B_4)$ calls $AddVertex(1, 0, B_5)$. Again, $BDDPool$ does not contain such a vertex, so a new entry, indexed by B_6, is created and is the desired result, namely, the ROBDD representation of the conjunction of the functions represented by B_3 and B_4:

$$Conj(B_3, B_4) \ = \ AddVertex(1, 0, B_5) \ = \ B_6.$$

Avoiding Recomputation

The recursive algorithm described so far may call the function $Conj$ repeatedly with the same two arguments. To avoid unnecessary computation, a table is

Input: bdd B, B'.
Output: bdd B'' such that $f(B'')$ is equivalent to $f(B) \wedge f(B')$.

table$[(\text{bdd} \times \text{bdd}) \times \text{bdd}]$ $Done$ $=$ EmptyTable

return $Conj(B, B')$.

bdd $Conj$ (bdd B, B')
   bddnode u, u'; bdd $B'', B_0, B_1, B_0', B_1'$; nat$[1, k]$ j, j'

  if $(B = 0 \vee B' = 1)$ then return B;
  if $(B = 1 \vee B' = 0)$ then return B';
  if $B = B'$ then return B;
  if $Done[(B, B')] \neq \bot$ then return $Done[(B, B')]$;
  if $Done[(B', B)] \neq \bot$ then return $Done[(B', B)]$;
  $u := BDDPool[B]$; $u' := BDDPool[B']$;
  $j := $ Label(u); $B_0 := $ Left(u); $B_1 := $ Right(u);
  $j' := $ Label(u'); $B_0' := $ Left(u'); $B_1' := $ Right(u');
  if $j = j'$ then $B'' := AddVertex(j, Conj(B_0, B_0'), Conj(B_1, B_1'))$;
  if $j < j'$ then $B'' := AddVertex\ (j, Conj(B_0, B'), Conj(B_1, B'))$;
  if $j > j'$ then $B'' := AddVertex\ (j', Conj(B, B_0'), Conj(B, B_1'))$;
  $Done[(B, B')] := B''$;
  return B''.

Figure 3.26: Algorithm for Taking Conjunction of ROBDDs

used to store the arguments and the corresponding result of each invocation of *Conj*. When *Conj* is invoked with input arguments B and B', it first consults the table to check whether the conjunction of $f(B)$ and $f(B')$ was previously computed. The actual recursive computation is performed only the first time, and the result is stored into the table.

A *table* data structure stores values that are indexed by keys. If the type of values stored is **value** and the type of the indexing keys is **key**, then the type of the table is table$[\text{key} \times \text{value}]$. This data type supports the retrieval and update operations like arrays: $D[k]$ is the value stored in the table D with the key k, and the assignment $D[k] := m$ updates the value stored in D for the key k. The constant table EmptyTable has the default value $\bot$ stored with every key. Tables can be implemented as arrays or hash-tables. The table used by the algorithm uses a pair of ROBDDs as a key and stores ROBDDs as values.

Let us analyze the time complexity of the algorithm of figure 3.26. Suppose the ROBDD pointed to by B has n vertices and the ROBDD pointed to by B' has n' vertices. Let us assume that the implementation of the set *BDDPool* supports constant time membership tests and insertions and the table *Done* supports constant-time creation, access, and update. Then within each invocation of *Conj*, all the steps, apart from the recursive calls, are performed in constant time. Thus, the time complexity of the algorithm is the same, within a constant factor, of the total number of invocations of *Conj*. For any pair of vertices, the function *Conj* produces two recursive calls only the first time *Conj* is invoked with this pair as input and zero recursive calls during the subsequent invocations. This gives an overall time-complexity of $O(n \cdot n')$.

Theorem 3.5 [ROBDD Conjunction] *Given two ROBDDs, B and B, the algorithm of figure 3.26 correctly computes the ROBDD for $f(B) \wedge f(B')$. If the ROBDD pointed to by B has n vertices and the ROBDD pointed to by B' has n' vertices, then the time complexity of the algorithm is $O(n \cdot n')$.* ∎

Similar algorithms can be developed to implement other operations such as `Disj` (logical disjunction), `Diff` (set difference), and `Exists` (existential quantification).

Symbolic Search Using ROBDDs

We now have all the machinery to implement the symbolic search algorithm using ROBDDs as a representation for regions. We have already discussed how to construct symbolic representation of transition systems as initialization and transition formulas φ_I and φ_T. If all the variables in the source description are Boolean variables, then the formulas φ_I and φ_T are Boolean formulas built using logical connectives and existential quantification. We can build ROBDDs corresponding to these formulas using the operations we have discussed.

If the formula is an atomic formula of the form $x = 1$, then the corresponding ROBDD is obtained by calling the *AddVertex* function: if the position of the variable x in the variable ordering is j, then the desired ROBDD is simply *AddVertex*$(j, 1, 0)$. If the formula f is of the form $f_1 \wedge f_2$, then we first build ROBDDs B_1 and B_2 corresponding to the (simpler) formulas f_1 and f_2, respectively, and then the ROBDD for f is obtained by invoking *Conj*(B_1, B_2). The ROBDD operations *Disj*, *Diff*, and *Exists* are used to process the corresponding operations of disjunction, negation, and existential quantification in the formulas. We can define the function *FormulaToBdd* that maps Boolean formulas to ROBDDs as follows:

$$
\begin{aligned}
FormulaToBdd(x_j = 1) &= AddVertex(j, 0, 1) \\
FormulaToBdd(x_j = 0) &= AddVertex(j, 1, 0) \\
FormulaToBdd(f_1 \wedge f_2) &= Conj(FormulaToBdd(f_1), FormulaToBdd(f_2)) \\
FormulaToBdd(f_1 \vee f_2) &= Disj(FormulaToBdd(f_1), FormulaToBdd(f_2)) \\
FormulaToBdd(\neg f) &= Diff(1, FormulaToBdd(f)) \\
FormulaToBdd(\exists X. f) &= Exists(FormulaToBdd(f), X)
\end{aligned}
$$

For symbolic invariant verification, given a property φ, we also need to build the ROBDD for the formula φ. Then we can use the algorithm of figure 3.18, where every region is a pointer into the global data structure storing ROBDD vertices.

While ROBDDs can be used directly when all system variables are of type `bool`, they can also be used for analysis of finite-state systems with variables of enumerated or other finite types by encoding a finite type using a sequence of Boolean variables. For example, consider the state variable *mode* of the `Train` component that takes three possible values `away`, `wait`, and `bridge` (see figure 3.4). We can encode the variable *mode* using two Boolean variables $mode_0$ and $mode_1$, using values 00, 01, and 10 to encode the three possibilities for *mode*. Expressions of the form *mode* = `away` are replaced by $mode_0 = 0 \land mode_1 = 0$; expressions of the form *mode* = `wait` are replaced by $mode_0 = 0 \land mode_1 = 1$; and expressions of the form *mode* = `bridge` are replaced by $mode_0 = 1 \land mode_1 = 0$.

The ROBDD representation of a Boolean formula can be exponential in the number of variables and is sensitive to the ordering of variables. Given a system with Boolean state variables S, to build the representation of the initialization and transition formulas, we need to choose an ordering $<$ of the variables in $S \cup S'$. Recall that one of the steps in the image computation is to rename all the primed variables to unprimed variables. This renaming step can be implemented by renaming the labels of the internal vertices of the ROBDD if the ordering of the primed variables is consistent with the ordering of the corresponding unprimed variables. This gives us our first rule for choosing $<$: for all variables $x, y \in S$, $x < y$ if and only if $x' < y'$. Another commonly used rule for choosing the ordering stipulates that a variable should appear only after all the variables it depends on. For example, when the update is specified using task graphs, if a task A writes a state variable x, then x' should appear after the variables read by the task A as well as the tasks preceding A according to the precedence constraints. Finally, the variables that are related to each other should be clustered together. In particular, instead of ordering all the primed variables after all the unprimed variables, we can try to minimize the distance between a primed variable and the unprimed variables it depends on.

The practical tools for analyzing systems using ROBDDs employ a wide array of techniques to combat the growth in the ROBDD size with the number of variables. As a result, symbolic invariant verification using ROBDDs has had significant success in analyzing industrial-scale hardware designs and embedded controllers. Yet it is not a *magic bullet*, and the performance of ROBDD-based tools remains unpredictable: sometimes they can reveal hitherto unknown bugs in complex designs, and sometimes the breadth-first search algorithm can finish only a few iterations before the number of ROBDD vertices created becomes too large compared to the available memory.

Exercise 3.20: Consider the Boolean formula

$$(x \lor y) \land (\neg x \lor z) \land (\neg y \lor \neg z).$$

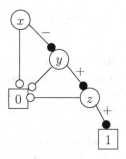

Figure 3.27: A Decision Graph with Complement Edges

Draw the ROBDD for this formula with respect to the variable ordering $x < y < z$. ∎

Exercise 3.21: Consider the Boolean formula

$$(x_1 \wedge x_2 \wedge x_3) \vee (\neg x_2 \wedge x_4) \vee (\neg x_3 \wedge x_4).$$

Choose a variable ordering for the variables $\{x_1, x_2, x_3, x_4\}$ and draw the resulting ROBDD. Can you reduce the size of the ROBDD by reordering the variables? ∎

Exercise 3.22: Let V be the set $\{x_0, x_1, y_0, y_1, z_0, z_1, c\}$. Choose an appropriate ordering of the variables and construct the ROBDD for the requirement that the output $z_1 z_0$, together with the carry bit c, is the sum of the inputs $x_1 x_0$ and $y_1 y_0$. ∎

Exercise 3.23 *: Give an algorithm for computing the existential quantification for ROBDDs: given a ROBDD B and a set X of variables, *Exists*(B, X) should return the ROBDD for the formula $\exists X. f(B)$. ∎

Exercise 3.24 *: An ordered binary decision diagram with *complement edges* (COBDD) is similar to an ordered binary decision diagram B with an additional component that classifies each right edge as positive $+$ or negative $-$. The function $f(u)$ for an internal vertex u is redefined so that $f(u)$ equals $(\neg label(u) \wedge f(left(u))) \vee (label(u) \wedge f(right(u)))$ if the right edge of u is positive and $(\neg label(u) \wedge f(left(u))) \vee (label(u) \wedge \neg f(right(u)))$ if the right edge of u is negative. Thus, when the right edge is negative, we negate the function associated with the right successor. For instance, in figure 3.27, the vertex labeled with y represents the function $y \wedge z$ while the root represents the function $(x \wedge \neg(y \wedge z))$.

(1) Is there a function with a COBDD representation that is smaller than its ROBDD representation for the same variable ordering? (2) Can we define a notion of *reduced* ordered binary decision diagrams with complement edges (RCOBDD) as a subclass of COBDDs such that every boolean function has a unique representation? ∎

Bibliographic Notes

The concept of invariants and inductive invariants was introduced in 1960s in early papers to formalize the notion of program correctness [Hoa69]. For an introduction to principles and tools for program verification, we refer the reader to [Lam02] and [BM07]. There have been prominent successes of software verification in industrial projects recently; see [BLR11, BBC$^+$10, IBG$^+$11, Hol13] as illustrative examples.

Efficient on-the-fly enumerative search for invariant verification was developed in the 1980s and forms the core analysis engine in the model checker SPIN [Hol04, Hol97] (see also the model checker MURPHI [Dil96]).

Bryant introduced ROBDDs as an efficient representation on Boolean functions [Bry86]. Symbolic search using ROBDDs was first introduced in the model checker SMV and was instrumental in the success of verification tools in analyzing hardware protocols [BCD$^+$92, McM93] (see also the model checkers VIS and NUSMV and the associated optimized implementations of ROBDD operations [BHSV$^+$96, CCGR00]).

Many of the illustrative examples in this chapter are borrowed from the draft textbook on *Computer Aided Verification* [AH99a].

We have briefly mentioned concepts in computability theory such as decidability and NP-completeness. For a comprehensive introduction to these topics, see [Sip13].

4

Asynchronous Model

We now shift our focus to the *asynchronous* model of computation that does not require concurrent activities to execute in lock-step. Such models naturally capture multi-processor machines and networked distributed platforms. In this chapter, we first formalize this model of computation and then study how to design protocols to achieve coordination necessary to solve computing problems in the presence of asynchrony.

4.1 Asynchronous Processes

Like a synchronous reactive component, an asynchronous process interacts with other processes via inputs and outputs and maintains an internal state. However, the execution does not proceed in rounds, and the speeds at which different processes execute are *independent*. Within a process, the reception of inputs is decoupled from the production of outputs, and this corresponds to the assumption that any internal computation takes an unknown but nonzero amount of time.

As an example, consider the `Buffer` process shown in figure 4.1 that models the asynchronous version of the synchronous reactive component `Delay` of figure 2.1. The input and output variables of a process are called *channels*. The process `Buffer` has a Boolean input channel *in* and a Boolean output channel *out*. The internal state of the process `Buffer` is a buffer of size 1, which can either be empty or contain a Boolean value. This is modeled by the variable x that ranges over the enumerated type $\{\texttt{null}, 0, 1\}$. Initially, the buffer is empty. The key difference between the synchronous component `Delay` and the asynchronous process `Buffer` lies in the specifications of their dynamics. The process `Buffer` has two possible types of actions. It can process an input value available in the input channel *in* by copying it into its buffer. Alternatively, if the buffer is non-empty, then the process can output the buffer state by writing it to the output channel *out* and then reset the buffer to empty. Each type of action is specified using a task, and *in one step, only one of the tasks is executed*.

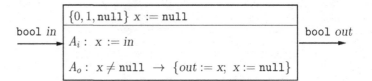

Figure 4.1: Asynchronous Process `Buffer`

4.1.1 States, Inputs, and Outputs

In general, an asynchronous process P has a set I of typed input channels, a set O of typed output channels, and a set S of typed state variables. All these three sets are finite and disjoint from one another so that there are no name conflicts.

As in the case of synchronous reactive components, a state of a process P is a valuation over the set S of its state variables, and the set of its states is the set Q_S of all possible valuations over S. The initialization *Init* assigns initial values to all the state variables in S. As before, we allow multiple initial values to capture situations where initial values are only partially known. A state q is called an initial state if, for every state variable x, the value $q(x)$ is consistent with the initialization of the variable x. The set of all initial states is denoted $[\![Init]\!]$.

In the asynchronous model of computation, when there are multiple input channels, the arrival of input values on different channels is not synchronized. Hence, an input of a process consists of a single input channel x along with a value v that belongs to the type of x. We denote such an input by $x\,?\,v$. Such an input can be interpreted as *receiving* the value v on the input channel x.

The modeling of outputs is symmetric. When there are multiple output channels, in one step, a process can produce a value for only one of the output channels. An output of a process consists of a single output channel y along with a type-consistent value v. We denote such an output by $y\,!\,v$. Such an output can be interpreted as *sending* the value v on the output channel y.

For the process `Buffer`, the set S of state variables is $\{x\}$, the set I of input variables is $\{in\}$, the set O of output variables is $\{out\}$, the set of states is $\{0, 1, \mathtt{null}\}$, the set of initial states is $\{\mathtt{null}\}$, the set of inputs is $\{in\,?\,0,\ in\,?\,1\}$, and the set of outputs is $\{out\,!\,0,\ out\,!\,1\}$.

4.1.2 Input, Output, and Internal Actions

For synchronous reactive components, execution during a round is specified using a set of tasks, where the execution of a single task captures an atomic unit of computation. For asynchronous processes, we also specify its computation using a set of tasks. As before, the update description of a task assigns values to variables in its write-set using the values of variables in its read-set and is usually given as a straight-line code consisting of conditional and assignment statements. In contrast to the synchronous case, during one step, instead of executing all the tasks, only one task is executed. To indicate whether a task is ready to be executed, we explicitly associate a *guard* condition with each task. This condition is given as a Boolean formula over the state variables, and the task is enabled in a given state if the state satisfies this formula. If multiple tasks are enabled, then one of them is chosen nondeterministically for execution. Precedence constraints among tasks are no longer meaningful since there is no need to order tasks within a round. In the synchronous model, a careful specification of the subset of state variables that a task reads and writes is necessary to identify potential write-conflicts, and we require that tasks with write-conflicts are ordered by precedence constraints for scheduling within a single round. This is not relevant in the asynchronous model, and we assume that each task reads and writes all the state variables. To ensure that a process either receives a single input value or sends a single output value in a step, we require that each task can either read at most one input channel or write at most one output channel.

Input Tasks

Processing of an input is called an *input action*. During an input action, the process can only update its state and does not produce outputs. Input actions are specified using *input tasks*, each of which is associated with one input channel. The description of an input task A associated with an input channel x is given as $Guard \rightarrow Update$, where $Guard$ gives the condition under which this task is willing to process inputs on the channel x and $Update$ describes how the task updates state variables based on the old values of the state variables together with the input value received on the channel x. Semantically, $Guard$ defines a set $[\![Guard]\!]$ of valuations over S, and $Update$ defines a relation $[\![Update]\!]$ from valuations over the read-set $S \cup \{x\}$ to valuations over the write-set S. An input task A is said to be enabled in a state s if the state s satisfies the guard condition $Guard$. Such a task defines the set of input actions of the form $s \xrightarrow{x\,?\,v} t$ such that the state s satisfies the guard $Guard$ and the state t can be obtained by executing the description $Update$ in state s given the value v for the input channel x, that is, if $s \in [\![Guard]\!]$ and $(s[x \mapsto v], t) \in [\![Update]\!]$.

For the process `Buffer` of figure 4.1, there is a single input task A_i that reads the input channel *in*. The task is always enabled, meaning that the process is always willing to accept an input on the channel *in*. In such a case, the guard equals the Boolean constant 1 and is omitted from the description. The task updates the state variable x using the assignment $x := in$. This task leads to six input actions: for each state $s \in \{0, 1, \mathtt{null}\}$, $s \xrightarrow{in\,?\,0} 0$ and $s \xrightarrow{in\,?\,1} 1$. Note that if the process is supplied an input value when the buffer is non-empty, then the old state is lost.

Usually each input channel x has one input task associated with it. If no input

$$\boxed{\begin{array}{l} \texttt{nat } x := 0;\ y := 0 \\ \hline A_x :\ x := x+1 \\ \hline A_y :\ y := y+1 \end{array}}$$

Figure 4.2: Asynchronous Process `AsyncInc`

task is associated with a channel, then the process cannot receive any inputs on this channel. We can associate multiple tasks with the same channel to specify different ways of processing input values on this channel. The set of all input tasks associated with an input channel x is denoted $\mathcal{A}_x$. In our example, $\mathcal{A}_{in} = \{A_i\}$.

Output Tasks

Producing an output is called an *output action*. Output actions are specified using *output tasks*, where each output task is associated with one output channel y. An output task A associated with an output channel y is described using a guard condition *Guard* that specifies the set of states in which the output task is ready to be executed and an update description *Update* that specifies how the task updates the state variables and the output value for y based on the values of the state variables it reads. Thus, for such a task, $[\![Update]\!]$ is a relation from valuations over the set S of state variables to valuations over the set $S \cup \{y\}$. Given a state s that satisfies the guard condition *Guard*, we can execute the update description *Update* to compute the new values of the state variables resulting in a state t, along with a value v to be issued on the output channel y. Thus, such a task defines the set of output actions $s \xrightarrow{y!v} t$, such that $s \in [\![Guard]\!]$ and $(s, t[y \mapsto v]) \in [\![Update]\!]$. As in the case of input tasks, multiple output tasks may be associated with the same channel, and the set of all tasks associated with an output channel y is denoted $\mathcal{A}_y$.

For the process `Buffer` of figure 4.1, there is a single output task A_o that produces an output on the channel *out*. The guard for this task is the condition $x \neq \texttt{null}$: the output task is enabled only when the buffer contains a non-null value. The update is described by the sequence of assignments $out := x;\ x := \texttt{null}$. This leads to the following two output actions: $0 \xrightarrow{out!0} \texttt{null}$ and $1 \xrightarrow{out!1} \texttt{null}$.

Internal Tasks

As a second example, consider the process `AsyncInc` of figure 4.2. The process does not have any input or output channels. It has two state variables x and y, both of type `nat` and initialized to 0. Since the process has no input and output channels, there are no input or output tasks.

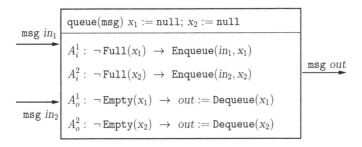

Figure 4.3: Asynchronous Process `Merge`

The internal computation of a process is described using *internal actions*. Such actions neither process inputs nor produce outputs but update internal state and are described using *internal tasks*. An internal task A has a Boolean guard condition *Guard* that describes the states in which the task is enabled and an update description *Update* that specifies how the task updates the state variables based on their old values. Given a state s, we evaluate the guard *Guard* to check whether the task is ready to be executed and, if so, execute the update description *Update* to compute the new values for the state variables leading to a state t. Thus, the task specifies the set of internal actions $s \xrightarrow{\varepsilon} t$ such that $s \in [\![Guard]\!]$ and $(s,t) \in [\![Update]\!]$. The label ε indicates that there is no observable communication during an internal action.

For the process `AsyncInc`, the state is updated by two internal tasks A_x and A_y. The task A_x is always enabled (that is, its guard condition is always satisfied) and increments the state variable x as specified by the update code $x := x + 1$. The task A_y is symmetric and increments the state variable y. The set of all internal tasks of a process is denoted $\mathcal{A}$ and equals $\{A_x, A_y\}$ for the process `AsyncInc`. A step of the process corresponds to executing one of these two tasks. Thus, the set of internal actions consists of $(i,j) \xrightarrow{\varepsilon} (i+1,j)$ and $(i,j) \xrightarrow{\varepsilon} (i,j+1)$ for every pair of natural numbers i and j.

Asynchronous Merge

As a third example, consider the process `Merge`, shown in figure 4.3, with two input channels in_1 and in_2, both of type `msg`. The process uses a buffer dedicated to each of the two input channels to store values received on that channel. We model a buffer using the type `queue`: `null` represents the empty queue, the operation `Enqueue`(v, x) updates the queue x by adding the value v as its last element, the operation `Dequeue`(x) returns the first element of the queue x while updating the queue by removing the first element, the operation `Front`(x) returns the first element of the queue x without removing it from the queue, the operation `Empty`(x) returns 1 if the queue x is empty and 0 otherwise, and the operation `Full`(x) returns 1 if the queue x is full and 0 otherwise.

The input task A_i^1 captures how the values received on the input channel in_1 are processed: if the queue x_1 is not full, then the value of in_1 is enqueued in the queue x_1. This is captured by the guard condition $\neg \texttt{Full}(in_1)$ and the update code $\texttt{Enqueue}(in_1, x_1)$. Compared to the process $\texttt{Buffer}$, this captures a different style of synchronization: the environment, or the process sending values on the channel in_1, is blocked if the process $\texttt{Merge}$ has its internal queue x_1 full. The input task A_i^2 corresponding to processing of the channel in_2 is similar.

The process $\texttt{Merge}$ has two output tasks. The task A_o^1 dequeues an element from the queue x_1 and transmits it on the output channel out. This is possible when the queue x_1 is not empty. Hence, the task is described by the guard condition $\neg \texttt{Empty}(x_1)$ and the update code $out := \texttt{Dequeue}(x_1)$. The task A_o^2 is symmetric and corresponds to transferring the front element of the queue x_2 to the output channel. Note that both the output tasks are associated with the same channel, and thus $\mathcal{A}_{out} = \{A_o^1, A_o^2\}$. When the queues x_1 and x_2 are non-empty, both output tasks are enabled, and either of them can be executed.

The definition is summarized below.

ASYNCHRONOUS PROCESS

An *asynchronous process* P has:

- a finite set I of typed input channels defining the set of inputs of the form $x\,?\,v$ with $x \in I$ and a value v for x;

- a finite set O of typed output channels defining the set of outputs of the form $y\,!\,v$ with $y \in O$ and a value v for y;

- a finite set S of typed state variables defining the set Q_S of states;

- an initialization $Init$ defining the set $[\![Init]\!] \subseteq Q_S$ of initial states;

- for each input channel x, a set $\mathcal{A}_x$ of input tasks, each described by a guard condition over S and an update from the read-set $S \cup \{x\}$ to the write-set S defining a set of input actions $s \xrightarrow{x\,?\,v} t$;

- for each output channel y, a set $\mathcal{A}_y$ of output tasks, each described by a guard condition over S and an update from the read-set S to the write-set $S \cup \{y\}$ defining a set of output actions $s \xrightarrow{y\,!\,v} t$; and

- a set $\mathcal{A}$ of internal tasks, each described by a guard condition over S and an update from the read-set S to the write-set S defining a set of internal actions $s \xrightarrow{\varepsilon} t$.

Exercise 4.1: We want to design an asynchronous adder process $\texttt{AsyncAdd}$ with input channels x_1 and x_2 and an output channel y, all of type **nat**. If the ith input message arriving on the channel x_1 is v and the ith input message arriving

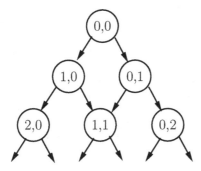

Figure 4.4: Executions of the Process `AsyncInc`

on the channel x_2 is w, then the ith value output by the process `AsyncAdd` on its output channel should be $v + w$. Describe all the components of the process `AsyncAdd`. ■

4.1.3 Executions

The operational semantics of an asynchronous process can be captured by defining its executions. An execution starts in an initial state. At every step, one of the tasks that is enabled in the current state is chosen and executed. This task may be an input task, an output task, or an internal task. Only one task is executed at every step, and the order in which different tasks are executed is totally unconstrained. Such a semantics for asynchronous interaction is called the *interleaving semantics*.

Figure 4.4 shows possible executions of the asynchronous process `AsyncInc` of figure 4.2. Each state is a pair (i, j) of natural numbers corresponding to the values of the variables x and y, respectively. The state $(0, 0)$ is the sole initial state, and each state has two possible transitions: one that increments the value of x corresponding to the execution of the internal task A_x, and one that increments the value of y corresponding to the execution of the task A_y. An execution is a (finite) path through the graph shown in figure 4.4 starting at the root. Note that for the process `AsyncInc`, every state of the form (i, j) is a reachable state. In particular, the state $(i, 0)$ is reachable via an execution that consists of executing the task A_x i times without ever executing the task A_y, corresponding to the left-most path in figure 4.4.

Formally, a finite *execution* of an asynchronous process P consists of a finite sequence of the form

$$s_0 \xrightarrow{l_1} s_1 \xrightarrow{l_2} s_2 \xrightarrow{l_3} s_3 \cdots s_{k-1} \xrightarrow{l_k} s_k$$

where for $0 \le j \le k$, each s_j is a state of P, s_0 is an initial state of P, and for $1 \le j \le k$, $s_{j-1} \xrightarrow{l_j} s_j$ is an input, an output, or an internal action P.

For instance, one possible execution of the process `Buffer` of figure 4.1 is:

$$\texttt{null} \xrightarrow{in\,?\,1} 1 \xrightarrow{out\,!\,1} \texttt{null} \xrightarrow{in\,?\,0} 0 \xrightarrow{in\,?\,1} 1 \xrightarrow{in\,?\,1} 1 \xrightarrow{out\,!\,1} \texttt{null}.$$

Note that the process `Buffer` may execute an unbounded number of input actions before it executes an output action, which issues the most recent input value received.

For the process `Merge` of figure 4.3, below is one possible execution, where the state lists the contents of the queues x_1 and x_2 in that order:

$$(\texttt{null},\texttt{null}) \xrightarrow{in_1\,?\,0} ([0],\texttt{null}) \xrightarrow{in_1\,?\,2} ([02],\texttt{null}) \xrightarrow{in_2\,?\,5} ([02],[5]) \xrightarrow{out\,!\,5}$$
$$([02],\texttt{null}) \xrightarrow{in_2\,?\,3} ([02],[3]) \xrightarrow{out\,!\,0} ([2],[3]) \xrightarrow{out\,!\,3} ([2],\texttt{null}) \xrightarrow{in_1\,?\,0} ([20],\texttt{null}).$$

In a state such as $([02],[5])$ where both the buffers are non-empty, assuming that the two input buffers are also not full, all the four tasks are enabled. For every possible value v of type `msg`, the possible input actions are: $([02],[5]) \xrightarrow{in_1\,?\,v}$ $([02v],[5])$ and $([02],[5]) \xrightarrow{in_2\,?\,v} ([02],[5v])$, obtained by executing the input tasks A_i^1 and A_i^2, respectively; and the possible output actions are $([02],[5]) \xrightarrow{out\,!\,0}$ $([2],[5])$ and $([02],[5]) \xrightarrow{out\,!\,5} ([02],\texttt{null})$, obtained by executing the output tasks A_o^1 and A_o^2, respectively.

Note that the sequence of values output by the process represents a merge of the sequences of input values received on the two input channels. The relative order of values received on the input channel in_1 is preserved in the output sequence, and so is the relative order of values received on the channel in_2, but an input value received on the channel in_1 before a value received on the channel in_2 may appear on the output channel later.

In this example, each individual task is deterministic: for each task, given a state in which the task is enabled, the execution of the task results in a unique update of the variables in its write-set. However, the asynchronous execution model is inherently nondeterministic: at each step, one of the enabled tasks is chosen and executed, and the order in which the tasks execute affects the outputs.

Exercise 4.2: We want to design an asynchronous process `Split` that is the dual of `Merge`. The process `Split` has one input channel *in* and two output channels out_1 and out_2. The messages received on the input channel should be routed to one of the output channels in a nondeterministic manner so that all possible splittings of the input stream are feasible executions. Describe all the components of the desired process `Split`. ∎

4.1.4 Extended-State Machines

In section 2.1.6, we explored the use of extended-state machines to specify the behavior of synchronous reactive components. Extended-state machines are

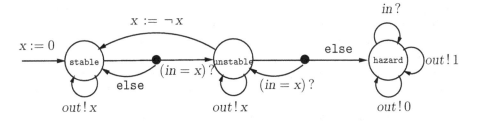

Figure 4.5: An Asynchronous Not Gate `AsyncNot`

used to describe the behavior of asynchronous processes also. In an extended-state machine description, there is an implicit state variable *mode* that ranges over a finite enumerated type. The behavior is described by a graph whose vertices correspond to the modes and whose edges correspond to mode-switches. In the asynchronous case, each mode-switch can access at most one input channel or at most one output channel, and each mode-switch corresponds to a task. We will illustrate the notation using the example of an asynchronous process modeling an *asynchronous Not* gate shown in figure 4.5.

In an asynchronous circuit, in contrast to synchronous circuits, there is no single global clock, and a change in the value of an output due to changes in the values of the inputs is delayed. An asynchronous logic gate is *stable* when its output is the desired function of the inputs and *unstable* otherwise. If the gate is stable and any of the inputs change in a way that violates the stability condition, then the gate turns unstable. The output of an asynchronous gate can change only if the gate is unstable, and when this happens, the gate becomes stable. The time it takes for the gate to update its output is assumed to be arbitrary so that a correctly designed asynchronous circuit is not dependent on concrete values of the delay parameters. If the gate is unstable and any of the inputs change without causing the stability condition to become true, then the gate remains unstable. However, if any of the inputs of an unstable gate change in a way that causes the stability condition to become true, then a hazard is encountered and the gate fails. If a gate has failed, then its output may change arbitrarily. Asynchronous gates and latches should be composed together to form an asynchronous circuit in a manner so as to ensure that no gate ever fails.

Figure 4.5 describes the asynchronous process `AsyncNot`. The process has an input channel *in* and an output channel *out* modeling the input and output wires of the gate. The extended-state machine has three modes `stable`, `unstable`, and `hazard` corresponding to the three modes of operation of the gate. The state variable x captures the value of the output, and this value is issued on the output channel *out*.

Initially, the gate is in the `stable` mode with the output x equal to 0. If the value received on the input channel *in* equals the current output, then this violates the logic of the Not gate, making it unstable, and if the value of the input channel is the negation of the output x, then the gate continues to stay stable. This is expressed by a *conditional* mode-switch that has multiple targets: the switch out of `stable` corresponding to processing inputs has no guard (that is, it is always enabled), and then if the condition $(in = x)$ holds the target of the mode-switch is `unstable`; otherwise the target of the mode-switch is `stable`.

In the unstable mode, the gate can switch back to the stable mode, toggling the value of x. Processing of an input value in the unstable mode causes the gate to ignore the input by either keeping the mode unchanged (this is the case if the input value is equal to the current output, thereby maintaining the validity of the pending toggling of the output) or switching to the mode `hazard` (this is the case if the input value is the negation of the current output implying a meaningful change in input values in rapid succession without giving the gate a chance to update its output appropriately). Processing of inputs is again expressed by a conditional mode-switch with two possible targets.

In the mode `hazard`, the gate ignores input values (that is, processing an input value has no effect on the state) and issues both output values in a nondeterministic manner.

Each mode-switch corresponds to a single task, and at each step, exactly one mode switch of the machine is executed. In our example, the self-loop on the mode `stable` contributes an output task with the guard condition $(mode = $ `stable`$)$ and the update code $out!x$. Each of the other three self-loops that involve the output channel contribute one output task each. The switch from the mode `unstable` to `stable` contributes the sole internal task with the guard condition $mode = $ `unstable` and the update code $x := \neg x; mode := $ `stable`. The conditional mode-switch out of `stable` contributes an input task with the guard condition $(mode = $ `stable`$)$ and the update code `if` $(in = x)$ `then` $mode := $ `unstable`. The input task corresponding to the conditional mode-switch out of the mode `unstable` is similar. Finally, the self-loop on the mode `hazard` labeled with $in?$ contributes the input task with the guard condition $(mode = $ `hazard`$)$ and empty update code (that is, the state stays unchanged). A possible execution of the process is shown below:

$$(\text{stable}, 0) \xrightarrow{out!0} (\text{stable}, 0) \xrightarrow{in?0} (\text{unstable}, 0) \xrightarrow{in?0} (\text{unstable}, 0) \xrightarrow{\varepsilon}$$

$$(\text{stable}, 1) \xrightarrow{out!1} (\text{stable}, 1) \xrightarrow{out!1} (\text{stable}, 1) \xrightarrow{in?1} (\text{unstable}, 1) \xrightarrow{in?0}$$

$$(\text{hazard}, 1) \xrightarrow{out!0} (\text{hazard}, 1) \xrightarrow{out!1} (\text{hazard}, 1) \xrightarrow{in?0} (\text{hazard}, 1).$$

Note that starting in the initial state, if the process is supplied the input $in?0$, followed by the input $in?1$, with no intervening output actions, then the resulting mode may be `hazard` or `unstable`. The latter is a possibility if in between the two input actions the process executes the internal action that toggles the state variable x. Note that the internal action of toggling x is decoupled from issuing the output on the output channel out. The only way to ensure that the

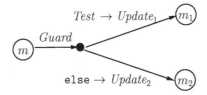

Figure 4.6: A Conditional Mode-switch in Extended-state Machines

gate does not enter the mode **hazard** is for its environment, after supplying the input $in?0$, to wait for the output $out!1$ before issuing the subsequent input $in?1$.

The execution semantics of processes specified using extended-state machines is intuitively simple and can be directly incorporated in the simulation and analysis tools for asynchronous processes. Alternatively, it is possible to translate the extended-state machine description to the task-based formal definition. The general form of a conditional mode-switch is shown in figure 4.6. The mode-switch can be executed if the current mode is m and the guard condition *Guard* holds. The update code corresponds to evaluating the condition *Test*, and if that holds, the code $Update_1$ is executed, and the mode variable is updated to m_1; otherwise the code $Update_2$ is executed, and the mode variable is updated to m_2. Such a mode-switch contributes a task with the guard condition

$$(mode = m) \ \wedge \ Guard$$

and update code

$$\text{if } Test \text{ then } \{Update_1; \ mode := m_1\} \text{ else } \{Update_2; \ mode := m_2\}.$$

The variables accessed in the two conditions *Guard* and *Test* and the two updates $Update_1$ and $Update_2$ should be such that the task can be classified as an internal task, an input task associated with a single input channel, or an output task associated with a single output channel. In particular, the key restriction is that the guard condition *Guard* should not refer to input values. That's why we cannot replace the conditional mode-switch out of the mode **stable** by two separate mode-switches, one from the mode **stable** to **unstable** with the guard condition $(in = x)$ and one self-loop with the negated guard condition $(in \neq x)$.

Exercise 4.3: Describe an asynchronous process **AsyncAnd** that models an asynchronous *And* gate with two Boolean input channels in_1 and in_2 and a Boolean output channel *out*. The process can be described as an extended-state machine with three modes as in the case of the process **AsyncNot** in figure 4.5 and with three Boolean state variables. ∎

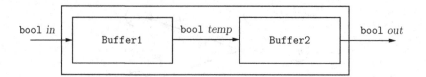

Figure 4.7: Block Diagram for DoubleBuffer from Two Buffer Processes

4.1.5 Operations on Processes

As discussed in chapter 2, block diagrams can be used to describe composition
of synchronous components to form systems in a hierarchical manner. The same
design methodology applies to asynchronous processes also. As an example, consider the block diagram of figure 4.7 that uses two instances of the asynchronous
process Buffer to form a composite process DoubleBuffer. The block diagram
is structurally identical to the block diagram of the synchronous component
DoubleDelay of figure 2.15. As before, the meaning of such diagrams can be
made precise using three operations: instantiation, parallel composition, and
output hiding. A textual description of the process DoubleBuffer using these
operations is:

$$(\texttt{Buffer}[\,out \mapsto temp\,] \mid \texttt{Buffer}[\,in \mapsto temp\,]) \setminus temp.$$

Input/Output Channel Renaming

The operation of input or output channel renaming is used to achieve the desired
communication pattern. In figure 4.7, the asynchronous process Buffer1 is
obtained by renaming the output channel out of the process Buffer to $temp$ and
corresponds to the renaming expression $\texttt{Buffer}[\,out \mapsto temp\,]$. Analogously, the
process Buffer2 is obtained by renaming the input channel in of the process
Buffer to $temp$ and corresponds to the process $\texttt{Buffer}[\,in \mapsto temp\,]$. When
these two processes are composed, the shared name $temp$ ensures that the output
issued by the process Buffer1 is consumed by the process Buffer2 as its input.

When composing processes, we assume that the names of state variables are
private, and state variables are implicitly renamed to avoid name conflicts. In
our example, we can assume that the state variable of Buffer1 is called x_1
instead of x, and the state variable of Buffer2 is called x_2.

The formal definition of the input/output channel renaming operation for processes is similar to the corresponding definition for synchronous components and
corresponds to syntactic substitution of channel names throughout its description.

Parallel Composition

The parallel composition operation combines two processes into a single process whose behavior captures the interaction between the two processes running concurrently so that an output action of one is synchronized with an input action of another with the common channel name, and remaining actions are interleaved. To differentiate the asynchronous composition with the synchronous composition (which is denoted $\|$), we use $P_1 \mid P_2$ to denote the composition of two processes P_1 and P_2.

As in the synchronous case, two processes can be composed only if their variable declarations are mutually consistent: there are no name conflicts concerning state variables, and the two sets of output channels are disjoint. These requirements capture the assumption that only one process is responsible for controlling the value of any given variable. An input channel of one can be an input or output channel of the other. Note that the problem of mutually cyclic await-dependencies discussed for the synchronous case does not arise in the asynchronous interaction. If x is an output channel of process P_1 and an input channel of process P_2, and y is an output channel of P_1 and an input channel of P_2, then we can compose P_1 and P_2 without any complications. This is because production of an output is a separate step from processing an input for each of the processes, and hence there can be no dependencies among variables within the same step.

The set of input channels, output channels, and state variables of the composite process are defined as in the synchronous case. Each state variable of a component process is a state variable of the composite process. Each output channel of a component process is an output channel of the composite process. Each input channel of a component process that is not an output of the other is an input channel of the composite process.

The state of the composite process is of the form (s_1, s_2), where s_1 is a state of the process P_1 and s_2 is a state of the process P_2. The two processes initialize their states independently, and thus a composite state (s_1, s_2) is initial if both states s_1 and s_2 are initial states of processes P_1 and P_2, respectively.

Tasks of the Composite Process

When an input channel x is a common input channel to both the processes, both consume an input value on channel x simultaneously, and a possible input action of the composite process corresponding to such an input is obtained by executing input actions of the two processes together. That is, $(s_1, s_2) \xrightarrow{x\,?\,v} (t_1, t_2)$ is an input action of the composite process precisely when $s_1 \xrightarrow{x\,?\,v} t_1$ is an input action of P_1 and $s_2 \xrightarrow{x\,?\,v} t_2$ is an input action of P_2. Consider an input task A_1 of P_1 associated with the channel x, and suppose its guard is $Guard_1$ and update description is $Update_1$. Similarly, suppose an input task A_2 of P_2 associated with the channel x has guard $Guard_2$ and update description

Update₂. Then by combining the tasks A_1 and A_2, we obtain an input task A_{12} for the composite process associated with the channel x: its guard condition is *Guard₁* $\land$ *Guard₂* and the update code is *Update₁*; *Update₂*. That is, the task A_{12} for processing input values on the channel x is enabled exactly when both the corresponding input tasks of the two component processes are enabled, and it updates the state variables of P_1 using the update code *Update₁* and then updates the state variables of P_2 by executing the update code *Update₂*. The order in which the two update descriptions are executed does not really matter as they update disjoint sets of variables. When the processes P_1 and P_2 have multiple input tasks associated with the channel x, the composite process has tasks corresponding to all possible pairings of such tasks of the two processes.

If a channel x is an output channel of one process, say process P_1, and an input channel of the other process P_2, then the two processes synchronize using this channel: when P_1 executes an output action sending a value on the channel x, the receiver P_2 executes a matching input action. The resulting joint action is an output action for the composite. That is, $(s_1, s_2) \xrightarrow{x\,!\,v} (t_1, t_2)$ is an output action of the composite process precisely when $s_1 \xrightarrow{x\,!\,v} t_1$ is an output action of P_1 and $s_2 \xrightarrow{x\,?\,v} t_2$ is an input action of P_2. If the description of an output task A_1 of P_1 associated with channel x is *Guard₁* $\rightarrow$ *Update₁* and the description of an input task A_2 of P_2 associated with channel x is *Guard₂* $\rightarrow$ *Update₂*, then the description for the task A_{12} of the composite process obtained by pairing these two tasks is: *Guard₁* $\land$ *Guard₂* $\rightarrow$ *Update₁*; *Update₂*. Thus, the task is enabled when the guard conditions of both the tasks are satisfied. The update description *Update₁* updates the state variables of the process P_1 and computes an output value for the channel x. This value is then used by the update code *Update₂* to update the state variables of P_2. It is worth emphasizing that the guard condition of an input task corresponding to a channel x refers only to the state variables: whether a process is willing to process an input on a channel x depends on its state but not on the value supplied on the channel x. Thus, in the synchronization between two processes P_1 and P_2 using the channel x, the willingness of both the processes to participate in the synchronization using the tasks A_1 and A_2 is captured by the condition *Guard₁* $\land$ *Guard₂*, which can be evaluated in a given composite state before the process P_1 has executed its update code to determine which output value is to be transmitted on the channel x. If P_1 has multiple output tasks associated with the channel x and/or P_2 has multiple input tasks associated with the channel x, the set of tasks associated with the channel x in the composite process is obtained by considering all possible pairings.

Now consider the case when P_1 has an input channel x that is not a channel of the other process P_2. In this case, to process an input value on the channel x, the composite process simply executes the input task of P_1 corresponding to the channel x, and the state of P_2 stays unchanged during such an input action. For every input action $s_1 \xrightarrow{x\,?\,v} t_1$ of process P_1 and every state s of process P_2, the composite process has an input action $(s_1, s) \xrightarrow{x\,?\,v} (t_1, s)$. For this purpose, we

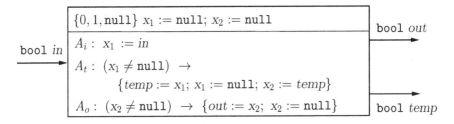

Figure 4.8: Asynchronous Parallel Composition of Two `Buffer` Processes

declare each input task of process P_1 associated with the channel x to be also an input task of the composite. Note that the guard condition and the update description of such a task stays unchanged and does not refer to the variables of P_2.

The same holds for output actions for a channel that involves only one process. If y is an output channel of the process P_1 and is not a channel of P_2, then each output task of P_1 associated with the output channel y is declared to be an output task of the composite process with the same guard condition and update description. Enabledness of such a task does not depend on the process P_2, and executing such a task leaves the state of P_2 unchanged. Thus, for every output action $s_1 \xrightarrow{y!v} t_1$ of process P_1 and every state s of process P_2, the composite process has an output action $(s_1, s) \xrightarrow{y!v} (t_1, s)$.

Finally, an internal action of the composite process is an internal action of exactly one of the two component processes, with the other process maintaining its state unchanged. Thus, every internal task of each of the two processes is declared to be an internal task of the composite with the same guard condition and update description.

The composition of processes `Buffer1` and `Buffer2` gives the process shown in figure 4.8. It has state variables $\{x_1, x_2\}$, output channels $\{temp, out\}$, and input channels $\{in\}$. For the composite process, the input task A_i is the same as the corresponding input task for the process `Buffer1`, and the output task A_o is the same as the corresponding output task for the process `Buffer2`. Since $temp$ is a common channel, the corresponding output task A_t is obtained by composing the specifications of the output task of the process `Buffer1` responsible for producing outputs on the channel $temp$, with the input task of the process `Buffer2` responsible for processing the inputs on the channel $temp$. The guard condition for this task then is the conjunction of the guard conditions of the two contributing tasks, which turns out to be only $(x_1 \neq \texttt{null})$ since the input task of `Buffer2` is always enabled with guard condition 1. The update description executes the update code of the output task of `Buffer1` followed by the input task of `Buffer2`. The composite has no internal tasks. Thus, only the process `Buffer1` participates in the processing of the channel in, the two processes synchronize on *temp*, and only the process `Buffer2` participates in producing the output on the channel *out*.

We now summarize the formal definition of parallel composition of asynchronous processes.

ASYNCHRONOUS PROCESS COMPOSITION

Let $P_1 = (I_1, O_1, S_1, Init_1, \{\mathcal{A}_x^1 \mid x \in I_1\}, \{\mathcal{A}_y^1 \mid y \in O_1\}, \mathcal{A}_1)$ and $P_2 = (I_2, O_2, S_2, Init_2, \{\mathcal{A}_x^2 \mid x \in I_2\}, \{\mathcal{A}_y^2 \mid y \in O_2\}, \mathcal{A}_2)$ be two asynchronous processes such that O_1 and O_2 are disjoint. Then the *parallel composition* $P_1 \mid P_2$ is the asynchronous process P defined by:

- the set S of state variables is $S_1 \cup S_2$;

- the set O of output channels is $O_1 \cup O_2$;

- the set I of input channels is $(I_1 \cup I_2) \setminus O$;

- the initialization is given by $Init_1; Init_2$;

- for each input channel $x \in I$, (1) if $x \notin I_2$, then the set of input tasks $\mathcal{A}_x$ is $\mathcal{A}_x^1$; (2) if $x \notin I_1$, then the set of input tasks $\mathcal{A}_x$ is $\mathcal{A}_x^2$; and (3) if $x \in I_1 \cap I_2$, then for each task $A_1 \in \mathcal{A}_x^1$ and $A_2 \in \mathcal{A}_x^2$, the set of input tasks $\mathcal{A}_x$ contains the task described by $Guard_1 \wedge Guard_2 \to Update_1; Update_2$, where $Guard_1 \to Update_1$ is the description of the task A_1 and $Guard_2 \to Update_2$ is the description of the task A_2;

- for each output channel $y \in O$, (1) if $y \in O_1 \setminus I_2$, then the set of output tasks $\mathcal{A}_y$ is $\mathcal{A}_y^1$; (2) if $y \in O_2 \setminus I_1$, then the set of output tasks $\mathcal{A}_y$ is $\mathcal{A}_y^2$; (3) if $y \in O_1 \cap I_2$, then for each task $A_1 \in \mathcal{A}_y^1$ and $A_2 \in \mathcal{A}_y^2$, the set of output tasks $\mathcal{A}_y$ contains the task described by $Guard_1 \wedge Guard_2 \to Update_1; Update_2$, where $Guard_1 \to Update_1$ is the description of the task A_1 and $Guard_2 \to Update_2$ is the description of the task A_2; and (4) if $y \in O_2 \cap I_1$, then for each task $A_1 \in \mathcal{A}_y^1$ and $A_2 \in \mathcal{A}_y^2$, the set of output tasks $\mathcal{A}_y$ contains the task described by $Guard_2 \wedge Guard_1 \to Update_2; Update_1$, where $Guard_1 \to Update_1$ is the description of the task A_1 and $Guard_2 \to Update_2$ is the description of the task A_2;

- the set $\mathcal{A}$ of internal tasks of the composite is $\mathcal{A}_1 \cup \mathcal{A}_2$.

Output Hiding

If y is an output channel of a process P, then the result of *hiding* y in P gives a process that behaves exactly like the process P, but y is no longer an output that is observable outside. This is achieved by removing y from the set of output channels and turning each output task associated with the channel y into an internal task by declaring y to be a *local* variable. Recall that a local variable is an auxiliary variable used in the description of the update code of a task and is not stored in the state.

Let us revisit the process `Buffer1 | Buffer2`. If we hide the intermediate output channel *temp*, we get the desired composite process `DoubleBuffer`: the set of state variables is $\{x_1, x_2\}$, the set of output channels is $\{out\}$, the set of input channels is $\{in\}$, and the initialization is given by $x_1 := $ `null`; $x_2 := $ `null`. The input task A_i and the output task A_o are unchanged from `Buffer1 | Buffer2`. The process `DoubleBuffer` has one internal task described by

$(x_1 \neq $ `null`$) \ \rightarrow$
  { `local bool` *temp*;
  temp $:= x_1$; $x_1 := $ `null`;
  $x_2 := $ *temp* }.

Exercise 4.4: Consider the asynchronous process

$$(\texttt{Merge}[\, out \mapsto temp\,] \mid \texttt{Merge}[\, in_1 \mapsto temp\,][\, in_2 \mapsto in_3\,]) \setminus temp$$

obtained by connecting two instances of the process `Merge`. Show the "compiled" version of this composite process similar to the description in figure 4.8. Explain the input/output behavior of this composite process. ∎

4.1.6 Safety Requirements

In chapter 3, we studied how to specify and verify safety requirements of transition systems. The same techniques apply to asynchronous processes also. Given an asynchronous process P, we can define an associated transition system T as follows:

- the state variables S of P are the state variables of T;

- the initialization specification *Init* of P is also the initialization of T; and

- the transition description of T corresponds to choosing either an internal, an input, or an output task A of P such that the guard of A is satisfied, and executing the corresponding update description. For output tasks, the corresponding output channel is converted into a local variable; and for input tasks, the corresponding input channel is converted into a local variable whose value is chosen nondeterministically at the beginning.

Thus, $s \rightarrow t$ is a transition of T precisely when the process P has either an input or an output or an internal action from state s to t.

A property φ over state variables of an asynchronous process is an invariant of the system if all reachable states of the corresponding transition system satisfy the property φ. For instance, consider the process `AsyncInc` of figure 4.4 and the requirement that the values of the two variables x and y remain at most c apart for a given constant c. This corresponds to checking whether the property

$|x - y| \leq c$ is an invariant of the system. It turns out that this is not the case for the process `AsyncInc`, no matter how large the constant c is.

Concepts such as inductive invariants can be used to prove safety requirements of asynchronous processes. For instance, to show that a state property φ is an inductive invariant, we need to show that it (1) holds initially, and (2) is preserved by every transition. Since a transition corresponds to executing exactly one task, we need to show that φ is preserved by the execution of every task.

Safety monitors can be used to capture safety requirements that cannot be directly stated in terms of state variables. In the asynchronous setting, a safety monitor for a process with input variables I and output variables O is another asynchronous process with internal state and $I \cup O$ as its input variables. Such a monitor synchronizes with the observed system P on the input and output actions of P. The monitor is described by an extended-state machine, such that an execution that ends up in an "error" mode of the monitor indicates a violation of the desired safety requirement.

Enumerative and symbolic reachability algorithms discussed in sections 3.3 and 3.4 also apply to verification of asynchronous processes.

Exercise 4.5: Consider the process `AsyncNot` of figure 4.5. In this exercise, we want to design an asynchronous process `AsyncNotEnv` that interacts with `AsyncNot`. The process `AsyncNotEnv` has a Boolean input channel *out* and a Boolean output channel *in*. It first outputs the value 0 and then is able to receive inputs. It waits until the received input equals 1 and proceeds to output the value 1, and then waits until the received input equals 0. This cycle is then repeated. Model the desired asynchronous process `AsyncNotEnv` as an extended-state machine. Consider the asynchronous composition `AsyncNot | AsyncNotEnv` and argue that (`AsyncNot.`*mode* $\neq$ `hazard`) is an invariant of the composite process. ∎

4.2   Asynchronous Design Primitives

4.2.1   Blocking vs. Non-blocking Synchronization

In the asynchronous model, exchange of information between two processes, and thus synchronization between them, occurs when the production of an output by one process is matched with the consumption of the corresponding input by another. Suppose x is an output channel of a process P_1 and an input channel of another process P_2. Let A_1 be an output task of P_1 corresponding to the channel x. Suppose the process P_1 is in a state s_1, in which this output task A_1 is enabled. Then the process P_1 is ready to send a value on its output channel x. Suppose s_2 is the current state of P_2. If some input task A_2 of P_2 associated with the channel x is enabled in the state s_2, then the process P_2 is willing to accept an input on the channel x, and the composite process can execute a synchronizing action on the channel x. However, if none of the input tasks of P_2 associated with the channel x is enabled in the state s_2, then the process P_2 is not willing to accept an input on the channel x, and effectively the process P_1 is blocked from executing its output task. This is a form of *blocking* communication where the producer P_1 needs the cooperation of the receiver P_2 to produce an output on the channel x. A process that is willing to accept every input in every state does not prevent the producer from producing outputs and is said to be *non-blocking*.

In our model, a process is willing to process inputs on a channel x precisely when the guard condition of one of the input tasks associated with the channel x is satisfied. For a process to be non-blocking, we require that the *disjunction* of the guards of all the tasks corresponding to an input channel be a valid formula, that is, equivalent to the Boolean constant 1.

NON-BLOCKING PROCESS

An asynchronous process P is said to be *non-blocking* if for every input channel x and for every state s, some task in the set $\mathcal{A}_x$ of tasks associated with the channel x is enabled in the state s.

The process `Buffer` of figure 4.1 is non-blocking: its environment can always supply a value on the input channel *in* even though some of these values are effectively lost. However, the process `Merge` of figure 4.3 is blocking: an input on the channel in_1 cannot be processed if the queue x_1 is full, and thus the producer of outputs on the channel in_1 has to wait until this queue becomes non-full. The process `AsyncNot` of figure 4.5 is non-blocking: it always accepts inputs even though supplying inputs in rapid succession can lead it to the hazardous state.

The process `DoubleBuffer` obtained by composing two `Buffer` processes is non-blocking. In fact, it is easy to verify that all the operations defined in section 4.1.5 preserve the property of being non-blocking: if all the component processes in a block diagram are non-blocking, then so is the composite process corresponding to the block diagram.

In designing asynchronous systems, both styles of synchronization, non-blocking and blocking, are common. In the non-blocking designs, if a process P_1 sends an output value to another process P_2, then typically an explicit acknowledgment from the process P_2 back to process P_1 is needed for P_1 to be sure that its output was examined by P_2. In the implementation of blocking synchronization, the run-time system must somehow ensure that the receiver is willing to participate in the synchronizing action.

4.2.2   Deadlocks

Deadlock is a commonly occurring error in asynchronous designs. In a system composed of multiple processes, a deadlock refers to a situation in which each process is waiting for some other process to execute a task, but no task is enabled, and thus there is no continuation of the execution.

To illustrate how deadlocks can arise, consider the system consisting of two processes P_1 and P_2 shown in figure 4.9. The process P_1 can generate requests of one type that are serviced by the process P_2. For our purpose, the data values exchanged are not particularly relevant. Hence, we model a request by process P_1 as a message with the value $\mathtt{req}_1$ sent on the channel x_1 from process P_1 to process P_2 and the corresponding response by process P_2 as a message with the value $\mathtt{resp}_1$ sent on the channel x_2 from process P_2 to process P_1. Similarly, the process P_2 can generate requests of another kind, and each such request is modeled as a message with the value $\mathtt{req}_2$ sent on the channel x_2 from process P_2 to process P_1. Each such request is serviced by process P_1, and the corresponding response is modeled as a message with the value $\mathtt{resp}_2$ sent on the channel x_1 from process P_1 to process P_2.

The description of the process P_1 is shown in figure 4.9. The process has an internal queue y_1 that is used to store messages received on its input channel x_2. The description of its tasks uses a mixture of two styles of specifications we have seen so far: the input task is listed explicitly, whereas the computation of the process corresponding to the internal and output tasks is described using an extended-state machine. The input task A_2 is always enabled and simply enqueues each message received on the input channel into the queue y_1. Initially, the mode is $\mathtt{idle}$. In this mode, if the process finds a message req_2 at the front of the queue y_1, then it dequeues this request and switches to the mode $\mathtt{busy}$. The mode $\mathtt{busy}$ captures the internal state of the process P_1, where the computation needed to service an incoming request occurs. The corresponding response is then issued on the output channel (captured by the update $x_1!\mathtt{resp}_2$), and the process returns to the $\mathtt{idle}$ mode. In the idle mode, the process can also generate a request on its own. This is modeled by the switch to the mode $\mathtt{wait}$ with the output action $x_1!\mathtt{req}_1$. In the mode $\mathtt{wait}$, the process P_1 is waiting for a response from the other process and is unwilling to process requests issued by P_2. Hence, the process can switch from the mode $\mathtt{wait}$ back to $\mathtt{idle}$ only when the first message in the queue y_1 is a response message $\mathtt{resp}_1$; if so, it is removed from the queue.

The description of the process P_2 is symmetric. Now consider the composed process $P_1 \,|\, P_2$. Suppose the process P_1 issues the request $\mathtt{req}_1$ on the channel x_1. This request gets stored in the internal queue y_2 of process P_2. This step is captured by the action shown below, where each state is described by listing the values of variables $P_1.\mathtt{mode}$, y_1, $P_2.\mathtt{mode}$, and y_2, in that order:

$$(\mathtt{idle}, \mathtt{null}, \mathtt{idle}, \mathtt{null}) \xrightarrow{x_1!\,req_1} (\mathtt{wait}, \mathtt{null}, \mathtt{idle}, [\mathtt{req}_1]).$$

At this point, two tasks are enabled: the internal task of the process P_2 corresponding to the mode-switch from the mode $\mathtt{idle}$ to mode $\mathtt{busy}$ and the output task of the process P_2 corresponding to the mode-switch from the mode $\mathtt{idle}$ to mode $\mathtt{wait}$. If the former task executes first, then the computation will progress as intended. However, if the latter task executes first, then the corresponding

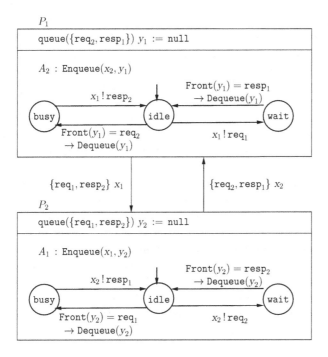

Figure 4.9: Illustrating Deadlocks

transition is:

$$(\texttt{wait}, \texttt{null}, \texttt{idle}, [\texttt{req}_1]) \overset{x_2 ! req_2}{\longrightarrow} (\texttt{wait}, [\texttt{req}_2], \texttt{wait}, [\texttt{req}_1]).$$

In the resulting state, no task is enabled: the process P_1 is expecting a response from P_2 and vice versa. Such a state is a deadlock and should be considered a bug in the design.

In general, a state s of an asynchronous process P is a *deadlock* state if (1) no task is enabled in the state s, and (2) the state s does not correspond to a *successful* termination of the system. The latter condition is specific to the design problem; for instance, in the leader election problem, a state in which all processes have already made a decision to be a leader or a follower is considered to be a successful terminal state. Except for such successful terminal states, we expect the system to continue executing. Thus, *absence of deadlocks* is a generic safety requirement that is expected to be an invariant of all asynchronous designs.

4.2.3 Shared Memory

In a shared memory architecture, processes communicate by reading and writing shared variables and, more generally, shared objects. In this section, we will

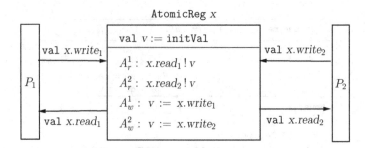

Figure 4.10: Atomic Register Supporting Read and Write Operations

illustrate how to model shared variables as asynchronous processes. In the asynchronous model, executions of different tasks are interleaved. A crucial design decision concerns how much computation can happen in one computation step of a task, that is, which operations are supported by shared objects as *atomic* operations that can be executed in a single step. We first discuss the model of atomic registers, where the only allowed operations are the most basic read and write operations.

Atomic Registers

Figure 4.10 shows the process `AtomicReg` that models a variable (or a register) x shared between two processes P_1 and P_2. The only atomic operations supported by this shared object are read and write, and such an object is called an *atomic register*. The description is parameterized by the set of values that the register can hold, denoted `val`, and the initial value of the register, denoted `initVal`.

The internal state variable v of the object holds its current value and is initialized to the value `initVal`. The channels $x.read_1$ and $x.read_2$ are used to model the read operations. The channel $x.read_1$ is an output channel for the atomic register and is an input channel for the process P_1. When the process P_1 wants to read the register, it executes the input action $y := x.read_1$, where y is a state variable of P_1, which is synchronized with the output action of the task A_r^1. Executing this action transmits the current state of the register x, and as a result, the updated value of the state variable y of P_1 is the current value of the register, whereas the state of the register stays unchanged. Thus, synchronization of the atomic register with the process P_1 on the channel $x.read_1$ transmits the value of the register to P_1. Note that the task A_r^1 of the process `AtomicReg` is always enabled, and thus whether the process P_1 can execute the task corresponding to reading the register depends solely on the guard condition of the task in P_1.

Analogously, the channels $x.write_1$ and $x.write_2$ are used to model the write operations. When the process P_1 wants to update the register by writing the value u, it executes the output action $x.write_1 ! u$, which is synchronized with the input action of the task A_w^1 and updates the internal state of the register x to the

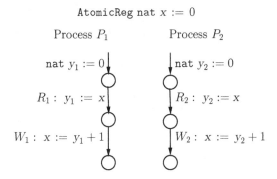

Figure 4.11: Data Race Example: Shared Counter

value received on the channel. If y is a state variable of the process P_1, then to update the shared register x with the current value of y, the process can execute the output statement $x.write_1 := y$. Since the task A_w^1 is always enabled, the enabledness of this joint activity depends solely on the guard condition of the corresponding task in P_1.

The communication pattern for the process P_2 is analogous.

Data Races

Consider the asynchronous system shown in figure 4.11. It consists of three processes. The shared register x is an instantiation of the `AtomicReg` process where the type **val** is **nat** and the initial value `initVal` is 0. In our description, such a shared register is declared using the familiar syntax for declaring variables. We will also refer to the shared variable without explicitly mentioning the associated read/write channels. For example, a read reference to the shared register x by process P_1 is an abbreviation for its input channel $x.read_1$, and an update of the shared register x by process P_2 is an update of its output channel $x.write_2$.

The two asynchronous processes P_1 and P_2 communicate by reading and writing the shared register x. The process P_1 has a state variable y_1, which is initialized to 0. The process first reads the value of x by executing the statement $y_1 := x$ (task R_1). Execution of this statement involves synchronization of the processes P_1 and x on the channel $x.read_1$. The process P_1 then writes the value $y_1 + 1$ back to the shared register x by executing the statement $x := y_1 + 1$ (task W_1), which involves synchronization of the processes P_1 and x on the channel $x.write_1$.

The process P_2 is symmetric: it reads the value of the shared register x in its internal state variable y_2 (task R_2) and writes the incremented value back to the shared register (task W_2).

Interleaving	x	y_1	y_2
$R_1; R_2; W_1; W_2$	1	0	0
$R_1; W_1; R_2; W_2$	2	0	1
$R_1; R_2; W_2; W_1$	1	0	0
$R_2; R_1; W_2; W_1$	1	0	0
$R_2; W_2; R_1; W_1$	2	1	0
$R_2; R_1; W_1; W_2$	1	0	0

Figure 4.12: All Possible Executions of Shared Counter of Figure 4.11

A step of the composed system corresponds to executing one of the tasks of the two processes. Executions of the composed system resulting from all possible interleavings of the four tasks are shown in figure 4.12. With each such execution, we list the values of the variables x, y_1, and y_2 at the end of the execution. Observe that when all the tasks have been executed once, the final value of the shared register x can be either 1 or 2. If the desired intent of each process is to increment the value of x, then a final value of 1 corresponds to a lost increment, a potential bug. Such a bug is caused by the other process accessing the shared register in between the execution of read and write access statements of one process. This type of interference between concurrent accesses to shared objects by asynchronous processes is called a *data race*.

Mutual Exclusion Problem

In the illustrative example of a shared counter of figure 4.11, suppose the process P_1 wants to ensure that the value of the shared object x stays unchanged between the execution of the read and write statements by P_1. Note that the shared object does not support both reading and writing in a single atomic step: the process P_1 cannot use the statement $x := x + 1$ to increment the counter atomically as it would involve two distinct synchronizations on two separate channels, $x.read_1$ and $x.write_1$. To ensure that the process P_1 has an *exclusive* access to the shared object as it reads and then updates the shared register, we need to design a protocol to solve the classical coordination problem of *mutual exclusion*.

Suppose we have two or more asynchronous processes that need access to a critical shared resource, such as the shared counter of our illustrative example. At any time, only one process should be using the shared resource. The allocation of the resource is not governed by a central coordinator, but processes need to coordinate among themselves to ensure such a mutually exclusive access. We assume that processes can communicate using atomic registers. Initially, a process starts in the mode `Idle`. It accesses the shared resource in the mode `Crit`, classically known as the *critical section*. In our example, once the process enters the critical section, it can read the value of the shared counter, add one to it, and write the updated value back to the shared register. We want to design the

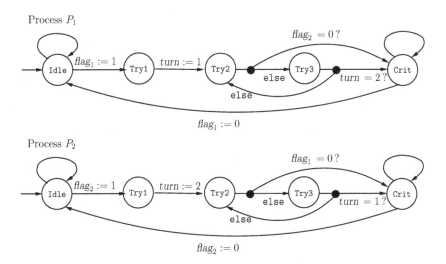

Figure 4.13: Peterson's Mutual Exclusion Protocol

entry code that the process should execute when it wants to switch from the mode `Idle` to the mode `Crit` and the exit code that the process should execute when it has finished its job in the critical section before returning to the idle mode. The safety requirement is *mutual exclusion*: no two processes should be in the critical section simultaneously. The other requirement is *deadlock freedom*: it should not be the case that one of the processes wants to enter the critical section but none is allowed to enter. Note that the safety requirement can be made precise using invariants, and formalization of the deadlock freedom as a liveness requirement is addressed in chapter 5.

Peterson's Mutual Exclusion Algorithm

Figure 4.13 shows Peterson's protocol, a classical solution to the mutual exclusion problem for the case of two processes. The processes communicate via three shared atomic registers: *turn*, $flag_1$, and $flag_2$. The process P_1 is initially in the mode `Idle`. The process has seven tasks, each corresponding to a mode-switch as described below.

1. The self-loop on the mode `Idle` indicates that the process may stay in this mode for arbitrarily many steps.

2. In the mode `Idle`, when the process needs access to the shared resource, it sets the Boolean register $flag_1$ to 1 and switches to the mode `Try1`.

3. In the mode `Try1`, the process updates the shared variable *turn* to its identifier, namely 1, and switches to the mode `Try2`.

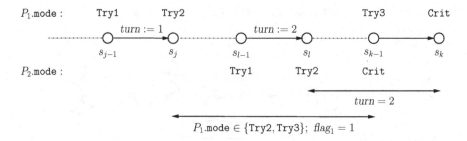

Figure 4.14: Analysis of Potential Counterexample Execution

4. In the mode **Try2**, the process reads the value of the variable $flag_2$. If $flag_2$ is 0, then it concludes that the other process does not need the resource and proceeds to the critical section. Otherwise, it switches to the mode **Try3**. This is formally modeled as a conditional mode-switch that involves the read channel between the process P_1 and shared register $flag_2$.

5. In the mode **Try3**, using a conditional switch, the process checks the value of the shared register $turn$. If $turn$ equals 2, then it concludes that the process P_2 updated $turn$ to 2 *after* the process P_1 updated $turn$ to 1, and in this case, the process P_1 proceeds to the critical section. If it finds $turn$ to be 1, then it concludes that the process P_2 updated $turn$ to 2 *before* the process P_1 updated $turn$ to 1 and returns to the mode **Try2** to check the value of $flag_2$ again.

6. The mode **Crit** corresponding to the critical section has a self-loop indicating that the process may spend an arbitrary number of steps in the critical section.

7. In the mode **Crit**, when the process no longer needs the shared resource, it updates the variable $flag_1$ back to 0 and returns to the initial mode **Idle**.

The process P_2 is symmetric.

We want to argue that Peterson's solution indeed satisfies the mutual exclusion requirement. First, observe that only the process P_1 writes to the shared register $flag_1$. The process sets $flag_1$ to 1 when it leaves the mode **Idle** and resets it to 0 when it returns to this mode. Hence, the value of $flag_1$ is 0 exactly in those states in which the mode of the process P_1 is **Idle**. Symmetrically, the value of the Boolean variable $flag_2$ is 0 exactly in those states in which the mode of the process P_2 is **Idle**.

To prove that there is no execution that leads both processes to be in the critical section simultaneously, let us assume to the contrary. Let $\rho = s_0, s_1 \ldots s_k$ be a shortest execution such that the modes of both processes equal **Crit** in the state s_k. In such an execution, the last step must correspond to some process, say P_1, switching its mode to `Crit` (if not, in the state s_{k-1}, both processes are already in their critical sections, and thus ρ is not the shortest counterexample demonstrating the violation of the desired requirement). The process P_1 can update its mode from `Try2` to `Crit` provided $flag_2$ is 0, or from `Try3` to `Crit` provided $turn$ is 2. In the state s_{k-1}, the process P_2 is already in the critical section, and hence $flag_2$ must be 1, and thus only the latter case is possible. Suppose the transition from the state s_{j-1} to s_j is the latest write to $turn$ by the process P_1, and the transition from the state s_{l-1} to s_l is the latest write to $turn$ by the process P_2 in this execution. Since $turn$ is 2 at the end of the execution, we conclude that $j < l$ (that is, the most recent update to $turn$ must be by the process P_2). Figure 4.14 depicts the scenario corresponding to this execution. The mode of the process P_1 must be either `Try2` or `Try3` in all the states $s_j, s_{j+1}, \ldots s_{k-1}$, and hence the value of $flag_1$ must be 1 in all these states. We can conclude that the value of $turn$ is 2 and $flag_1$ is 1 in all the states $s_l, s_{l+1}, \ldots s_{k-1}$. This implies that the switching conditions for the two possible ways for the process P_2 to enter its critical section ($(flag_1 = 0)$ from the mode `Try2`, and $(turn = 1)$ from the mode `Try3`) are false during this interval. Since the mode of the process P_2 is `Try2` in the state s_l and `Crit` in the state s_k, we obtain a contradiction (that is, the postulated execution ρ witnessing a violation of the safety requirement cannot exist).

We can also show that Peterson's protocol does not deadlock: it cannot happen that one of the processes wants to enter the critical section but none is allowed to enter. If only one process, say P_1, wants to enter the critical section, then the other process P_2 is in the mode `Idle`, and the variable $flag_2$ is 0. In this case, the process P_1 will succeed when it checks the value of $flag_2$ in the mode `Try2`. If both processes are trying to enter the critical section, then it cannot happen that both get stuck in the cycle between the modes `Try2` and `Try3`: once both are past their updates to the variable $turn$, its value does not change, and depending on its value, one of them must succeed in the test in the mode `Try3`.

Test&Set Registers

Figure 4.15 shows the process `Test&SetReg` that models a shared object that stores a Boolean value but supports the primitive operations of *test&set* and *reset*. The test&set operation sets the shared register to 1 while returning the old value, and the reset operation updates the shared register to 0. The state variable v, initialized to 0, stores the current value of the register. When the process P_1 wants to execute the test&set operation, it executes an input action on the channel $x.t\&s_1$ and synchronizes with the output action executed by the register. The output task A_{ts0}^1 is enabled when the value of the register is 0 and it transmits 0 while updating the state to 1. The output task A_{ts1}^1 is enabled when the value of the register is 1, and it transmits 1 while keeping the state unchanged. Note that the transmission of the current value and its update happen

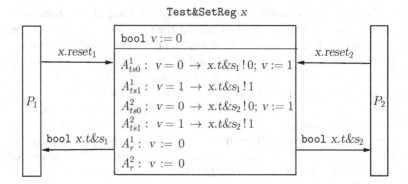

Figure 4.15: Boolean Register Supporting *test&set* and *reset* Operations

atomically within a single step, and this results in a more powerful communication scheme. If both processes P_1 and P_2 are attempting to synchronize with the `Test&SetReg` process, then the first process to synchronize will receive the value 0, and this will set the state of the register to 1, causing the subsequent process to receive the response 1.

Whenever the process P_1 wants to reset the register, it executes the output action for the channel $x.reset_1$, which gets synchronized with the execution of the input task A_r^1 which updates the register to 0. Note that no value needs to be associated with the reset operation.

In the presence of `Test&Set` registers, it is easy to implement a solution to the mutual exclusion problem. Figure 4.16 shows a solution using a single shared `Test&Set` register *free*. When the value of this shared object is 0, the critical section is unoccupied. When a process wants to enter the critical section, it simply executes the *test&set* operation on the shared register. If the operation returns 0, then the process proceeds to the critical section, and if the operation returns 1, the process tries again. Upon leaving the critical section, the process resets the value of the shared register to 0. Note that both processes are identical. It is easy to verify that the protocol satisfies the mutual exclusion requirement and is also deadlock-free.

The specification of shared objects such as `AtomicReg` and `Test&SetReg` can be generalized so that the object is shared among multiple processes instead of two.

Exercise 4.6 : Consider the transition system corresponding to Peterson's mutual exclusion protocol. The set of state variables for this system contains the variables P_1.mode, P_2.mode, *turn*, *flag*$_1$, and *flag*$_2$. Draw the reachable subgraph of this transition system. How many states are reachable? ∎

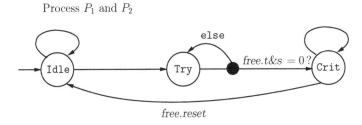

Figure 4.16: Mutual Exclusion Using **Test&Set** Register

Exercise 4.7: In an attempt to "optimize" the two-process mutual exclusion protocol of figure 4.13, someone proposes that the shared register *turn* is not necessary. Consider the modified solution of figure 4.17 that uses only the Boolean shared registers $flag_1$ and $flag_2$. Does this solution satisfy the mutual exclusion requirement? If your answer is yes, then give an informal argument of correctness or else show a counterexample execution. Is this revised protocol a satisfactory solution to the mutual exclusion problem? ∎

Exercise 4.8: In Peterson's mutual exclusion protocol (see figure 4.13), the process P_1, when it wants to enter the critical section, first sets the register $flag_1$ to 1 and then sets the register *turn* to 1. Suppose we switch the order in which these two steps are executed. That is, consider a modified version of Peterson's protocol in which the process P_1, when it wants to enter the critical section, first sets the register *turn* to 1 and then sets the register $flag_1$ to 1; symmetrically, the process P_2, when it wants to enter the critical section, first sets the register *turn* to 2 and then sets the register $flag_2$ to 1. Everything else stays the same. Does the modified protocol satisfy the requirement of mutual exclusion? If yes, give a brief justification; if no, show a counterexample. ∎

Exercise 4.9*: Consider two asynchronous processes P_1 and P_2 that communicate using a shared atomic register x of type **nat** with initial value 1. The process P_1 reads the shared register and stores the value in its internal state variable u_1, reads it again and stores the value in another state variable v_1, updates the shared register value with the sum of u_1 and v_1, and repeats this sequence of read, read, and write. The process P_2 is symmetric: it reads the shared register and stores the value in its internal state variable u_2, reads it again and stores the value in another state variable v_2, updates the shared register value with the sum of u_2 and v_2, and repeats this sequence. Let us say that a value n is reachable if there is an execution of the system in which the value of the shared register x is n at the end of the execution. Which values are reachable? Hint: try to find executions that demonstrate the reachability of values 5, 6, 7, and 8. ∎

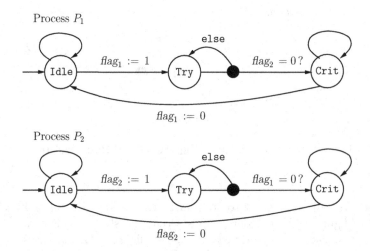

Figure 4.17: Modified Peterson's Mutual Exclusion Protocol

4.2.4   Fairness Assumptions [*]

The execution of a process in the asynchronous model is obtained by interleaving executions of different tasks. At every step of the execution, if multiple tasks can be executed, there is a choice. For example, for the `Buffer` process (figure 4.1), at every step, one can obtain the next state by executing either the input task A_i or the output task A_o (provided the state is non-null). We do not want to make assumptions about the relative frequency at which the two tasks are executed, but we would like to rule out an execution where the output task is *never* executed. Similarly, for the `Merge` process (figure 4.3), while the exact order in which the values arriving on the two input channels are merged is arbitrary by design, it is natural to assume that all these values eventually appear on the output channel. For the shared objects such as `AtomicReg` and `Test&SetReg`, if multiple processes are competing to write to them, the asynchronous model of computation allows them to succeed in an arbitrary order, possibly one process executing multiple writes before another process gets to execute a single write. However, if a process is denied a chance to write successfully forever, then no meaningful computation can occur. Hence, we would like to assume that a read or write operation by a process on a shared register is not delayed forever.

Infinite Executions

The standard mathematical framework for capturing the informal assumption that *execution of a task can be delayed arbitrarily long, but not forever*, requires us to consider *infinite* executions. An infinite execution, also called an

ω-execution, of a process P starts in one of the initial states and has an infinite sequence of states such that every state in this sequence is obtained from the previous one by executing one of the actions of the process.

Let us revisit the process AsyncInc of figure 4.2: it has two tasks A_x and A_y that are always enabled and increment the variables x and y, respectively. Consider the following ω-execution of the process AsyncInc:

$$(0,0) \xrightarrow{A_x} (1,0) \xrightarrow{A_x} (2,0) \xrightarrow{A_x} (3,0) \xrightarrow{A_x} (4,0) \cdots$$

where we have labeled each internal action by the task that was executed. In this ω-execution, the next state is always obtained by executing the internal task A_x. We will say that such an ω-execution is *unfair* to the task A_y: at every step the task A_y is enabled, but it is never executed. It is reasonable to assume that no implementation produces such unfair executions. When we state requirements, we will only require that all fair executions should satisfy the requirements. Consider the process AsyncInc and a correctness requirement that *the value of y should not always be zero*. Even though the ω-execution in which the task A_y is never executed violates this requirement, we still want to conclude that the process AsyncInc meets this requirement since in all fair executions y is guaranteed to be incremented.

Consider a finite execution of the process AsyncInc, say consisting of 1000 steps. Even if all the actions in this execution correspond to incrementing x, and thus the task A_y is enabled at every step without being executed, this is considered a plausible or valid execution of the process. If the desired correctness requirement states that "the value of y should be non-zero when the value x is 1000," we want to conclude that the AsyncInc is buggy since the finite execution consisting of executing the task A_x 1000 times demonstrates the reachability of the state $(1000, 0)$ that violates the requirement. If we put a *concrete quantitative bound* on the number of steps for which it is acceptable to ignore an enabled task, but not beyond this bound, then no matter what specific number we choose, it would seem to be an arbitrary assumption about the implementation of the process. That's why fairness is defined to be an assumption about infinite executions: every finite prefix of the unfair execution of AsyncInc illustrated above is considered legal, but the infinite execution is unfair. In a sense that can be made mathematically precise, fairness is a property of *limits* of finite executions. As we will study in chapter 5, even though infinite executions is an abstract mathematical concept and seems difficult to reason about at first glance, effective analysis algorithms exist for reasoning about such executions since such reasoning can typically be reduced to analyzing cycles in the graph of reachable states.

Consider an ω-execution of the process AsyncInc in which the tasks A_x and A_y are executed in an alternate manner, say 1000 times, but after that only the task A_x is executed indefinitely. In such an infinite execution, the value of y is "stuck" at 1000, but x keeps increasing in an unbounded manner. The fairness assumption with respect to the execution of the task A_y rules out this execution also. For an ω-execution of the process `AsyncInc` to be fair with respect to the task A_y, it must contain infinitely many actions that increment y. Symmetrically, an ω-execution is considered fair with respect to the task A_x only if it contains infinitely many actions that increment x. In the infinite tree of figure 4.4, a fair ω-execution is an infinite path through the tree that zigzags along left and right branches in an arbitrary manner but is guaranteed to take both left and right branches repeatedly. In particular, for every number n, along every fair execution, we are guaranteed that the value of x will exceed n and the value of y will also exceed n.

For the process `Buffer`, an ω-execution where only the input task A_i is executed at every step is unfair to the output task A_o, and we want to rule out such an execution by assuming fairness with respect to the task A_o. For an ω-execution of `Buffer` to be fair with respect to its output task, it must contain infinitely many output actions. Again, when we state requirements such as *a message is eventually delivered*, we will demand that all fair executions should satisfy the requirements. For the process `Buffer`, we don't make any fairness assumptions about the execution of the input task A_i. Notice that for this particular process, every infinite execution must contain infinitely many input actions: this is because every time `Buffer` executes an output action, the buffer becomes empty, and the next output cannot be produced until another input is received.

Now consider the following infinite execution of the process `Merge` of figure 4.3. It receives a value on the input channel in_1. Then it repeatedly executes the loop in which it receives a value on the input channel in_2 and transfers it to the output channel by executing the task A_o^2. That is, the infinite sequence of tasks it executes is A_i^1, followed by the periodic execution of $A_i^2; A_o^2$. This clearly starves the output task A_o^1 that can transfer the element from the queue x_1 to the output channel, which is enabled at every step but never executed. We again want to rule out such an ω-execution as unfair.

Before we define the notion of fair ω-executions precisely, note that we can require a task to be executed only when it is enabled. An unfair ω-execution is one in which, after a certain point, a task is always enabled but never executed.

Consider another infinite execution of the process `Merge`: it repeatedly executes the loop in which it receives a value on the input channel in_2 and transfers it to the output channel by executing the task A_o^2. We consider this to be a fair execution. The input task A_i^1 is never executed, but this is a plausible scenario, and a bug revealed in such an execution may be a real bug. Demanding repeated execution of an input task would mean that we are making implicit assumptions about the environment. Thus, fairness is assumed only for the tasks that the process controls. If the process `Merge` is composed with another process P whose output channel is in_1, then the fairness with respect to an output task of P corresponding to in_1 can force actions involving the channel in_1. The ω-execution that repeatedly executes A_i^2 and A_o^2 in a loop is also (vacuously) fair

$$\boxed{\begin{array}{l} \texttt{nat } x := 0;\ y := 0 \\ \hline A_x:\ x := x + 1 \\ A_y:\ \texttt{even}(x)\ \rightarrow\ y := y + 1 \end{array}}$$

Figure 4.18: Asynchronous Process `AsyncEvenInc`

with respect to the output task A_o^1. This is because the queue x_1 is always empty, and thus the task A_o^1 is never enabled.

Strong Fairness

The notion of fairness we have discussed so far corresponds to what is known as *weak fairness*. Weak fairness for a task assumes that if the task is continuously enabled, then it is eventually executed. A stronger assumption is *strong fairness*, which demands that a task that is repeatedly enabled should eventually be executed.

An an illustrative example, consider the process `AsyncEvenInc` shown in figure 4.18. Similar to the process `AsyncInc` of figure 4.2, it has two state variables x and y that are incremented by the internal tasks A_x and A_y, respectively, but now the task A_y is enabled only when the value of x is an even number. Consider the following ω-execution of the process `AsyncEvenInc`:

$$(0,0) \xrightarrow{A_x} (1,0) \xrightarrow{A_x} (2,0) \xrightarrow{A_x} (3,0) \xrightarrow{A_x} (4,0) \cdots$$

During this execution, the status of the task A_y switches between enabled and disabled. As a result, this execution satisfies the condition "if continuously enabled then eventually taken" for both the tasks, and it is weakly fair with respect to both the tasks. However, this execution is not strongly fair with respect to the task A_y: the task A_y is enabled infinitely often but never taken. If we assume that the implementation platform ensures only weak fairness, then such an execution is a possible execution, and it is not guaranteed that y gets incremented. If we assume that the implementation platform ensures strong fairness, then such an execution is not a possible execution, and it is guaranteed that y gets incremented.

Modeling an Unreliable FIFO Link

As another illustrative example, consider the unreliable FIFO buffer modeled by the process `UnrelFIFO` shown in figure 4.19. The input task A_i simply transfers the input message to the internal queue x. The transfer of messages from the queue x to the output channel is done by three tasks. The task A_o^1 transfers a message from the queue x to the output channel correctly dequeuing a message

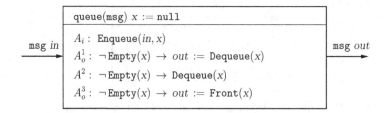

Figure 4.19: Asynchronous Process `UnrelFIFO` for Unreliable Link

and sending it on the output channel. The (internal) task A^2 models a loss of message and simply removes a message from the queue x without transferring it. The task A_o^3 models duplication of messages: it transmits the message at the front of the queue x to the output channel without removing it from the queue. The process thus models a communication link that may lose some messages and may duplicate others. However, it preserves the order and does not reorder messages. The fairness assumptions should ensure that an input message will eventually appear on the output channel.

Consider the following execution of the process `UnrelFIFO`. A message arrives on the channel *in* and is enqueued in the queue x. This message is removed by the task A^2. Since the execution of this task models loss of a message, it does not transmit it on the output channel. Suppose the tasks A_i and A^2 are repeated forever in an alternating manner. This ω-execution is weakly fair with respect to the task A_o^1 that models the correct transfer of messages. This is because every time the input task enqueues the input message in the queue x, the task A_o^1 is enabled, but every time the internal task A^2 removes this message, the task A_o^1 is disabled. Since it does not stay continuously enabled, the weak fairness assumption does not ensure its eventual execution. However, this infinite execution is not strongly fair with respect to the task A_o^1: the task is repeatedly enabled but is never executed. Thus, to capture the informal assumption that repeated attempts to transfer a message will eventually succeed, we should restrict attention to ω-executions that are strongly fair with respect to the task A_o^1.

Fairness Specification

The specification of the process `UnrelFIFO` also highlights that we do not have to assume fairness with respect to all the tasks. In particular, an infinite execution in which the task A_o^3 that duplicates a message or the task A^2 that loses a message is never executed is an acceptable and realistic execution. Losing or duplicating a message is not an active task to be executed and does not need to be executed repeatedly. While the correct functioning of the system could rely on fairness with respect to the task A_o^1, it should not rely on fairness with respect to A^2: a protocol that works correctly *only when* the underlying network repeatedly loses messages should not be considered correct.

This suggests that the description of an asynchronous process should annotate its output and internal tasks: for some strong fairness is assumed, for some weak fairness is assumed, and some do not have any fairness assumption.

FAIRNESS ASSUMPTION

An ω-execution of an asynchronous process P consists of an infinite sequence of the form $s_0 \xrightarrow{l_1} s_1 \xrightarrow{l_2} s_2 \xrightarrow{l_3} s_3 \cdots$ where each s_j is a state of P, s_0 is an initial state of P, and for each $j > 0$, $s_{j-1} \xrightarrow{l_j} s_j$ is an input, an output, or an internal action of P. A task A is *taken* at step j if the transition $s_{j-1} \xrightarrow{l_j} s_j$ corresponds to the execution of the task A. The ω-execution is *weakly fair* with respect to an internal or an output task A, if for all positions j, if the task A is enabled in the state s_j, then there exists a later position $l > j$ such that the task A is either taken at step l or not enabled in state s_l. The ω-execution is *strongly fair* with respect to an output or an internal task A, if for infinitely many indices j, the task A is enabled in state s_j, then for infinitely many indices l, the task A is taken at step l. A *fairness assumption* for an asynchronous process P consists of a subset *SF* of its internal and output tasks demanding strong fairness and a subset *WF* of its internal and output tasks demanding weak fairness. Given such a specification, a *fair ω-execution* of P is an ω-execution that is strongly fair with respect to every task in *SF* and is weakly fair with respect to every task in *WF*.

In the formal definition above, the weak fairness assumption is *if enabled then eventually either taken or disabled*, which is equivalent to *repeatedly disabled or repeatedly taken*. Similarly, the strong fairness assumption is *if repeatedly enabled then repeatedly taken*, which is equivalent to *continuously disabled or repeatedly taken*. Note that any execution that is strongly fair with respect to a task is also weakly fair with respect to that task, but the converse may not hold. In chapter 5, we will specify fairness assumptions using temporal logic, which can help in gaining more insight into the subtle distinction between weak and strong assumptions.

Fairness Assumptions for Mutual Exclusion

To illustrate how to augment a process description with fairness assumptions, let us revisit the solutions to the mutual exclusion problem. First, let us consider Peterson's protocol described in figure 4.13. Each mode-switch corresponds to a task, and let us examine all the tasks of process P_1 (the assumptions for the tasks of the process P_2 are symmetric). There are no fairness assumptions regarding the mode-switches out of the mode Idle. This means that the protocol does not assume how long a process waits in the mode Idle and does not rely on whether a process requests to enter the critical section repeatedly. The mode-switch out of the mode `Try1` represents an output action by the process P_1, and we assume weak fairness for this task. This rules out an infinite execution in which the process P_1 is waiting in the mode `Try1` to execute the statement $turn := 1$ while only the process P_2 is being executed repeatedly. The conditional mode-switches out of the modes `Try2` and `Try3` correspond to testing the values of the shared registers. Such read actions are output actions of the corresponding shared register process `AtomicReg` shown in figure 4.10. We assume weak fairness for the output tasks A_r^1 of the registers $flag_2$ and $turn$. This ensures that the process P_1 cannot just stay in the mode `Try2` waiting to read $flag_2$ while only the process P_2 is executed repeatedly. There are no fairness assumptions about the self-loop on the mode `Crit`, as we don't want to rely on the process P_1 staying in the critical section for a specific duration. But we do want to assume that it eventually does leave the critical section (otherwise the process P_2 will be blocked forever), and hence we assume weak fairness for the mode-switch from the mode `Crit` to `Idle`. Note that in each case, weak fairness suffices since each task, once enabled, stays enabled until it is executed.

Now let us consider the protocol of figure 4.16. As in the case of Peterson's protocol, we don't make any fairness assumptions about the self-loops on the modes `Idle` and `Crit` and the switch from the mode `Idle` to `Crit`. Weak fairness is assumed for the switch from the mode `Crit` to `Idle` to capture the assumption that a process eventually does exit its critical section. Regarding the conditional switch out of the mode `Try` that tests the value of the shared register $free$, note that this corresponds to output actions of the shared register, and thus the fairness assumptions should be added to the description of the process corresponding to $free$ (see the description of the `Test&SetReg` process in figure 4.15). We assume *strong fairness* for the four output tasks A_{ts0}^1, A_{ts1}^1, A_{ts0}^2, and A_{ts1}^2. Note that each task has a guard, and as the value of the shared register changes, each of the tasks can switch between being enabled and being disabled. Strong fairness with respect to the task A_{ts0}^1 ensures that if it is the case repeatedly that the process P_1 is in the mode `Try` and the register is 0, then the task A_{ts0}^1 must eventually be executed, resulting in a synchronization that returns the value 0 to the process P_1. Thus, it cannot happen that the process P_1 waits in the mode `Try` indefinitely while the process P_2 enters and exits the critical section repeatedly. Note that weak fairness for the task A_{ts0}^1 will not rule out such a scenario.

Correctness under Fairness Assumptions

When proving liveness requirements of an asynchronous process with fairness assumption, we can restrict attention only to fair ω-executions. For example, for the process `Merge`, weak fairness is assumed for the output tasks A_o^1 and A_o^2. With such a fairness assumption, if it processes the input $in_1 ? v$ at any step, then it is guaranteed that at some future step it will produce the output $out ! v$. This is because in every fair execution, if the ith action processes the input $in_1 ? v$, then the message v will be added to the queue x_1. Once the queue x_1

has a message, the output task A_o^1 stays enabled at least until this message has been transmitted on the output channel. The weak-fairness assumption for the output task A_o^1 ensures that it will be executed: if the queue x_1 contains multiple messages ahead of v, then they will all be eventually transferred, finally along with v itself. The desired requirement does not hold for unfair executions, but such executions are merely an artifact of modeling definitions and not indicative of a violation in a real implementation.

For the mutual exclusion protocols of figures 4.13 and 4.16, in every fair execution, if the process P_1 wants to enter the critical section, then it will eventually enter the critical section.

The choice of fairness assumptions clearly affect the requirements that an asynchronous process satisfies. Let us illustrate this by considering different requirements for the processes `AsyncInc` and `AsyncEvenInc`:

- The requirement φ_1 states that "the value of x eventually exceeds 10." The process `AsyncInc` does not satisfy this requirement in absence of fairness assumptions but does satisfy this requirement if weak fairness for the task A_x is assumed. Similarly, the process `AsyncEvenInc` does not satisfy this requirement in absence of fairness assumptions but does satisfy this requirement if weak fairness for the task A_x is assumed.

- Consider the requirement φ_2 that "the value of y eventually exceeds 10." The process `AsyncInc` does not satisfy this requirement in absence of fairness assumptions but does satisfy this requirement if weak fairness for the task A_y is assumed. The process `AsyncEvenInc`, in contrast, does not satisfy this requirement in absence of fairness assumptions, and does not satisfy this requirement if only weak fairness is assumed, but it does satisfy this requirement if strong fairness for the task A_y is assumed.

- The requirement φ_3 states that "the value of y eventually exceeds the value of x." The process `AsyncInc` does not satisfy this requirement. The requirement is still not satisfied even if we assume fairness for the two tasks. In particular, the infinite execution where we first execute the task A_x, then execute the task A_y, and repeat this pattern is fair to both the tasks, and along this execution, the condition $y \leq x$ holds in every state (and thus the requirement φ_3 is violated). By a similar argument, the process `AsyncEvenInc` does not satisfy the requirement φ_3 with or without any form of fairness assumptions.

Fairness assumptions only ensure eventual execution of tasks and cannot enforce any specific pattern in relative frequencies of executions of different tasks. If such a pattern is required, as is the case in the requirement φ_3, then the coordination logic within the system must be modified to meet this requirement.

Exercise 4.10: Let us revisit the asynchronous process `Split` that you designed in exercise 4.2. Suppose we want to capture the assumption that the distribution of messages among the two output channels should be, while un-specified, fair in the sense that if infinitely many messages arrive on the input channel *in*, then both output channels out_1 and out_2 should have infinitely many messages transmitted. How would you add fairness assumptions to your design to capture this? If you are using strong fairness, then argue that weak fairness would not be enough (that is, describe an infinite execution that is weakly fair but the split of messages is not fair as desired). ■

Exercise 4.11: By modifying the description of the process `UnrelFIFO` of fig-ure 4.19, construct a precise specification of the process `VeryUnrelFIFO`, which, in addition to losing and duplicating messages, can also reorder messages. What would be natural fairness assumptions for the modified process? ■

Exercise 4.12: Consider the modified version of Peterson's mutual exclusion protocol shown in figure 4.17. What fairness assumptions should be added to this description? With these fairness assumptions, does the protocol satisfy the requirement that *if a process wants to enter the critical section, then it eventually will enter the critical section?* ■

Exercise 4.13: Consider an asynchronous process P with two variables x and y, both of type nat, with x initialized to 0 and y initialized to 2. The behavior of the process is described by two tasks. The task A_1 is always enabled, and its update code $x := x + 1$. The task A_2 is always enabled, and its update code is $y := x + y$. Answer each of the questions below with a brief justification. When adding fairness assumptions, clearly specify whether you are using strong fairness or weak fairness and for which task.

1. Is it guaranteed that the value of x eventually exceeds 5? If not, is there a suitable fairness assumption for the two tasks under which this guarantee holds?

2. Is it guaranteed that the value of y eventually exceeds 5? If not, is there a suitable fairness assumption for the two tasks under which this guarantee holds?

3. Is it guaranteed that at some step in the execution the values of x and y become equal? If not, is there a suitable fairness assumption for the two tasks under which this guarantee holds?

■

4.3   Asynchronous Coordination Protocols

In a network of processes communicating asynchronously, in each step a sin-gle process executes a computation step, and such a step can either receive an input value on an incoming channel or send an output value on an outgo-ing channel. As a result, algorithms for solving coordination problems cannot

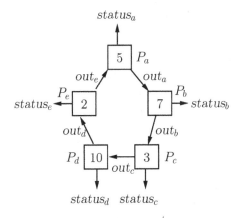

Figure 4.20: An Asynchronous Network with Ring Topology

proceed in lock-step rounds as in the synchronous case. We illustrate some of
the design challenges using three classical problems: electing a leader in a ring
of processes, implementing reliable communication using unreliable links, and
reaching consensus among two processes using shared objects.

4.3.1 Leader Election

Let us revisit the coordination problem of leader election discussed in section 2.4.3, now in the asynchronous setting. Let us assume that the underlying
network connects the nodes in a unidirectional ring (see figure 4.20 for an example ring with five nodes). Each node has a unique identifier, and the protocol
consists of a strategy for nodes to exchange messages so that eventually a single node declares itself to be the leader, with the remaining nodes declaring
themselves to be followers.

We model each network node as an asynchronous process P. The input channel
in receives identifiers sent by the unique predecessor of P in the ring, and the
output channel *out* sends identifiers to the unique successor of P in the ring.
An internal queue x is used to store messages received on the channel *in*, and
the queue y is used to store messages to be sent, which get delivered by the
output task on the channel *out* one by one. When the process concludes that it
is either the leader or one of the followers, the decision is issued on the output
channel *status*. The description of the process is parameterized by the identifier
of the corresponding network node, denoted myID. We will assume that each
identifier is a positive number. To form a ring, we create multiple instances of
the process P and compose them together using the asynchronous composition
operation. For example, the system corresponding to the ring of five processes

in figure 4.20 is $P_a \mid P_b \mid P_c \mid P_d \mid P_e$, where

$$
\begin{aligned}
P_a &= P[status \mapsto status_a][in \mapsto out_e][out \mapsto out_a][\texttt{myID} \mapsto 5], \\
P_b &= P[status \mapsto status_b][in \mapsto out_a][out \mapsto out_b][\texttt{myID} \mapsto 7], \\
P_c &= P[status \mapsto status_c][in \mapsto out_b][out \mapsto out_c][\texttt{myID} \mapsto 3], \\
P_d &= P[status \mapsto status_d][in \mapsto out_c][out \mapsto out_d][\texttt{myID} \mapsto 10], \\
P_e &= P[status \mapsto status_e][in \mapsto out_d][out \mapsto out_e][\texttt{myID} \mapsto 2].
\end{aligned}
$$

Our goal is to complete the description of the process P so that when multiple instances of this process are composed to form a ring, the following requirements are met: (1) every process eventually terminates, that is, there is no infinite execution of the protocol; and (2) in every terminating execution, exactly one process has output the value **leader** on its output channel *status*, and the remaining processes have output the value **follower** on their output channels *status*.

Recall that in the synchronous solution, if N is the total number of nodes in the network, then assuming the network to be strongly connected, a node could infer that its identifier has reached all the nodes in the network within N rounds. In the asynchronous case, no such inference can be drawn, as nodes are executing at independent speeds, and there is no concept of a *round* that involves all the processes. A node can infer that the message it sent to its successor on the output channel *out* has propagated to all the processes in the ring only when it receives an appropriate input message from its predecessor. Consequently, a process does not need to know the number of processes.

One possible solution to the asynchronous leader election in a ring is obtained by adopting the flooding algorithm of section 2.4.3 that elects the process with the highest identifier. We describe a more interesting algorithm that reduces the number of messages that are exchanged. If the ring contains N processes, then the algorithm to be discussed will generate only about $N \log N$ messages, as opposed to N^2 messages that the flooding algorithm can generate. As it turns out, $N \log N$ is also a *lower bound* on the number of messages that have to be exchanged in order to elect a leader among processes communicating asynchronously over a ring network.

The algorithm is shown in figure 4.21. The input task A_i is always enabled and simply stores each input message in the internal queue x. The output task A_o outputs pending messages from the queue y to the output channel *out* and can be executed at any time provided the queue y is not empty.

The core computation of the process is described as an extended-state machine. Initially, the mode is undecided. Once a decision is reached, the process switches to the leader or the follower mode, and during this switch, the decision is output on the channel *status*. In the follower mode, the process simply relays messages from its input queue x to its output queue y, and this is captured by the internal task A_7.

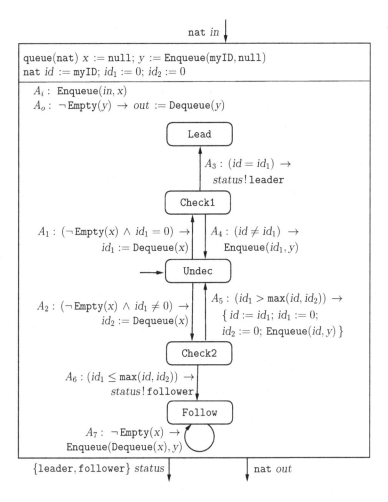

Figure 4.21: Asynchronous Leader Election in a Ring

The execution of the algorithm progresses in phases, and in each phase, the number of undecided processes decreases at least by a factor of 2, until only one process remains undecided, which then becomes the leader.

Initially, each process sends its identifier to two successive processes along the ring. To achieve this, each process first sends its identifier, as well as the first input message it receives, on the output channel. When a process receives two messages on the input channel, it knows its own identifier, captured by the variable id, the identifier of its predecessor, captured by the variable id_1, and the identifier of the predecessor's predecessor, captured by the variable id_2.

Initially, the variable id is set to $\texttt{myID}$, the unique positive number associated with the process. The variables id_1 and id_2 are set to 0. The identifier is enqueued in the outgoing queue to be transmitted to the next process. Then the process waits until there is a message to be processed in the incoming queue x. When the value of id_1 is 0, the next message to be processed, which is at the front of the queue x, is the value from the predecessor, and id_1 is set to this value by dequeuing x, and the process switches to the mode Check1 (see task A_1). If the value of id_1 is non-zero, the next input message to be processed is the identifier of the predecessor's predecessor. The task A_2 dequeues this message, stores it in the variable id_2, and switches to the mode Check2.

In the mode Check1, the process checks the value of the predecessor's identifier stored in id_1. When this value equals the current value of id, the process has won the election. In this case, the process outputs the value **leader** on the channel *status*, and the mode is updated to Lead (see task A_3). Otherwise, when the predecessor's identifier is different from the current value of id, this value is enqueued in the outgoing queue y to be sent to the successor process, and the mode switches to Undec (see task A_4).

Once the process has received the values of both id_1 and id_2, in the mode Check2, it compares these two identifiers with the value of id. If id_1 is the highest among these three identifiers, then the process continues to remain undecided, adopting the value of id_1 as its own identifier, and initiates a new phase starting in the mode Undec (see task A_5). If id_1 is not the highest among these three identifiers, then the process outputs the decision **follower** on the channel **status** and switches to the follower mode Follow (see task A_6). Subsequently, this process will only relay messages without examining them.

Note that every process is repeating the same computation. Suppose for a process P, $id = m_0$, $id_1 = m_1$, and $id_2 = m_2$. The process P will continue to stay undecided if both $m_1 > m_0$ and $m_1 > m_2$. Consider the predecessor P' of P. Then for the process P', its own identifier, that is, the value of its id variable is m_1, and the identifier of its predecessor, that is, the value of its id_1 variable, is m_2. This guarantees that *if P decides to stay undecided adopting m_1 as its identifier, P' will become a follower*. Consequently, the number of processes that continue to stay undecided is at most half of the current number of undecided processes. Furthermore, the number of processes that continue to stay undecided is at least 1: among all the undecided processes, the successor of the process with the highest identifier is guaranteed to stay undecided.

For the example network shown in figure 4.20, for the process P_c with the original identifier 3, the values of id, id_1, and id_2 in the first phase will be 3, 7, 5, respectively, and it will continue to the next phase as an undecided process, with 7 as its identifier. For the process P_b with the original identifier 7, the values of id, id_1, and id_2 in the first phase will be 7, 5, and 2, respectively, and it will become a follower. After the first phase, only processes P_c and P_e will be undecided, with modified identifiers 7 and 10, respectively.

When a process continues to stay undecided, it repeats the protocol again. It sends its current identifier (which was adopted from its predecessor in the preceding round) and the next input message on its output channel. After receiving two input messages, it examines the relative ordering of its identifier and the identifiers of its two (undecided) predecessors, making decisions as before. That is, in every subsequent phase, the current ring with the reduced number of undecided processes repeats the same protocol, thereby again reducing the number of undecided processes by at least half. The presence of follower processes does not influence the logical argument since they are simply relaying messages.

When an undecided process receives an input message that is equal to its current identifier, it can conclude that it is the only undecided process and proceeds to declaring itself as the leader. Note that even though this identifier is guaranteed to be the highest among all the original identifiers, it is not the original identifier of this leader process.

Continuing our example from figure 4.20, during the second phase, for the process P_c, the values of id, id_1, and id_2 will be 7, 10, and 7, respectively, and it will continue to the next phase as the only undecided process adopting the identifier 10. In the third phase, the first message it sends will come back to it as the predecessor's identifier, with all other processes simply relaying this message. This will cause the process P_c to declare itself as the leader.

The formal correctness argument is complicated by the fact that the phases are not synchronized, and at any given time, neighboring processes may be executing different phases. In each phase, each process sends at most two messages. If the ring contains N processes, then each phase contributes at most $2N$ messages, and the number of phases is at most $\log N$, leading to an overall bound of $2N \log N$ messages.

In this protocol, no process ever sends messages repeatedly. Thus, no infinite execution is possible, and as a result, correctness does not require any fairness assumptions.

Exercise 4.14: For the leader election protocol of figure 4.21, consider a ring with 16 nodes where the identifiers of the processes in order are: 25, 3, 6, 15, 19, 8, 7, 14, 4, 22, 21, 18, 24, 1, 10, 23. Which process will be elected as the leader? ■

Exercise 4.15 *: For the leader election protocol of figure 4.21, describe the best- and worst-case scenarios: (a) describe the scenario in which only one node will stay active after the first phase, and (b) describe the scenario in which the protocol will need $\log N$ phases before the election. ■

4.3.2 Reliable Transmission

Given an unreliable communication medium, how can we implement a reliable FIFO link that delivers each message exactly once in the order received? More specifically, we want to design processes P_s and P_r so that the composite system shown in figure 4.22 acts as a reliable FIFO buffer with respect to its input and output channels using two instances of the unreliable communication link

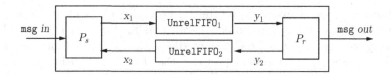

Figure 4.22: The Block Diagram for Reliable Communication

UnrelFIFO (see figure 4.19). The process P_s acts as an interface for the sender, and the process P_r acts as an interface for the receiver. The unreliable link UnrelFIFO$_1$ transfers messages from the process P_s to P_r, and the unreliable link UnrelFIFO$_2$ transfers messages from the process P_r to P_s.

Alternating Bit Protocol

To deliver a message that the process P_s receives on its input channel *in*, it may need to send the message repeatedly to the process P_r since the link UnrelFIFO$_1$ may lose messages, and the process P_r needs to send an explicit acknowledgment back to the process P_s notifying successful delivery. The acknowledgment also needs to be sent repeatedly to ensure eventual successful delivery to account for lost messages. A key design challenge is to match messages with acknowledgments, in the presence of potential duplication of messages as well as duplication of acknowledgments. One classical solution for this purpose is the *alternating bit protocol* that synchronizes the sender and the receiver processes using a Boolean tag bit that alternates.

The sender interface process P_s is shown in figure 4.23. It maintains a queue x of messages that it receives on its input channel *in*, and it is processed by the input task A_i. The state variable *tag* is a Boolean variable that is initially 1. When the process P_s sends the message at the front of its internal queue x to the receiver process P_r using the unreliable FIFO link on the channel x_1, it augments the message with the current value of *tag* and does not remove the message from the queue x. The output task A_1 may get executed repeatedly. When the sender process P_s gets an acknowledgment on the channel x_2 in the form of a tag bit from the receiver, it checks whether the received tag matches its own tag; if this check succeeds, it removes the message from its queue x and toggles the tag. The processing of acknowledgment tags is modeled by the input task A_2. The toggling of the tag will cause the next message in the queue x to be sent, possibly repeatedly, on the output channel x_1 augmented with this updated tag. Note that the task A_2 is always enabled, but it does not modify the state if the incoming acknowledgment tag does not match the expected tag. The fairness assumption consists of weak fairness for the output task: once a message is enqueued in the queue x, the task A_1 stays enabled and should eventually be executed sending the first message on the channel x_1.

The receiver process P_r is shown in figure 4.24. The messages it receives are

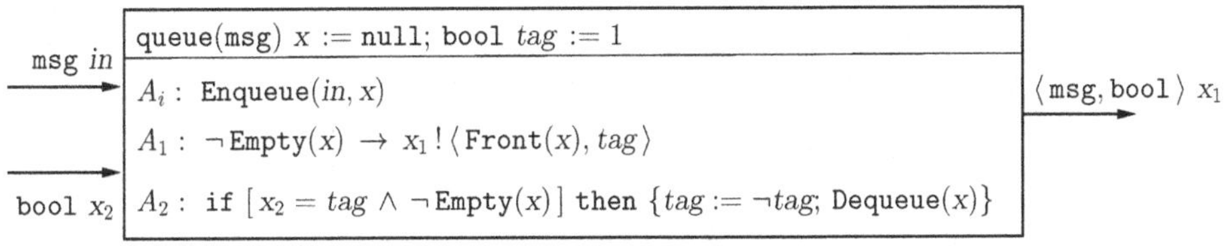

```
                    ┌────────────────────────────────────────────────────────────────┐
         msg in     │ queue(msg) x := null; bool tag := 1                             │
        ──────────▶ ├────────────────────────────────────────────────────────────────┤
                    │ A_i : Enqueue(in, x)                                            │    ⟨msg, bool⟩ x₁
                    │ A₁ : ¬Empty(x) → x₁ ! ⟨Front(x), tag⟩                           │  ──────────────▶
        ──────────▶ │                                                                 │
         bool x₂    │ A₂ : if [ x₂ = tag ∧ ¬Empty(x) ] then {tag := ¬tag; Dequeue(x)} │
                    └────────────────────────────────────────────────────────────────┘
```

Figure 4.23: The Sender Process for the Alternating-bit Protocol

stored in the internal queue y. The process also maintains a Boolean-valued tag state variable, which is initially 0. Note that the initial values of the tag bits of the sender P_s and the receiver P_r are complements of each other: initially and at every step, the sender P_s expects the incoming tag from the receiver to be the same as its internal tag, whereas the receiver expects the incoming tag from the sender to be the complement of its internal tag. When the receiver process P_r receives a message on the input channel y_1, it checks whether the tag of the incoming message is the complement of its own tag. If so, the incoming message is considered a new message, and it is added to the queue y. This is captured by the input task A_1, where the primitives First and Second are used to retrieve the two fields of the incoming message. Note again that the task A_1 is always enabled, and if the incoming tag is not what it expects, the incoming message is simply ignored. Messages in the queue y are transmitted on the output channel by the output task A_o. The receiver P_r also repeatedly sends the current value of its tag to the sender process P_s as an acknowledgment on the channel y_2 (captured by the output task A_2). To ensure eventual delivery on both the output channels, the fairness assumption consists of weak fairness for both the output tasks A_2 and A_o. Since these tasks are not disabled by competing actions, we don't need strong fairness.

The following scenario describes how the protocol executes. Suppose the process P_s receives a message, say m_1, on its input channel *in*. Then it will repeatedly send the message $(m_1, 1)$ to the process P_r using the unreliable channel. Each such message may be lost or duplicated. Meanwhile, the process P_r can repeatedly send the tag bit 0 to the process P_s, but P_s will ignore all such acknowledgments. The first time the message $(m_1, 1)$ is successfully delivered to the process P_r on the channel y_1, the process P_r will change its tag to 1 and enqueue m_1 in its output queue y. The message m_1 will eventually be transmitted on the output channel *out*. Additional copies of the message $(m_1, 1)$ received on the channel y_1 will be ignored by the process P_r since its tag is now 1: it will recognize the next message as a fresh message only when the message is tagged with 0. The process P_r will repeatedly send the tag 1 to P_s as an acknowledgment on the channel y_2. Each such message again may be lost or duplicated, but eventually the process P_s will receive the tag 1 on the channel x_2. At this point, the process P_s will remove the message m_1 from its internal queue x and toggle its *tag* variable to 0. If additional messages are received on

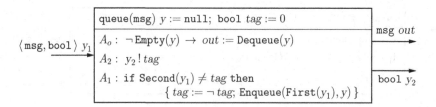

Figure 4.24: The Receiver Process for the Alternating-bit Protocol

the channel *in* during this period, then they get enqueued in the queue x, and if m_2 is the next pending message, then the process P_s will start sending the message $(m_2, 0)$ to P_r on the channel x_1. If the process P_s receives additional tag messages 1 on the channel x_2, then it will ignore them. The message m_2 will be dequeued by the process P_s from its queue x only when it receives the tag 0.

Exercise 4.16: Suppose we know that the communication link from the receiver back to the sender is reliable. How would you modify the alternating-bit protocol to take advantage of this? That is, design simplified versions of the processes P_s and P_r so that the composite system shown in figure 4.22 acts a reliable FIFO buffer when the process UnrelFIFO$_2$ is replaced by the process Buffer. ∎

Exercise 4.17 *: Consider the description of the process VeryUnrelFIFO designed in exercise 4.11 of an unreliable link that may lose messages, duplicate messages, and reorder messages. First, show that the alternating-bit protocol does not work correctly if we replace each instance of UnrelFIFO with a corresponding instance of the process VeryUnrelFIFO. How would you modify the processes P_s and P_r so that reliable communication is guaranteed even in the presence of this added complication of reordering? Argue that the modified protocol works correctly. Hint: a Boolean-valued tag is not enough, and messages need to be tagged with a counter variable of type nat. ∎

4.3.3   Wait-Free Consensus *

To see how the choice of atomic primitives supported by shared objects impacts the ability to solve distributed coordination problems, let us consider the classical problem of *wait-free two-processes consensus*. Each process starts with an initial preference that is known only to itself. The processes want to communicate and arrive at a consensus decision value. This problem has been posed in many different forms, for instance, requiring two Byzantine Generals in charge of collaborating armies separated by the enemy army to exchange messengers to arrive at a mutually agreed time of attack. The core coordination problem of reaching agreement in the presence of unpredictable delays is central to many distributed computing problems.

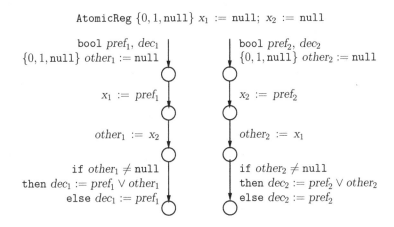

AtomicReg $\{0, 1, \text{null}\}$ $x_1 := \text{null}$; $x_2 := \text{null}$

bool $pref_1$, dec_1
$\{0, 1, \text{null}\}$ $other_1 := \text{null}$

$x_1 := pref_1$

$other_1 := x_2$

if $other_1 \neq \text{null}$
then $dec_1 := pref_1 \vee other_1$
else $dec_1 := pref_1$

bool $pref_2$, dec_2
$\{0, 1, \text{null}\}$ $other_2 := \text{null}$

$x_2 := pref_2$

$other_2 := x_1$

if $other_2 \neq \text{null}$
then $dec_2 := pref_2 \vee other_2$
else $dec_2 := pref_2$

Figure 4.25: First Solution to Two-Process Consensus Using Atomic Registers

Problem Description

More specifically, we have two asynchronous processes, say P_1 and P_2, each of which has an initial Boolean value, denoted v_1 and v_2, respectively, unknown to the other. The processes want to arrive at Boolean decision values d_1 and d_2, respectively, so that the following three requirements are met: (1) the decision values d_1 and d_2 of the two processes are identical, (2) the decision value must be equal to one of the initial values v_1 or v_2, and (3) at any time, if tasks involving only one of the processes are repeatedly executed, this process should reach a decision. The first requirement, called *agreement*, means that the two processes should come to a common decision even if they start with different preferences. The second requirement, called *validity*, says that if both prefer the same value, then they must decide on that value. This requirement rules out input-oblivious solutions such as: *both decide on 0 no matter what the initial preferences are.* The third requirement, called *wait freedom*, ensures that a process can decide on its own without having to wait indefinitely for the other.

Suppose we want to design the processes P_1 and P_2 so that they communicate using shared objects such as atomic registers and test&set registers. In the composite system, every action then will be either an internal action of one of the processes or will be a primitive operation by one of the processes involving one of the shared objects.

Incorrect Solutions Using Atomic Registers

Correctness requirements for the problem and challenges in designing a correct solution can be best illustrated using protocols that meet only some of the requirements. As a first attempt, consider the solution shown in figure 4.25. The solution uses two shared atomic registers x_1 and x_2, each of which is an

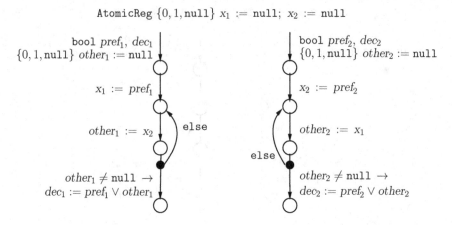

Figure 4.26: Second Solution to Two-Process Consensus Using Atomic Registers

instance of the `AtomicReg` process shown in figure 4.10. Here, the set `val` of values is $\{0, 1, \mathtt{null}\}$, and the initial value `initVal` for each of the shared registers is `null`.

The initial preference of the process P_1 is stored in its state variable $pref_1$, and its objective is to set the decision variable dec_1 to the decision value on termination. The process P_1 first writes its preference to the shared register x_1, and analogously, the process P_2 writes its preference (captured by the initial value of its state variable $pref_2$) to the shared register x_2. The process P_1, after writing its preference to x_1, reads the shared register x_2 into its internal state variable $other_1$. Since the execution of the two processes proceeds asynchronously, when the process P_1 executes the action of reading x_2, there is no guarantee that the process P_2 has already written its preference to x_2. To account for this possibility, the process P_1 checks whether the value it reads is `null`. If it is `null`, then it decides on its own preference; otherwise, it knows the preference of the process P_2 and decides on the logical disjunction of the two preferences. The process P_2 is symmetric.

The protocol, however, is buggy. Suppose the initial preferences of P_1 and P_2 are 0 and 1, respectively. Consider the execution in which we first execute only the tasks involving the process P_1 until it finishes and then execute all the actions of the process P_2. In this scenario, when the process P_1 reads x_2, its value is still `null`, and hence P_1 decides on its own preference, 0. However, when the process P_2 reads x_1, it receives the value 0, and it decides on the disjunction of the two initial preferences, namely, 1. Thus, the protocol violates the *agreement* requirement. Observe that the protocol does meet the requirements of validity (if both initial preferences are 0, both will decide 0; and if both initial preferences are 1, both will decide 1 no matter in which order the two processes execute

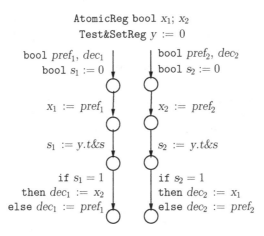

AtomicReg bool x_1; x_2
Test&SetReg $y := 0$

bool $pref_1$, dec_1 bool $pref_2$, dec_2
bool $s_1 := 0$ bool $s_2 := 0$

$x_1 := pref_1$ $x_2 := pref_2$

$s_1 := y.t\&s$ $s_2 := y.t\&s$

if $s_1 = 1$ if $s_2 = 1$
then $dec_1 := x_2$ then $dec_2 := x_1$
else $dec_1 := pref_1$ else $dec_2 := pref_2$

Figure 4.27: Solution to Two-Process Consensus Using a Test&Set Register

their respective actions) as well as wait freedom (each process executes exactly three actions before terminating and enabling of each of these actions does not depend on the other process).

We can try to "fix" the protocol by requiring each process to wait until it knows the preference of the other process. Figure 4.26 shows the revised protocol: after the process P_1 reads the shared register x_2, if its value is `null`, it loops back and reads x_2 again. In the revised version, the requirement of agreement is satisfied since both processes decide only after they know both the preferences. The requirement of validity also holds. However, the requirement of wait freedom is violated. The reason is that if, say the process P_2, has not yet executed its write to x_2, then the process P_1 will repeatedly read x_2 and will not be able to reach a decision on its own.

Solution Using Test&Set Registers

It is possible to solve consensus using a single test&set register. Consider the following protocol that uses two Boolean atomic registers x_1 and x_2 and a test&set register y (see figure 4.27). The initial values of the registers x_1 and x_2 do not matter, and y is initially 0. The process P_1 executes the following sequence of actions, with the process P_2 following a symmetric protocol. The process P_1 first writes its own preference to the atomic register x_1. Then it executes a *test&set* operation on the register y. If the value returned (stored in the state variable s_1) is 0 (implying that the register y was 0 before the process P_1's *test&set* operation was executed), then P_1 goes ahead and decides on its own preference. If the value received by the *test&set* operation on y is 1, then the process P_1 concludes that the other process P_2 had already executed its *test&set* operation successfully, and hence the register x_2 must contain the preference of the process P_2. The process P_1 then proceeds to read x_2 and decides on the value it contains. In summary, each process publishes its preference in a shared atomic register, executes *test&set* to resolve contention, and, based on the result of this test, decides whose preference to adopt. Each process executes a fixed number of actions and thus can decide without waiting for the other.

Impossibility of Solving Consensus Using Atomic Registers

The key to the correct solution of figure 4.27 is the use of the atomic operation *test&set* that updates the register and returns its old value without interference from other processes. If we are required to use only atomic registers, where the operations of reading and writing a register are decoupled, then no matter how many shared registers the protocol employs and how many values each such register can hold, there is no solution that satisfies all three requirements of agreement, validity, and wait-freedom.

For the protocol of figure 4.27, consider the initial state in which $pref_1 = 0$ and $pref_2 = 1$. Such a state is called *uncommitted* in the sense both decisions are still feasible: starting from such a state, there is a possible execution in which both processes end up deciding on 0, and there is another possible execution that results in both processes deciding on 1. The first step of the proof below is to establish that such an uncommitted state must exist in every protocol that correctly solves the consensus problem. For the protocol of figure 4.27, starting with $pref_1 = 0$ and $pref_2 = 1$, consider the state after both the processes have taken one step each and have written their respective preferences to the variables x_1 and x_2. This state is still uncommitted, but if the next step is by P_1, then the final decision is guaranteed to be 0 (irrespective of how the execution proceeds subsequently), while if the next step is by P_2, then the final decision is guaranteed to be 1. A key part of the proof is to establish that in every consensus protocol, there must be such an uncommitted reachable state such that the next step is the deciding factor for the final decision. For the protocol of figure 4.27, this critical step involves the *test&set* operation on a shared register. The proof concludes by showing that if all that a process can do in one step is either read or write an atomic register, then such a step cannot be the critical deciding factor, and thus the problem cannot be solved using only atomic registers.

Theorem 4.1 [Impossibility of Consensus using Atomic Registers] *There is no protocol for two-process consensus such that (1) the processes communicate using only atomic registers as shared objects, and (2) the protocol satisfies all three requirements of agreement, validity, and wait-freedom.*

Proof. Suppose there exists a solution to the two-process consensus problem using only atomic registers. Consider the transition system T that corresponds to the system obtained by composing the two processes P_1 and P_2 and all the atomic registers that the protocol uses. A state s of T consists of the internal states of the two processes and the states of all the shared atomic registers. A single transition of T is either an internal action of one of the two processes or a read action involving one of the processes and one shared register, or a write action involving one of the processes and one shared register.

Starting from a given state s, many executions are possible, but each one is finite and ends in a state where both processes have decided. Let us call a state s *uncommitted* if both decisions 0 and 1 are still possible: there is an execution starting in the state s in which both processes decide 0, and there is another execution starting in the state s in which both processes decide 1. A state is called 0-*committed* if in all executions starting in the state s both processes decide 0 and 1-*committed* if in all executions starting in the state s both processes decide 1.

Let us call two states s and t P_2-*indistinguishable* if the internal state of the process P_2 is the same in both states s and t, and the state of each of the shared registers is also the same in both states s and t. That is, the states s and t look the same from the perspective of the process P_2: if P_2 can execute an action in the state s, then it can execute the same action in the state t.

As a first step toward the proof, we first establish the following:

> Lemma 1. If two states s and t are P_2-indistinguishable, then it cannot be the case that the state s is 0-committed and the state t is 1-committed.

The wait-freedom requirement means that starting in any state, if we execute actions involving only one of the two processes, then it must reach a decision. Consider two states s and t that are P_2-indistinguishable such that the state s is 0-committed. In the state s, if we let only the process P_2 execute actions, then it will eventually reach a decision, and this must be 0 by the assumption that all executions starting in the state s lead to the decision 0. Now consider what happens if we let only the process P_2 take steps starting in the state t. Since the states s and t look the same as far as the process P_2 can tell, it can execute the same sequence of actions and reach the same decision 0. Thus, the state t cannot be 1-committed.

The next step in the proof is the following lemma:

> Lemma 2: There exists an uncommitted initial state.

Consider an initial state s in which the preferences v_1 and v_2 of the two processes are different, say 0 and 1, respectively. We claim that this state must be uncommitted. If not, suppose it is 0-committed. In the state s, the process P_2 has preference 1. Neither the initial values of the shared registers nor the initial values of the state variables belonging to P_2 reflect the initial preference of the process P_1. Consider the initial state t in which the initial values for the shared registers and the state variables of P_2 are identical to those in the state s but the initial state of the process P_1 is chosen so that its initial preference is 1. That is, the only difference in the states s and t is in the initial preference of the process P_1. The states s and t are P_2-indistinguishable by construction. By

Lemma 1, we can conclude that the state t cannot be 1-committed. But this is a contradiction to the *validity* requirement: in the state t, both preferences are 1, and thus every execution starting in the state t must lead to the decision 1 (otherwise the protocol does not satisfy the validity requirement).

We now proceed to establish that:

> Lemma 3: There exists an uncommitted reachable state s such that all successor states of the state s are committed.

The proof of the lemma is by contradiction. First observe that the protocol cannot terminate in an uncommitted state since both processes are required to reach a common decision upon termination. Then if Lemma 3 does not hold, we can assume that every reachable uncommitted state has an uncommitted successor state. Consider an uncommitted initial state s_0 guaranteed by Lemma 2. Clearly, the state s_0 is reachable, and by assumption, it must have an uncommitted successor, say state s_1. We can repeat this argument again: at every step j, we have a reachable uncommitted state s_j, and by assumption, we can find an uncommitted successor state s_{j+1} extending the execution by one more step. This means that there is an infinite execution in which processes have not reached a decision, a violation of the correctness requirement to reach agreement. It follows that Lemma 3 must hold.

Consider a state s promised by Lemma 3. The state s is uncommitted, that is, both decisions are still possible, but executing one more step by either of the processes commits the protocol to the eventual decision. Without loss of generality, we can assume that there exist actions $s \to s_1$ by the process P_1 using the task A_1 and $s \to s_2$ by the process P_2 using the task A_2, such that every execution starting in the state s_1 ends up with the decision 0, and every execution starting in the state s_2 results in the decision 1 (see figure 4.28). Each action can be an internal action, a reading of a shared register, or a writing of a shared register. To complete the proof, we consider all possible types of tasks for A_1 and A_2 and arrive at a contradiction in each case.

Suppose the task A_1 of the process P_1 corresponds to a reading of a shared atomic register. The execution of such a task does not modify the state of any of the shared objects and does not modify the internal state of the process P_2. Thus, the states s and s_1 are P_2-indistinguishable. Since the two states look the same to the process P_2, it can execute the task A_2 in the state s_1 also, and let the resulting state be t (see figure 4.28). The states t and s_2 are P_2-indistinguishable. But the state s_2 is 1-committed, while the state t is 0-committed, a contradiction to Lemma 1.

The cases when one of the tasks is an internal task and when the task A_2 involves a read action are similar. The interesting remaining case is when both the tasks A_1 and A_2 execute write actions. There are two sub-cases: they both write to the same register and they write to different registers. We will consider the former, leaving the latter as an exercise.

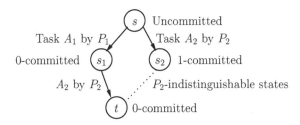

Figure 4.28: Impossibility Result for Consensus Using Atomic Registers

Consider the case when both the processes write to the same atomic register, say x. That is, in the state s, the process P_1 writes some value m_1 to the register x leading to the state s_1, and the process P_2 writes some value m_2 to the same register x leading to the state s_2. Note that in the state s_1, even though the value of the register x is different from its value in the state s, the internal state of the process P_2 is the same in both the states s and s_1. A key observation is that the execution of the task corresponding to writing a register is not influenced by the current value of the register. Thus, in the state s_1, the process P_2 can write the same value m_2 to the register x leading to the state t (see figure 4.28). The writing of the value m_1 by the process P_1 to x has been effectively lost and did not influence what the process P_2 was about to do in the state s. In the states s_2 and t, the internal states of the process P_2 are identical, and so are the states of all the shared registers. Thus, the states s_2 and t are P_2-indistinguishable, the state s_2 is 1-committed, and the state t is 0-committed: a contradiction to Lemma 1. ∎

Exercise 4.18: Complete the proof of Theorem 4.1 by considering the remaining case where the task A_1 writes to a shared register x, and the task A_2 writes to a different shared register y. ∎

Exercise 4.19: Consider the following solution to the two-process consensus problem in the asynchronous model. The processes use a shared atomic register x and a shared test-and-set register y. The possible values for the register x are *null*, 0, and 1, and the initial value is *null*. The possible values for the register y are 0 and 1, and its initial value is 0. Each process executes the following sequence of steps:

1. Write its initial preference to the register x.

2. Execute a test-and-set operation on the register y.

3. If step (2) returns 0, then decide on its own initial preference.

4. If step (2) returns 1, then read the register x and decide on the value read.

Consider the three requirements for consensus: validity, agreement, and wait-freedom. Which of these requirements are satisfied by this protocol? Justify your answer. ∎

Exercise 4.20 *: Consider the generalization of the consensus problem to multiple processes in which each process starts with an initial preference bit and wants to decide on a common Boolean value. The protocol must satisfy the requirements of agreement (all decide on the same value), validity (the decision value must be a preference of one the processes), and wait freedom (if a process takes steps all by itself, then it should reach a decision in finitely many steps without having to wait for the others). Assume that the description of atomic registers and test&set registers described in section 4.2.3 is suitably generalized so that a register can be accessed by multiple processes. Explain why the strategy described in the two-process protocol based on a single `Test&SetReg` register to resolve contention (see figure 4.27) does not generalize to three processes. Try to design a solution to the three-process consensus problem using two `Test&SetReg` registers and show that your attempts fail (note: when the number of processes is three (or more), there is no solution to the consensus problem using only atomic and test&set registers). ∎

Exercise 4.21 *: Consider the shared object `StickyBit` that supports read and write operations as in the case of an atomic register, with some modifications. The internal state of a `StickyBit` process can be `null`, 0, or 1 and is initially `null`. The read operation outputs the current value. The write operation has a Boolean (0 or 1) input value associated with it: if the current state is `null`, then the state is updated to the value of write, but if not (that is, if the state is already 0 or 1), then the value stays unchanged. Describe a protocol for solving two-process consensus using a single `StickyBit` object (you may use any number of additional atomic registers as you need). Can you solve consensus for three (or more generally, n) processes using multiple `StickyBit` and `AtomicReg` objects? ∎

Exercise 4.22 *: This exercise describes a classical puzzle that requires design of an asynchronous coordination strategy. There are N prisoners who get together initially to decide on a strategy. Then each prisoner is taken to her own isolated cell. A prison guard goes to a cell and takes its prisoner to a room where there is a switch. The switch can either be up or down. The prisoner is allowed to inspect the state of the switch and then has the option of flicking the switch. The prisoner is then taken back to her cell. The prison guard repeats this process infinitely often. The order in which he brings the prisoners to the cell is arbitrary, in particular, there is no bound on how many times one prisoner visits the room with the switch before some other prisoner gets to visit. However, the prison guard guarantees fairness: every prisoner will visit the room infinitely often. At any time, any prisoner can exclaim "I have concluded that every prisoner has visited the room with the switch at least once." Upon such a declaration, if the statement is indeed correct, all prisoners are set free; if the statement is not correct, all prisoners are immediately executed. What strategy should the prisoners use to ensure their eventual freedom? Note that the initial state of the switch is unknown to the prisoners, but as a warm-up, you may consider the same problem but with a known initial state of the switch. ∎

Bibliographic Notes

There is a rich history of formal models and distributed algorithms for asynchronous concurrent processes dating back to Dijkstra [Dij65]. The formal model described here is based on the model of I/O automata [LT87, Lyn96].

Different notions of fairness are discussed in [Fra86], and the literature on the model of fair transition systems contains many examples of specification and verification of asynchronous systems with weak and strong notions of fairness [MP91].

The coordination problems of mutual exclusion, consensus, leader election, and reliable communication in the presence of unreliable channels have been core research problems in both distributed computing and formal verification for decades (see for instance, [CM88], [Lyn96], and [Lam02]). The specific results we have discussed include Peterson's two-process mutual exclusion protocol [Pet81], leader election in a ring using $O(N \cdot \log N)$ messages [Pet82], alternating bit protocol for reliable communication [BSW69], and impossibility of consensus using atomic registers [FLP85, Her91].

5

Liveness Requirements

As discussed in chapter 3, requirements can be classified into two broad categories: *safety* requirements assert that "nothing bad ever happens," and *liveness* requirements assert that "something good eventually happens." For instance, in the leader election problem, the central safety requirement is that no two nodes should ever declare themselves to be the leaders, and the central liveness requirement is that each node should eventually make a decision. In chapter 3, we studied how to specify and verify safety requirements. Now we turn our attention to liveness requirements. Such requirements are specified using a formalism called *temporal logic*. The problem of checking whether a model satisfies its specification expressed in temporal logic is known as *model checking*.

5.1 Temporal Logic

Let us revisit our example of the system of traffic lights for a railroad from section 3.1.2. Given a model of the trains and the desired requirements, the design problem is to construct a controller so that the system composed of the trains and the controller satisfies the requirements. One basic requirement is that the two trains should not be on the bridge simultaneously. This safety requirement is captured by the property

$$\texttt{TrainSafety}: \ \neg\left[(mode_W = \texttt{bridge}) \ \wedge \ (mode_E = \texttt{bridge})\right]$$

and we require this property to be an invariant of the composite system. Obviously, this is not a *complete* specification for the desired controller: a controller that keeps both traffic lights to be red all the time is safe with respect to the property `TrainSafety` but is not an acceptable solution as it would never let any train onto the bridge. We need to augment the safety requirement with a liveness requirement that asserts that the controller should allow the trains onto the bridge. For resource allocation problems exemplified by our railroad system, while there is usually a canonical safety requirement, the liveness requirements can make differing demands. For instance, we may require that "if one of the trains wants to enter the bridge, then eventually some train should be allowed to enter," or we may demand a stronger requirement that "if a train wants to enter the bridge, then eventually that specific train should be allowed to enter."

Violation of a safety requirement is demonstrated by a finite execution that leads the system from an initial state to an erroneous state. For instance, the counterexample of figure 3.7 is a finite execution demonstrating that our first attempt at designing the controller for the railroad system is incorrect. Violation of a liveness requirement, in contrast, is not exhibited by such a finite execution. Instead, it consists of a cycle of states such that the cycle is reachable from an initial state, and if the cycle is executed repeatedly, the demand made by the liveness requirement is unmet. Hence, the mathematical formalization of liveness requirements considers infinite executions of the system. A natural formalism for specifying requirements about infinite executions is *temporal logic*. Temporal logics come in many varieties and can express safety as well as liveness requirements. We will study the classical temporal logic LTL, which stands for Linear Temporal Logic. This logic forms the core of the *Property Specification Language* (PSL), which has been standardized by IEEE and supported by commercial simulation and verification tools used in the electronic design automation industry.

5.1.1   Linear Temporal Logic

Let V be a set of typed variables, and suppose we are writing requirements to constrain the values these variables are allowed to take. Given a valuation q over V, that is, a type-consistent assignment of values to V and a Boolean expression e over V, $q(e)$ denotes the result of evaluating the expression e using the values assigned by the valuation q. When $q(e)$ equals 1, we say that the valuation q *satisfies* the expression e. We can thus interpret a Boolean expression e to express a constraint or requirement on individual valuations: the requirement e is satisfied when the expression evaluates to 1 according to the values assigned by the valuation. For example, suppose the set V contains two Boolean variables x and y. Then the expression $(x = y)$ expresses the requirement that both these variables should take the same value: a valuation q satisfies the requirement precisely when the value $q(x)$ is same as the value $q(y)$.

While an expression over variables V is evaluated with respect to a valuation for V, a temporal logic formula over V is evaluated with respect to an *infinite* sequence of valuations. That is, to interpret a temporal logic formula, we need to consider an infinite sequence $q_1 q_2 \ldots$ where each element q_i of the sequence is a valuation. For example, when the set V contains two Boolean variables x and y, each valuation is an assignment to the Boolean variables x and y, and the temporal logic formulas are evaluated with respect to an infinite sequence $\rho = (x_1, y_1)(x_2, y_2) \cdots$. We call such an infinite sequence of valuations a *trace* over V.

Boolean expressions are used to express constraints over individual valuations,

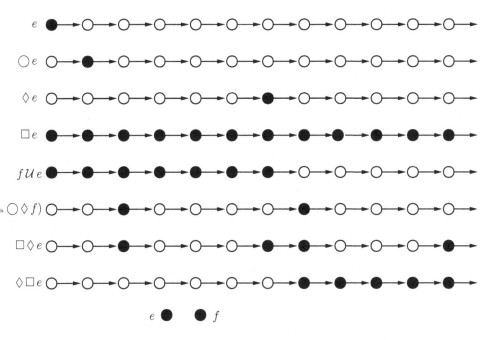

Figure 5.1: Illustrating Temporal Operators of LTL

and such expressions are combined using temporal operators that capture requirements for the sequence of valuations in a trace. Thus, a Boolean expression e over V is the simplest form of a temporal logic formula. We say that a trace ρ satisfies the Boolean expression e if the *first* valuation in the trace satisfies e. In our example, the trace $\rho = (x_1, y_1)(x_2, y_2) \cdots$ satisfies the LTL-formula $(x = y)$ precisely when the first valuation in the trace satisfies the expression, that is, when x_1 equals y_1.

Temporal Operators

Let us consider the temporal operator *always*, denoted $\Box$. The LTL-formula $\Box e$ is satisfied by a trace when every valuation in the trace satisfies e. For example, the trace $\rho = (x_1, y_1)(x_2, y_2) \cdots$ satisfies the LTL-formula $\Box (x = y)$ precisely when *every* valuation in the trace satisfies the expression $(x = y)$, that is, when $x_j = y_j$ for every j. Thus, the formula $\Box (x = y)$ expresses the requirement that the variable x should always be equal to y. Figure 5.1 illustrates the requirements imposed by different temporal logic formulas.

The dual of the *always* operator is the temporal operator *eventually*, denoted $\Diamond$. The LTL-formula $\Diamond e$ is satisfied by a trace when *some* valuation in the trace satisfies e. For example, the trace $\rho = (x_1, y_1)(x_2, y_2) \cdots$ satisfies the LTL-formula $\Diamond (x = y)$ precisely when some valuation in the trace satisfies the expression $(x = y)$, that is, when $x_j = y_j$ for some j. Thus, the formula

$\Diamond (x = y)$ expresses the requirement that eventually at some step, the values of the variables x and y coincide.

The temporal operator *next*, denoted $\bigcirc$, is used to assert requirements for the *next* valuation in the trace. The LTL-formula $\bigcirc e$ is satisfied by the trace $q_1 q_2 \ldots$ when the valuation q_2 satisfies the expression e. For example, the trace $\rho = (x_1, y_1)(x_2, y_2) \cdots$ satisfies the LTL-formula $\bigcirc (x = y)$ precisely when $x_2 = y_2$.

The final temporal operator we consider is called the *until* operator, denoted $\mathcal{U}$, that takes two formulas as arguments. The LTL-formula $f \, \mathcal{U} \, e$ is satisfied in a trace if the expression f is satisfied in every valuation in the sequence until we encounter a valuation that satisfies e. That is, the trace $q_1 q_2 \ldots$ satisfies the formula $f \, \mathcal{U} \, e$ precisely when there exists a position j such that the valuation q_j satisfies the expression e and each of the valuations from q_1 to q_{j-1} satisfies the expression f. For example, the trace $\rho = (x_1, y_1)(x_2, y_2) \cdots$ satisfies the LTL-formula $(x = 0) \, \mathcal{U} \, (y = 1)$ precisely when there is some position j such that $y_j = 1$ and $x_k = 0$ for all positions $1 \leq k < j$. This expresses the requirement that eventually y should become 1, and until then x should stay 0.

Temporal logic formulas can be combined using the standard logical operators: conjunction ($\wedge$), disjunction ($\vee$), implication ($\rightarrow$), and negation ($\neg$). For example, if φ_1 and φ_2 are two LTL-formulas, then we can combine them using the conjunction operator to obtain the LTL-formula $\varphi_1 \wedge \varphi_2$. A trace ρ satisfies the conjunction $\varphi_1 \wedge \varphi_2$ precisely when it satisfies both φ_1 and φ_2. Thus, a trace ρ satisfies the LTL-formula $\Box (x = y) \wedge \Diamond (x = 0)$ if in every valuation of the trace, the value of x is equal to the value of y, and there is a valuation in the trace which assigns the value 0 to x.

So far we have considered LTL-formulas in which the arguments to the temporal operators are expressions constraining individual valuations. In general, temporal operators can be nested, that is, arguments of temporal operators may be complex temporal logic formulas. For example, consider the LTL-formula $\bigcirc \Box (x = y)$. This formula says that in the next step, always $(x = y)$ holds: the trace $\rho = (x_1, y_1)(x_2, y_2) \cdots$ satisfies this formula at the first position precisely when it satisfies the formula $\Box (x = y)$ at position 2, that is, $x_j = y_j$ for all positions $j \geq 2$.

To formalize the meaning of LTL-formulas with nested temporal operators, we will define what it means for a trace to satisfy an LTL-formula at a given position: for a trace ρ, a position $j \geq 1$, and an LTL-formula φ, the notation $(\rho, j) \models \varphi$ stands for "the trace ρ satisfies the formula φ at position j." The trace $\rho = q_1 q_2 \ldots$ satisfies a Boolean expression e (without any temporal operators) at position j if the valuation q_j satisfies e. The trace ρ satisfies the next formula $\bigcirc \varphi$ at position j, where φ may be an arbitrary LTL-formula, if the trace ρ satisfies the formula φ at position $j + 1$. That is, "next φ" holds at a position if φ holds at the next position. Similarly, the trace ρ satisfies the eventually formula $\Diamond \varphi$ at position j, where φ may be an arbitrary LTL-formula, if the trace

ρ satisfies the formula φ at some position k with $k \geq j$. That is, "eventually φ" holds at a position if φ holds at some later (or future) position. Similarly, "always φ" holds at a position if φ holds at every subsequent position.

Syntax and Semantics

Now we can define the logic LTL precisely. The definition below defines both the *syntax* of the logic—what are the syntactically correct formulas of the logic and the *semantics* of the logic—the meaning of the formulas given by the rules to evaluate the formulas over traces. The definition is inductive; for instance, it describes the rule for evaluating the formula $\Box\varphi$ assuming we have already defined how to evaluate the simpler formula φ.

LINEAR TEMPORAL LOGIC

Given a set V of typed variables, the set of formulas of *linear temporal logic* (LTL) is defined inductively by the rules below:

- If e is a Boolean expression over V, then e is an LTL-formula.

- If φ is an LTL-formula, then so are $\neg\varphi$, $\bigcirc\varphi$, $\Diamond\varphi$, and $\Box\varphi$.

- If φ_1 and φ_2 are LTL-formulas, then so are $\varphi_1 \wedge \varphi_2$, $\varphi_1 \vee \varphi_2$, $\varphi_1 \rightarrow \varphi_2$, and $\varphi_1 \mathcal{U} \varphi_2$.

Given a trace $\rho = q_1 q_2 \ldots$ (that is, an infinite sequence of valuations over V), a position $j \geq 1$, and an LTL-formula φ, the *satisfaction* relation, $(\rho, j) \models \varphi$, meaning that the trace ρ satisfies the LTL-formula φ at position j, is defined inductively by the following rules:

- $(\rho, j) \models e$ if the valuation q_j satisfies the Boolean expression e.

- $(\rho, j) \models \neg\varphi$ if it is not the case that $(\rho, j) \models \varphi$.

- $(\rho, j) \models \varphi_1 \wedge \varphi_2$ if both $(\rho, j) \models \varphi_1$ and $(\rho, j) \models \varphi_2$.

- $(\rho, j) \models \varphi_1 \vee \varphi_2$ if either $(\rho, j) \models \varphi_1$ or $(\rho, j) \models \varphi_2$.

- $(\rho, j) \models \varphi_1 \rightarrow \varphi_2$ if either $(\rho, j) \models \neg\varphi_1$ or $(\rho, j) \models \varphi_2$.

- $(\rho, j) \models \bigcirc\varphi$ if $(\rho, j + 1) \models \varphi$.

- $(\rho, j) \models \Box\varphi$ if for every position $k \geq j$, $(\rho, k) \models \varphi$.

- $(\rho, j) \models \Diamond\varphi$ if for some position $k \geq j$, $(\rho, k) \models \varphi$.

- $(\rho, j) \models \varphi_1 \mathcal{U} \varphi_2$ if for some position $k \geq j$, $(\rho, k) \models \varphi_2$, and for all positions i such that $j \leq i < k$, $(\rho, i) \models \varphi_1$.

The trace ρ satisfies the LTL-formula φ if $(\rho, 1) \models \varphi$.

Notice that the satisfaction of a formula at a position j of a trace $\rho = q_1 q_2 \ldots$

depends only on the *suffix* of the trace starting at position j, that is, on the the sequence of valuations $q_j q_{j+1} \cdots$. This is because all the temporal operators refer to the *future* positions. It is worth emphasizing that the current position is considered part of the future: satisfaction of "eventually φ" at a position j demands φ to be satisfied at some position $k \geq j$. Thus, if a formula φ is satisfied at a position, then so is "eventually φ" satisfied at that position. Also note that for the until-formula $\varphi_1 \, \mathcal{U} \, \varphi_2$ to be satisfied at a position j, we demand that the formula φ_2 is satisfied at some position $k \geq j$, and the formula φ_1 holds at all positions following j, including j itself, and *strictly preceding k*. The eventuality operator is just a special case of the until-operator: $\Diamond \varphi$ has the same meaning as the until-formula $1 \, \mathcal{U} \, \varphi$, where 1 is the Boolean constant that is satisfied by every valuation.

Illustrative Temporal Patterns

Nesting of temporal operators give interesting and useful formulas. We highlight some typical patterns (see figure 5.1).

Sequencing: Nested applications of eventually operators can be used to require a sequence of events in a particular order. For instance, consider two events that correspond to satisfaction of formulas φ_1 and φ_2. Then a trace satisfies the LTL-formula $\Diamond (\varphi_1 \wedge \bigcirc \Diamond \varphi_2)$ if there are two positions i and j with $i < j$, such that the formula φ_1 is satisfied at position i and the formula φ_2 is satisfied at position j. For example, the trace $\rho = (x_1, y_1)(x_2, y_2) \cdots$ satisfies the LTL-formula $\Diamond ((x = 1) \wedge \bigcirc \Diamond (y = 1))$ precisely when we can find two positions i and j such that $i < j$ and $x_i = 1$ and $y_j = 1$. Note that the use of the next operator requires the two events to occur at *distinct* positions. The modified formula $\Diamond (\varphi_1 \wedge \Diamond \varphi_2)$ is satisfied by a trace if there are two positions i and j with $i \leq j$, such that the formula φ_1 is satisfied at position i and the formula φ_2 is satisfied at position j.

Recurrence Formulas: Consider the *always-eventually* formula $\square \Diamond \varphi$. A trace ρ satisfies the formula $\square \Diamond \varphi$ at the initial position if $\Diamond \varphi$ is satisfied at every position i. This condition holds if for every position i there exists a future position $j \geq i$ such that the trace ρ satisfies φ at position j. With a little bit of reasoning, convince yourself that this condition can be reformulated as: there exists an infinite sequence of positions $j_1 < j_2 < j_3 < \cdots$ such that φ is satisfied at each of these positions. In other words, $\square \Diamond \varphi$ is satisfied if φ is satisfied in a *recurrent* or repeating manner. For example, the trace $\rho = (x_1, y_1)(x_2, y_2) \cdots$ satisfies the recurrence formula $\square \Diamond (x = 0)$ precisely when for infinitely many positions j, $x_j = 0$. This expresses the requirement that x is assigned the value 0 repeatedly.

Persistence Formulas: The dual of the recurrence requirement expressed by the always-eventually formula is the *eventually-always* formula $\Diamond \square \varphi$. It is satisfied if there exists a position j where the always-formula $\square \varphi$ is satisfied, that is, φ is satisfied in every position following j. In other words, the requirement is that eventually the formula φ is satisfied and continues to hold in a persistent manner. For example, the trace $\rho = (x_1, y_1)(x_2, y_2) \cdots$ satisfies the persistence formula $\Diamond \Box (x = 0)$ precisely when for some position j, for every $k \geq j$, $x_k = 0$ (equivalently, if x is non-zero only at finitely many positions).

Let us consider another example to understand nested temporal formulas. Suppose there is a single variable x of type **nat**. Consider the expressions **even**(x), **odd**(x), and **prime**(x) that are satisfied when the value of x is an even number, an odd number, and a prime number, respectively. Consider the trace $\rho = 1, 2, 3, \cdots$; that is, the jth valuation in the trace assigns the value j to the variable x. Then the trace ρ satisfies the following formula

$$\Box \, [\, \mathtt{even}(x) \; \rightarrow \; (\bigcirc \, \mathtt{odd}(x) \; \wedge \; \bigcirc \bigcirc \, \mathtt{even}(x)\,)\,]$$

which asserts that at every position, if x is even, then in the next position, x is odd, and in the next-to-next position, x is even. The trace ρ also satisfies the recurrence formula

$$\Box \Diamond \, \mathtt{prime}(x)$$

which asserts that the trace contains infinitely many prime numbers.

As another example, suppose there is a single variable x of type **real**, and consider the trace $\rho = 1, 1/2, 1/4, 1/8, \cdots$ (that is, the jth valuation in the trace assigns the value 2^{-j} to the variable x. Then the trace ρ does *not* satisfy the eventuality formula $\Diamond (x = 0)$ but for every $\epsilon > 0$, however small, satisfies the persistence formula $\Diamond \Box (x \leq \epsilon)$.

Temporal Implications and Equivalences

An LTL-formula φ is said to be *valid* if every trace ρ satisfies φ. A valid LTL-formula is also called a *temporal tautology*: it holds no matter how we choose to assign values to the variables at every step. For two LTL-formulas φ_1 and φ_2, if the implication $\varphi_1 \rightarrow \varphi_2$ is valid, then whenever a trace ρ satisfies the formula φ_1, we are guaranteed that the trace ρ also satisfies the formula φ_2. In such a case, the requirement expressed by φ_1 is *stronger* than the requirement expressed by φ_2 since the satisfaction of one implies the satisfaction of the other. Two LTL-formulas φ_1 and φ_2 are *equivalent*, written $\varphi_1 \leftrightarrow \varphi_2$, if both the implications $\varphi_1 \rightarrow \varphi_2$ and $\varphi_2 \rightarrow \varphi_1$ are valid. For two equivalent formulas, a trace satisfies either both or none of them.

Let us consider a few valid implications and equivalences aimed at gaining a better intuition about the meaning of temporal operators.

If a trace ρ satisfies the LTL-formula $\Box \varphi$, then it satisfies φ at all positions, in particular at the initial position, and thus ρ satisfies φ. This argument is independent of the choice of φ. Thus, for every LTL-formula φ, the implication $\Box \varphi \rightarrow \varphi$ is valid, that is, the formula $\Box \varphi$ is a stronger requirement than the formula φ itself. The converse is not true: it is easy to find instances where a trace satisfies an LTL-formula φ but not the formula $\Box\varphi$. The following equivalence, however, is valid for every LTL-formula φ:

$$\Box\varphi \leftrightarrow [\varphi \wedge \bigcirc\Box\varphi].$$

It says that the always-formula $\Box\varphi$ is satisfied precisely when the current position satisfies φ and the next position satisfies the always-formula $\Box\varphi$. This equivalence can be viewed as an "inductive" definition of the always operator. Similar inductive definitions of the eventually and until operators can be obtained (see exercise 5.1).

Observe that if a trace satisfies the recurrence formula $\Box\Diamond\varphi$ at a particular position, say the first position, then it satisfies the same recurrence formula $\Box\Diamond\varphi$ at the next position also. The converse also holds. In fact, for every trace ρ, for all positions i and j, $(\rho, i) \models \Box\Diamond\varphi$ if and only if $(\rho, j) \models \Box\Diamond\varphi$. As a result, for any LTL-formula φ, all the following three formulas are equivalent:

$$\Box\Diamond\varphi \leftrightarrow \bigcirc\Box\Diamond\varphi \leftrightarrow \Diamond\Box\Diamond\varphi.$$

Laws for Temporal and Logical Operators

The interplay between temporal and logical operators can be understood by considering how the logical and temporal operators distribute with respect to one another. Let φ_1 and φ_2 be two LTL-formulas, and let ρ be a trace. Then the trace ρ satisfies the always-formula $\Box(\varphi_1 \wedge \varphi_2)$ precisely when $(\rho, j) \models (\varphi_1 \wedge \varphi_2)$ for every position j. This holds precisely when $(\rho, j) \models \varphi_1$ and $(\rho, j) \models \varphi_2$ for every position j. This is equivalent to saying that the trace ρ satisfies $\Box\varphi_1$ as well as $\Box\varphi_2$, which holds precisely when the trace ρ satisfies the conjunction $\Box\varphi_1 \wedge \Box\varphi_2$. Thus, we have established that the always-operator distributes over conjunction. Thus, for any two LTL-formulas φ_1 and φ_2, the following equivalence is valid:

$$\Box(\varphi_1 \wedge \varphi_2) \leftrightarrow (\Box\varphi_1 \wedge \Box\varphi_2).$$

Let us examine if a similar distributivity property holds for disjunction: do the formulas $\Box(\varphi_1 \vee \varphi_2)$ and $(\Box\varphi_1 \vee \Box\varphi_2)$ mean the same? Consider a trace ρ. Suppose ρ satisfies $\Box\varphi_1$. Then for every position j, $(\rho, j) \models \varphi_1$. By the semantics of disjunction, we have that for every position j, $(\rho, j) \models (\varphi_1 \vee \varphi_2)$. It follows that ρ satisfies the always-formula $\Box(\varphi_1 \vee \varphi_2)$. Symmetric reasoning allows us to conclude that if a trace ρ satisfies $\Box\varphi_2$, then it also satisfies $\Box(\varphi_1 \vee \varphi_2)$. A trace satisfies the disjunction $\Box\varphi_1 \vee \Box\varphi_2$ precisely when it satisfies either $\Box\varphi_1$ or $\Box\varphi_2$, and in either case, we have established that it must then satisfy $\Box(\varphi_1 \vee \varphi_2)$. Thus, the following implication is valid:

$$(\Box\varphi_1 \vee \Box\varphi_2) \rightarrow \Box(\varphi_1 \vee \varphi_2).$$

However, the converse implication is not valid. If we know that a trace ρ satisfies the always-formula $\Box(\varphi_1 \vee \varphi_2)$, then we know that at each position, either φ_1

or φ_2 is satisfied. But this does not necessarily mean that either all positions satisfy φ_1 or all positions satisfy φ_2. As a concrete counterexample, suppose the valuation assigns a Boolean value to a variable x. Consider the trace $\rho = 010101\cdots$ in which the values 0 and 1 are assigned to x in an alternate manner. This trace satisfies $\Box(x = 0 \lor x = 1)$ but satisfies neither $\Box(x = 0)$ nor $\Box(x = 1)$.

To conclude this section, let us note the interplay between the temporal operators and logical negation. First note that $\neg\Box\varphi$ is equivalent to $\Diamond\neg\varphi$ (and similarly, $\neg\Diamond\varphi$ is equivalent to $\Box\neg\varphi$). This results in a duality between the recurrence and persistence formulas: a property φ is not recurrent precisely when the negated property $\neg\varphi$ is persistent. That is, $\neg\Box\Diamond\varphi$ is equivalent to $\Diamond\Box\neg\varphi$. The dual of the next operator is itself: a trace does not satisfy the next-formula $\bigcirc\varphi$ at a position j, precisely when the trace does not satisfy the formula φ at position $(j+1)$, precisely when the trace satisfies the formula $\bigcirc\neg\varphi$ at position j. Thus, the LTL-formulas $\neg\bigcirc\varphi$ and $\bigcirc\neg\varphi$ are equivalent.

Exercise 5.1: We saw that the always-formula $\Box\varphi$ is equivalent to $\varphi \land \bigcirc\Box\varphi$. Find analogous formulas equivalent to the eventually-formula $\Diamond\varphi$ and to the until-formula $\varphi_1\,\mathcal{U}\,\varphi_2$. Justify your answers. ■

Exercise 5.2: For each of the pair of formulas below, say whether the two are equivalent and if not whether one of them is a stronger requirement than the other. In each case, justify your answer.

1. $\Diamond(\varphi_1 \land \varphi_2)$ and $(\Diamond\varphi_1 \land \Diamond\varphi_2)$.

2. $\Diamond(\varphi_1 \lor \varphi_2)$ and $(\Diamond\varphi_1 \lor \Diamond\varphi_2)$.

3. $\Box\Diamond(\varphi_1 \land \varphi_2)$ and $(\Box\Diamond\varphi_1 \land \Box\Diamond\varphi_2)$.

4. $\Box\Diamond(\varphi_1 \lor \varphi_2)$ and $(\Box\Diamond\varphi_1 \lor \Box\Diamond\varphi_2)$.

■

Exercise 5.3: Are the LTL-formulas $\neg(\varphi_1\,\mathcal{U}\,\varphi_2)$ and $(\neg\varphi_2)\,\mathcal{U}\,(\neg\varphi_1)$ equivalent? If not, is one of them a stronger requirement than the other? Justify your answer. ■

Exercise 5.4: Are the two LTL formulas $\Box\Diamond(\varphi_1 \land \Diamond\varphi_2)$ and $\Box\Diamond(\varphi_2 \land \Diamond\varphi_1)$ equivalent? Justify your answer clearly. ■

5.1.2   LTL Specifications

We can use LTL formulas to specify requirements for both synchronous and asynchronous systems. Let us first focus on synchronous systems.

Consider a synchronous reactive component C with input variables I and output variables O. The natural choice of observable variables for such a component is the set $I \cup O$ of input and output variables. An LTL specification for the component C is an LTL-formula φ over the set $I \cup O$ of observable variables. As the component executes, the infinite sequence of inputs and outputs it produces is a trace of the component. Formally, an infinite execution of the component C consists of an infinite sequence of the form

$$s_0 \xrightarrow{i_1/o_1} s_1 \xrightarrow{i_2/o_2} s_2 \xrightarrow{i_3/o_3} s_3 \cdots$$

such that s_0 is an initial state of C, and for each $j > 0$, $s_{j-1} \xrightarrow{i_j/o_j} s_j$ is a reaction C. Given such an execution, the infinite sequence $(i_1, o_1)(i_2, o_2)(i_3, o_3) \cdots$ of inputs and outputs is a *trace* of C. The component C satisfies the specification φ if *every* trace of C satisfies φ. An infinite execution is called a counterexample to the specification φ if the corresponding trace does not satisfy φ. The problem of checking whether a component satisfies a temporal logic specification is known as model checking.

For example, our very first component `Delay` (figure 2.1) has input variable *in* and output variable *out*. LTL-formulas over these variables can be used to express constraints over the desired temporal behavior. In particular, consider the specification:

$$\Box \, [\, (in = 0) \; \rightarrow \; \bigcirc \, (out = 0) \,] \; \wedge \; \Box \, [\, (in = 1) \; \rightarrow \; \bigcirc \, (out = 1) \,],$$

which says that at every position of a trace, if the value of *in* is 0, then in the next position the value of *out* is 0, and if the value of *in* is 1, then in the next position the value of *out* is 1. Indeed, every trace of the component `Delay` satisfies this specification, so we will say that the component satisfies the specification.

As another example, consider the component `ClockedCopy` (figure 2.6) with input variables *in* and *clock* and output variable *out*. Consider the following LTL formula:

$$\Box \, [\, (out = 0) \; \rightarrow \; (\, out = 0\,) \, \mathcal{U} \, clock? \,] \wedge \Box \, [\, (out = 1) \; \rightarrow \; (\, out = 1\,) \, \mathcal{U} \, clock? \,].$$

It says that if the value of the output variable *out* is 0 (or 1) in a given round, then it is guaranteed to stay 0 (or 1, respectively) until the event *clock* is present. It captures the requirement that the output should not change in rounds in which the event *clock* is absent. The component `ClockedCopy` does satisfy this requirement.

Requirements for Leader Election

Let us recall the leader election problem discussed in section 2.4.3. The decision of each node is captured by the output variable *status* that ranges over the enumerated type $\{$`unknown`, `leader`, `follower`$\}$. While the nodes use the variables *in* and *out* for exchanging messages with one another, the design requirements

of the problem specify which traces of values of the *status* variables of different processes are acceptable. The requirement that a node n should eventually
make a decision is expressed by the formula

$$\Diamond\, [\; status_n \;\neq\; \texttt{unknown}\;].$$

The formula states that for a given node n, eventually the value of the status
variable of the instance of the process `SyncLENode` corresponding to this node
should be different from **unknown**. The safety requirement that two nodes should
not consider themselves to be leaders can be expressed by the following formula,
which states that for every pair of distinct nodes m and n, either m is never a
leader or n is never a leader:

$$\Box\, (\, status_m \;\neq\; \texttt{leader}\,) \;\vee\; \Box\, (\, status_n \;\neq\; \texttt{leader}\,).$$

Requirements for Railroad Controller

Let us revisit the railroad controller system of section 3.1.2. Let us consider
the observable variables of the system to be $signal_W$ and $signal_E$ capturing the
traffic lights and the variables $mode_W$ and $mode_E$ capturing the train states.
While the latter are not modeled as output variables, it is acceptable to write
requirements that refer to the state of the models capturing the environment
since the modeling of the environment is part of the specification of the design
problem.

When an LTL-formula refers to a state variable x of a component, the valuation
at each position of a trace should specify the value for x also: in a trace corresponding to an infinite execution, the value of x at position j is the value of x
at the beginning of the jth round.

The basic safety requirement that the two trains should not be on the bridge
simultaneously is expressed by the always-formula:

$$\Box\; \neg\, [(mode_W = \texttt{bridge}) \;\wedge\; (mode_E = \texttt{bridge})].$$

Consider the following liveness requirement, which asserts that the west train
should enter the bridge repeatedly:

$$\Box\, \Diaond\; (\, mode_W = \texttt{bridge}\,).$$

For the given model of the trains, no controller can satisfy this requirement since
the model does not require the train to arrive at the bridge: a train could always
stay in the mode **away** forever. Indeed, this is not an appropriate requirement
for resource allocation problems. The granting of the response—setting the
signal green by the controller should be preconditioned on the request—waiting
at the bridge by the train. Consider the following revised liveness requirement,
which asserts that if the west train is waiting then eventually the west traffic
signal should turn green:

$$\Box\, [\; (mode_W = \texttt{wait}) \;\rightarrow\; \Diamond\; (signal_W = \texttt{green})\;].$$

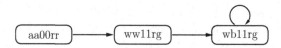

Figure 5.2: Cyclic Counterexample Illustrating Liveness Violation

This LTL-formula says that at every step, if the condition $mode_W = \texttt{wait}$ holds, then at some later step, the condition $signal_W = \texttt{green}$ must hold. This is a typical pattern for LTL formulas: *Always (Request implies Eventually Response)*. We want to check if every trace of the system `RailRoadSystem2` (see figure 3.8) satisfies this specification. It turns out that this is not the case. The counterexample, which consists of an initial sequence of steps followed by a cyclic execution that repeats, is illustrated in figure 5.2. As in figure 3.7, each state is denoted by listing the values of the variables $mode_W$, $mode_E$, $near_W$, $near_E$, *west*, and *east*, in that order, and a, w, b, g, and r are abbreviations for **away**, **wait**, **bridge**, **green**, and **red**, respectively. The cycle in the counterexample corresponds to the case when the east train is on the bridge refusing to leave, while the west train keeps waiting. We can conclude that if the controller lets the east train on the bridge and the train does not ever leave the bridge, a scenario consistent with the given model of the train, then the controller cannot possibly let the west train in. Since no controller can satisfy the requirement as specified, we need to modify our specification of the requirement.

An alternative standard form of liveness requirement is captured by the revised formula φ_{df}, which states that if the west train is waiting, then eventually either the corresponding signal is green or the east train is on the bridge:

$$\Box[\,(mode_W = \texttt{wait}) \;\rightarrow\; \Diamond\,[(signal_W = \texttt{green}) \vee (mode_E = \texttt{bridge})]\,].$$

This form of requirement is called *deadlock freedom*: while it does not ensure that the controller is responsive to the west train when it requests an entry, it does ensure utilization of the resource. In particular, a controller that keeps both the traffic lights red all the time would violate this requirement, and so would a controller that keeps both trains waiting for one another in a deadlocked manner due to a buggy design. The controller `Controller2` of figure 3.8 is free of such deadlocks and meets the specification φ_{df}.

The more stringent requirement that every request should be fulfilled by granting the resource to the requester is called *starvation freedom*. In our example, if the west train wants to enter, then starvation freedom requires that it should be allowed to enter. As discussed already, this is feasible only when the east train is well behaved in the sense that it does not stay on the bridge forever. The formula φ_{sf} below asserts that under the assumption that the east train is repeatedly off the bridge, if the west train is waiting, then eventually the west traffic signal should turn green:

$$\Box\Diamond\,(mode_E \neq \texttt{bridge}) \;\rightarrow\; \Box[\,(mode_W = \texttt{wait}) \;\rightarrow\; \Diamond\,(signal_W = \texttt{green})\,].$$

Note that the requirement expressed by φ_{sf} is stronger than φ_{df}: any trace that satisfies φ_{sf} also satisfies φ_{df} but not vice versa. The controller `Controller2` of figure 3.8 is in fact starvation-free and meets the specification φ_{sf}. In particular, the counterexample of figure 5.2 is ruled out: since the precondition $\square \lozenge \, (mode_E \neq \texttt{bridge})$ is violated by this trace, it satisfies φ_{sf}.

Exercise 5.5: Consider the design of the synchronous three-bit counter from section 2.4.1. Write an LTL-formula to express the requirement that if the input signal *inc* is repeatedly high, then it is guaranteed that the counter will be repeatedly at its maximum value (that is, all the three output bits out_0, out_1, and out_2 are 1). Does the circuit `3BitCounter` of figure 2.27 satisfy this specification? ∎

Exercise 5.6: Recall the synchronous design of a cruise controller system from section 2.4.2. Consider the following requirement: when the cruise-controller is "on," assuming the driver does not issue any further input events, eventually the speed becomes equal to the desired cruising speed and stays equal. Express this requirement in LTL using the variables *on*, *speed*, *cruiseSpeed*, *cruise*, *inc*, and *dec*. ∎

5.1.3   LTL Specifications for Asynchronous Processes *

LTL-formulas can be used to specify constraints on executions of asynchronous processes also. Consider an asynchronous process P with state variables S, input channels I, and output channels O. Then LTL-formulas over the set $I \cup O$ can be used to specify desired requirements on sequences of inputs and outputs. We can associate infinite traces with infinite executions of P in the same manner we associated traces with executions of synchronous components with the following two changes. First, in the asynchronous model, each action is either an internal action that does not involve any of the input or output channels, an input action involving a single input channel x, or an output action involving a single output channel y. To interpret an LTL-formula over the set $I \cup O$ of variables, we need an infinite sequence of valuations, where each valuation needs to assign values to all the input and output variables. To interpret a single action of P as a valuation for all the input and output variables, we can assign the undefined value $\bot$ to each input and output channel not involved in the action. Second, since an asynchronous process has (weak or strong) fairness assumptions associated with its tasks, to check whether the process satisfies an LTL-specification, we consider the traces corresponding only to the fair executions: the asynchronous process P satisfies the LTL-specification φ if the trace corresponding to every *fair* infinite execution of P satisfies φ.

Suppose for the reliable communication buffer with input channel *in* and output channel *out*, we want to specify that a message sent on the input channel eventually appears on the output channel (see section 4.3.2). The following LTL-formula specifies this requirement for a given value v of type `msg`:

$$\square \, (\, (in = v) \; \rightarrow \; \lozenge \, (out = v) \,).$$

The alternating-bit protocol discussed in section 4.3.2 satisfies this requirement under the fairness assumptions discussed there.

In some problems, the requirements for the asynchronous solutions are no different from the corresponding requirements for the synchronous designs. One such instance is the leader election problem. We have already studied LTL-formulas corresponding to the requirements that every node n eventually decides, and for every pair of nodes m and n, either the node m is never a leader or the node n is never a leader. The same formulas can be used as requirements for the asynchronous case.

The difference in the requirements for the synchronous and asynchronous cases is highlighted by the specification of logical gates. Consider an inverter with input in and output out. The natural specification in the synchronous case is the always formula

$$\square\,(\,out = \neg\,in\,),$$

which says that the output is always equal to the negation of the input. In the asynchronous case, this specification cannot be satisfied due to the decoupling of the changes in the output in response to the changes in the input. We can demand that if the input is 0, then we expect the output to eventually become 1, provided the input is maintained unchanged at 0. This is expressed by the following formula:

$$\square\,[\,(\,in = 0\,) \rightarrow (\,in = 0\,)\,\mathcal{U}\,(\,in = 1 \vee out = 1\,)\,].$$

A symmetric formula can express the requirement that if the input is 1, then unless the input is changed back to 0, and the output will eventually become 1.

We can also write LTL-requirements that refer to state variables as well as input and output channels. To interpret such a formula, the trace corresponding to an ω-execution retains the values of state variables also.

Fairness Assumptions

In section 4.2.4, we discussed how to annotate tasks of an asynchronous process with fairness assumptions so that we consider only those infinite executions in which enabled tasks are not delayed forever. To check whether an asynchronous process meets its specification given as an LTL-formula, we check if every fair execution satisfies the specification. Now we discuss how to capture fairness assumptions within LTL-specifications. LTL-formulas corresponding to fairness assumptions can be useful for a better understanding of the distinction between weak and strong fairness and also suggest how an analysis tool designed for checking whether all executions satisfy a given LTL-formula can easily be adapted to check whether all fair executions satisfy an LTL-specification.

Consider an asynchronous process P with state variables S, input channels I, and output channels O. To express fairness requirements in LTL, at every step of the execution, we need to be able to express whether a task is enabled and

nat $x := 0$; $y := 0$; $\{A_x, A_y\}$ *taken*
$A_x :\ x := x + 1$; *taken* $:= A_x$
$A_y :\ \mathbf{even}(x)\ \rightarrow\ \{y := y + 1$; *taken* $:= A_y\}$

Figure 5.3: Modified Version of `AsyncEvenInc`

whether a task is taken. For each output and internal task A of P, let $Guard(A)$ be the guard condition for the task A, which is a Boolean-valued expression over the state variables whose value in a state indicates whether the task A is enabled in that state. Thus, an infinite execution satisfies the LTL-formula $\Box \Diamond\, Guard(A)$ exactly when the task A is enabled at infinitely many steps of the execution. While a state of an asynchronous process contains enough information about whether a task is enabled, to refer to whether a task is taken, we introduce an additional variable, *taken*, that ranges over the set of tasks: its value at each step indicates the most recent task executed. The update code of a task A is modified so that it sets this variable to A.

To illustrate this, recall the asynchronous process `AsyncEvenInc` from section 4.2.4 (see figure 4.18). Figure 5.3 shows the corresponding process with the additional variable *taken*. An infinite execution of this modified process satisfies the recurrence formula $\Box \Diamond\, (\textit{taken} = A_y)$ exactly when the task A_y is executed infinitely often.

With this modification, for a given output or an internal task A, consider the following formula

$$\text{wf}(A) :\ \Diamond \Box\, Guard(A)\ \rightarrow\ \Box \Diamond\, (\textit{taken} = A).$$

This formula expresses the requirement that if the task A is persistently enabled, then it must be repeatedly taken. Thus, a trace satisfies this formula precisely when the trace corresponds to an infinite execution that is weakly fair with respect to the task A. To check if the process `AsyncEvenInc` guarantees the value of x to eventually exceed 10 assuming weak fairness for the task A_x, for which the guard condition is always true, we check if the modified process of figure 5.3 satisfies the LTL-formula

$$\Box \Diamond\, (\textit{taken} = A_x)\ \rightarrow\ \Diamond\, (x > 10).$$

Indeed this LTL-formula is satisfied along every infinite execution of the process. To check if the value of y is guaranteed to eventually exceed 10 assuming weak fairness for the task A_y, we check if the process of figure 5.3 satisfies the LTL-formula

$$[\Diamond \Box\, \mathbf{even}(x) \rightarrow \Box \Diamond\, (\textit{taken} = A_y)]\ \rightarrow\ \Diamond\, (y > 10).$$

This requirement does not hold: the infinite execution in which only the task A_x is executed at every step does not satisfy this formula, and thus weak fairness assumption for the task A_y does not suffice to ensure satisfaction of the eventuality $\Diamond\, (y > 10)$.

Note that the formula $\mathit{wf}(A)$ is equivalent to the following formula, which asserts that if the task A is enabled at a given step, then at a later position it is either taken or disabled:

$$\mathit{wf}(A):\ \Box\, [\, \mathit{Guard}(A)\ \rightarrow\ \Diamond\, (\, (\mathit{taken} = A)\ \lor\ \neg\, \mathit{Guard}(A)\,)\,].$$

The strong fairness assumption for a given task A is expressed by the formula:

$$\mathit{sf}(A):\ \Box\Diamond\, \mathit{Guard}(A)\ \rightarrow\ \Box\Diamond\, (\mathit{taken} = A).$$

This formula asserts that if the task is repeatedly enabled, then it must be repeatedly executed. Thus, a trace satisfies this formula precisely when the trace corresponds to an infinite execution that is strongly fair with respect to the task A.

To check if the process `AsyncEvenInc` guarantees the value of y to eventually exceed 10 assuming strong fairness for the task A_y, we check if the process of figure 5.3 satisfies the LTL-formula

$$[\Box\Diamond\ \mathbf{even}(x)\ \rightarrow\ \Box\Diamond\, (\mathit{taken} = A_y)]\ \rightarrow\ \Diamond\, (y > 10).$$

Indeed this LTL-formula is satisfied along every infinite execution of the process.

Note that, independent of exactly how a trace assigns values to the expressions $\mathit{Guard}(A)$ and $(\mathit{taken} = A)$ at each step, the following temporal implication is valid:

$$\mathit{sf}(A)\ \rightarrow\ \mathit{wf}(A).$$

This explains that "strong fairness" is indeed a stronger requirement than "weak fairness."

Instead of requiring that all fair executions of an asynchronous process P satisfy an LTL-formula φ, we can require all executions of the process P to satisfy the conditional LTL-formula $\varphi_{\mathit{fair}} \rightarrow \varphi$, where φ_{fair} is the conjunction of $\mathit{wf}(A)$ and $\mathit{sf}(A)$ formulas for tasks A for which weak and strong fairness assumptions are made. For example, for the process `UnrelFIFO` of figure 4.19, we assume strong fairness for the internal task A_1 that correctly transfers an element from the queue x to the queue y and weak fairness for the output task A_{out} that transmits the elements from the internal queue y to the output channel. To demand that all fair executions satisfy an LTL-formula φ is equivalent to requiring that all executions satisfy the formula

$$(\, \mathit{sf}(A_1)\ \land\ \mathit{wf}(A_{out})\,)\ \rightarrow\ \varphi.$$

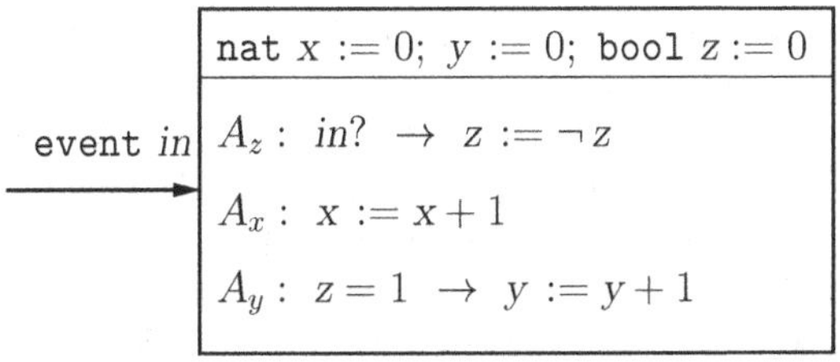

$$\boxed{\begin{array}{l} \texttt{nat x := 0; y := 0; bool z := 0} \\ \hline A_z :\ in?\ \rightarrow\ z := \neg z \\ A_x :\ x := x+1 \\ A_y :\ z = 1\ \rightarrow\ y := y+1 \end{array}}$$

event *in*

Figure 5.4: Exercise: Satisfaction Under Fairness Assumptions

Exercise 5.7: Consider the two specifications of weak fairness:

$$\varphi_1 :\ \Diamond\Box\ Guard(A)\ \rightarrow\ \Box\Diamond\ (taken = A)$$

and

$$\varphi_2 :\ \Box\ [Guard(A)\ \rightarrow\ \Diamond\ ((taken = A)\ \lor\ \neg\ Guard(A))].$$

Prove that these two LTL-formulas are equivalent. ∎

Exercise 5.8: Consider an asynchronous process P shown in figure 5.4 with the input task A_z and internal tasks A_x and A_y. For each of the LTL-formulas below, does the process P satisfy the formula? If not, is there a suitable fairness assumption regarding execution of tasks under which the process satisfies this specification? When adding fairness assumptions, clearly specify whether you are using strong fairness or weak fairness and for which tasks with a justification.

(1) $\Diamond\ (x > 5)$;
(2) $\Diamond\ (y > 5)$;
(3) $\Box\Diamond\ (z = 1)\ \rightarrow\ \Diamond\ (y > 5)$.

∎

5.1.4 Beyond LTL*

We conclude this section by noting some of the limitations of the logic LTL as a specification language for writing requirements. These limitations have spawned a number of extensions and variations of LTL. While a detailed study of various temporal logics and their comparative merits is beyond the scope of this textbook, the discussion below is a glimpse into the rich variety of alternative temporal logics.

Branching-Time Temporal Logics

An LTL-formula is evaluated over a trace corresponding to a single execution of a system, and a system satisfies an LTL-formula when *all* executions of the system satisfy the formula. With this interpretation, there is no way to demand

that some executions satisfy one type of a requirement and some others satisfy
another kind. In particular, for the consensus problem discussed in section 4.3.3,
consider the requirement that "if the preferences of the two processes P_1 and P_2
are different initially, then both decisions are possible." This requirement cannot
be specified in LTL but can be stated in the so-called "branching-time" temporal
logics. Recall that the variables $pref_1$ and $pref_2$ capture the initial preferences of
the two processes, and the variables dec_1 and dec_2 are assigned the decision values on termination. Then the following formula of the branching-time temporal
logic CTL (Computation Tree Logic) captures the desired requirement:

$$(pref_1 \neq pref_2) \; \rightarrow \; [\, \exists \Diamond \, (\, dec_1 = dec_2 = 0\,) \; \wedge \; \exists \Diamond \, (\, dec_1 = dec_2 = 1\,)\,].$$

The logic CTL, in addition to the logical and temporal operators, allows existential ($\exists$) and universal ($\forall$) quantifiers over executions. The formulas are
interpreted over the tree of all executions where nodes correspond to states and
branching corresponds to possible choices of the successor state at each node.
A quantified branching-time formula $\exists \varphi$ is satisfied at a node if there is an
execution ρ starting at the corresponding state such that ρ satisfies the formula
φ, which may contain temporal operators.

Stateful Temporal Logics

Given a Boolean variable e, consider the following requirement: the value of e is
1 in every even position. One can prove that no LTL-formula exactly captures
this requirement. Note that the LTL-formula

$$\bigcirc (e = 1) \; \wedge \; \square[(e = 1) \rightarrow \bigcirc \bigcirc (e = 1)]$$

expresses a much stronger requirement: for a trace to satisfy this formula, not
only is it necessary that at every even position the value of e must be 1, but if
the value of e happens to be 1 at some odd position, then the formula can be
satisfied only when the value of e is 1 in all subsequent odd positions. Intuitively,
the desired requirement "the value of e is 1 in every even position" requires the
specification logic to maintain an internal state variable that captures whether
a position is odd or even, and LTL-formulas cannot maintain such a state. This
shortcoming has led to extensions of LTL with regular expressions (or equivalently deterministic finite automata) to express *stateful* temporal constraints.
The IEEE standard Property Specification Language PSL allows a combination
of temporal operators and regular expressions.

Interpretation over Finite Traces

In our formalization of LTL, LTL-requirements specify constraints only on infinite executions of a system. As a result, if a synchronous reactive component
C has no infinite executions (this may happen if the component C is not input enabled), then no matter which LTL-formula φ we consider, the component
C satisfies the requirement φ vacuously. Also, since not every reachable state

necessarily appears on some infinite execution, given a state property φ, the system may satisfy the always-formula $\Box \, \varphi$, even though the property φ is not an invariant of the system. This anomaly can be avoided if we redefine the semantics of LTL-formulas so that a formula can be evaluated on a finite trace also. If $\rho = q_1 q_2 \cdots q_m$ is a finite trace and $1 \leq j \leq m$ is a position in the trace, then

$$(\rho, j) \models \Box \varphi \text{ if } (\rho, k) \models \varphi \text{ for all positions } k \text{ with } j \leq k \leq m; \text{ and}$$
$$(\rho, j) \models \bigcirc \varphi \text{ if } j < k \text{ and } (\rho, j+1) \models \varphi.$$

Thus, the main difference is that $\bigcirc \varphi$ now means that it's not yet the end of the trace and the next position satisfies φ. Now, a component C satisfies an LTL-formula φ if every infinite trace as well as every finite trace corresponding to a maximal execution of C satisfies φ. Here, a maximal execution is a finite execution of C that ends in a state that has no successors (that is, the execution cannot be extended by an additional state). Intuitively, maximal traces correspond to terminating (or deadlocked) executions, and including such executions ensures that while evaluating an LTL-formula, we examine all the reachable states of the component. Note that evaluating an LTL-formula on *all* finite executions is not meaningful: if a system satisfies an eventuality φ after, say five rounds, then executions of the system of length less than five do not satisfy the eventually formula $\Diamond \varphi$ but should not be considered as counterexamples (however if there is some maximal execution that does not contain a state satisfying φ, then it does indicate a violation of the requirement $\Diamond \varphi$). All the analysis techniques for establishing that a system satisfies its LTL-specification can be easily modified to account for such a revised interpretation.

5.2 Model Checking

In chapter 3, we saw that the canonical safety verification problem is the invariant verification problem: given a transition system T and a property φ over its state variables, we want to check if all the reachable states of the system T satisfy the property φ. For automated verification, we reduced the invariant verification to the reachability problem: to check whether the property φ is an invariant of the transition system T, we check whether a state violating φ is reachable and, if so, the corresponding execution is a counterexample to the invariant verification question. We then studied both enumerative and symbolic algorithms for solving the reachability problem.

Repeatability Problem

In model checking, given the system described as a synchronous reactive component, or as an asynchronous process, and an LTL-specification, we want to check if every execution of the system satisfies the given LTL-specification. The core computational problem for verification of liveness requirements turns out to be the *repeatability* problem: given a transition system T and a property φ

over its state variables, does there exist an infinite execution of the system T that repeatedly encounters states satisfying the property φ (that is, whether the recurrence formula $\Box \Diamond \varphi$ is satisfied by some infinite execution of the system)? This form of repeated reachability is also known as *Büchi reachability*, named after the logician J. Richard Büchi who studied finite automata over infinite words resulting in an elegant theory of ω-regular languages that mirrors the classical theory of regular languages over finite words. We will show that the LTL model-checking problem, namely, checking whether every trace of a given system satisfies a given LTL-specification, can be reduced to the repeatability problem for the composition of the system and a monitor derived from the LTL-specification. In the safety case, the finite execution that demonstrates the reachability of certain error states of the monitor corresponds to a counterexample indicating a violation of the requirement. Similarly, in model checking, the infinite execution that demonstrates the repeated reachability of certain error states of the monitor is a counterexample indicating a violation of the liveness requirement.

REPEATABILITY PROBLEM FOR TRANSITION SYSTEMS

An *infinite execution* of a transition system T consists of an infinite sequence of the form $\rho = s_0, s_1, \ldots$ such that s_0 is an initial state of T and for each $j > 0$, (s_{j-1}, s_j) is a transition of T. A property φ over the state variables of T is said to be *repeatable* if there exists *some* infinite execution ρ of T such that the execution ρ satisfies the recurrence LTL-formula $\Box \Diamond \varphi$. The *repeatability problem* is to check, given a transition system T and a property φ over its state variables, whether φ is repeatable.

When the answer to the repeatability problem, for a given transition system T and property φ, is positive, we want to demonstrate an infinite execution in which the property φ is repeatedly satisfied. Typically such an execution is illustrated by a state s, such that (1) the state s is reachable from some initial state, (2) the state s is reachable from itself using one or more transitions, and (3) the state s satisfies the property φ. That is, as evidence, we will produce a cycle that is reachable from some initial state and contains a state satisfying φ.

Recall the transition system $\text{GCD}(m, n)$ of figure 3.1 capturing the program for computing the greatest common divisor of two numbers m and n. Note that as long as both the variables x and y have positive values, the system stays in the mode loop updating the variables. Suppose we want to check the liveness requirement that the loop always terminates. This corresponds to checking repeatability of the property $(mode = \text{loop})$: an infinite execution where this condition is satisfied repeatedly corresponds to a nonterminating execution.

5.2.1 Büchi Automata

Now we describe how to compile LTL-formulas into a special kind of monitor, called Büchi automata, so that the model checking problem can be reduced to the repeatability problem for the composition of the system and the monitor.

Definition

Given a set V of Boolean variables, a Büchi automaton over V is a synchronous reactive component M with the set V as input variables, described as an extended-state machine. The only state variable for the automaton is its mode, and thus it has only finitely many states. It has no outputs. Thus, a mode-switch, or an edge, is completely described by the source and the target states of the switch and the guard condition, which is a Boolean expression over the input variables. Given an infinite sequence of inputs, that is, a trace over V, an execution of the machine produces an infinite sequence of states. A subset F of the states is declared as *accepting*. The execution corresponding to an input trace is accepting if some accepting state repeats infinitely often along this execution. The automaton can be nondeterministic: from a given state and a given input, the guard conditions of multiple outgoing switches can be true simultaneously. Thus, for a given input trace, multiple executions are possible. The automaton M accepts an infinite trace ρ over V if there exists an accepting infinite execution when supplied with the sequence ρ of inputs. The formal definition is summarized below:

BÜCHI AUTOMATON

A Büchi automaton M consists of

- a finite set V of Boolean input variables,

- a finite set Q of states,

- a set $Init \subseteq Q$ of initial states,

- a set $F \subseteq Q$ of accepting states, and

- a finite set E of edges, where each edge is of the form $(q, Guard, q')$ consisting of a source state $q \in Q$, a target state $q' \in Q$, and a Boolean expression $Guard$ over the input variables V.

For two states q and q' and an input $v \in Q_V$, $q \xrightarrow{v} q'$ is a transition of the automaton if there exists an edge $(q, Guard, q')$ such that the input v satisfies the expression $Guard$. Given a trace $\rho = v_1 v_2 \cdots$ over the input variables, an execution of the automaton over the input trace ρ is an infinite sequence of the form $q_0 \xrightarrow{v_1} q_1 \xrightarrow{v_2} \cdots$ such that q_0 is an initial state, and for each $i \geq 1$, $q_{i-1} \xrightarrow{v_i} q_i$ is a transition. The Büchi automaton M *accepts* the input trace ρ if there exists an execution over ρ such that for infinitely many indices i, $q_i = q$ for some accepting state $q \in F$.

The Büchi automaton M is said to be deterministic if from every state on a given input only one edge can be chosen: for every state q and every pair of edges $(q, Guard_1, q_1)$ and $(q, Guard_2, q_2)$, the conjunction $Guard_1 \wedge Guard_2$ is unsatisfiable (ensuring that an input v that satisfies the guard $Guard_1$ cannot satisfy the guard $Guard_2$, and vice versa).

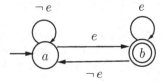

Figure 5.5: Büchi Automaton for $\Box \Diamond e$

Figure 5.6: Büchi Automaton for $\Diamond \Box e$

Examples

Figure 5.5 shows the Büchi automaton M that accepts only those traces that satisfy the recurrence formula $\Box \Diamond e$. It has a single Boolean input variable e and is a state machine with two states a and b. The state a is initial, and the state b is accepting (indicated by a double circle). In each round, if the input e is 1, the automaton transitions to the state b, and if the input e is 0, the automaton transitions to the state a. Given a trace $v_1 v_2 \cdots$ of inputs, the automaton M has a unique execution, and the state b repeats infinitely often along this execution precisely when the input sequence contains infinitely many 1 s. Thus, the automaton M accepts an input trace ρ precisely when the input trace contains infinitely many 1 s, that is, when the trace ρ satisfies the LTL-formula $\Box \Diamond e$.

Figure 5.6 shows another Büchi automaton. It also has a single Boolean input variable e and is a state machine with two states a and b, of which the state a is initial and the state b is accepting. This automaton is nondeterministic. In the initial state, in each round, independent of whether the input value is 0 or 1, the automaton can either stay in the state a or switch to the state b. Once the automaton switches to the state b, if the input is 0, then no transition is enabled, and the automaton gets stuck. Since the only accepting state is b, and once in the state b, the automaton can generate an infinite execution only when all the subsequent input values are 1 s, the automaton has an infinite execution with the state b repeating precisely when from some position onward the input trace contains only 1 s. That is, the automaton accepts exactly those traces that satisfy the LTL-formula $\Diamond \Box e$. There is no deterministic Büchi automaton corresponding to the LTL-formula $\Diamond \Box e$, and thus nondeterminism can be crucial for constructing Büchi automata corresponding to LTL formulas.

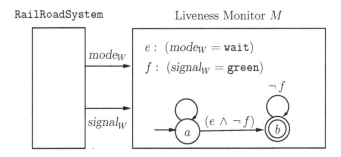

Figure 5.7: Monitoring Liveness Violations Using a Büchi Automaton

Monitoring for Violations of Liveness Requirements

To illustrate how Büchi automata can be used as monitors to detect violations of liveness requirements, let us revisit the railroad controller example and the liveness requirement, which asserts that if the west train is waiting, then eventually the west traffic signal should turn green:

$$\Box [\, (mode_W = \texttt{wait}) \; \rightarrow \; \Diamond \; (signal_W = \texttt{green}) \,].$$

This is a a commonly occurring pattern for liveness requirements of the form $\varphi : \Box (e \rightarrow \Diamond f)$, where e and f are expressions over the observable variables of the system. To check whether every trace of the system satisfies φ, we first *negate the specification* and check if there is some execution of the system that satisfies the negated specification. The negated specification is $\neg \Box (e \rightarrow \Diamond f)$, which is equivalent to the formula $\Diamond (e \land \Box \neg f)$. Thus, a violation of the requirement is an infinite execution in which the property e holds at some position, and then onward the property f never holds. Consider the nondeterministic Büchi automaton M shown in figure 5.7 that accepts exactly those traces that satisfy this negated formula. The automaton can switch to the accepting state b only when it encounters an input that satisfies the property e, and once in the state b, it can continue execution only when the input at every step does not satisfy the property f. In the composed system $\texttt{RailRoadSystem} \| M$, there is an infinite execution in which the automaton state is repeatedly b if and only if the component $\texttt{RailRoadSystem}$ can produce a trace that satisfies $\neg \varphi$. Thus, the model checking problem of verifying that every trace of the system satisfies the LTL-formula φ has been reduced to checking the repeatability of the property $(M.mode = b)$ for the composed system. As already discussed, $\texttt{RailRoadSystem2}$ (see figure 3.8) does not satisfy the specification: in the system $\texttt{RailRoadSystem2} \| M$, the property $(M.mode = b)$ is repeatable since the state with $mode_W = \texttt{wait}$, $mode_E = \texttt{bridge}$, $near_W = near_E = 1$, $signal_W = \texttt{red}$, $signal_E = \texttt{green}$, and $M.mode = b$ is reachable and has a self-loop.

As another example of monitoring for violations of liveness requirements, consider the LTL specification $\varphi : \Box \Diamond e \rightarrow \Box \Diamond f$, where e and f are expressions

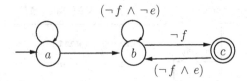

Figure 5.8: Büchi Automaton for Detecting Violation of $\square \lozenge\, e \rightarrow \square \lozenge\, f$

over the observable variables of the component C. It asserts conditional recurrence demanding that if the property e is recurrent, then so should be the property f. To check whether every trace of the component C satisfies φ, we first negate the specification and check if there is some execution of the component C that satisfies the negated specification. The negated specification is equivalent to $\square \lozenge\, e \wedge \lozenge \square \neg f$, stating that the property e is recurrent and the property $\neg f$ is persistent. Thus, we want to find an infinite execution in which after a certain position, $\neg f$ holds continuously and e holds repeatedly. This is captured by the Büchi automaton M shown in figure 5.8 with three states. Initially the state is a. The automaton loops in this initial state for an arbitrary number of steps and then nondeterministically switches to the state b. Subsequently, every time the condition e is satisfied, it transitions to the state c and switches back to the state b in the next step. In both states b and c, the execution can continue only if the condition $\neg f$ holds. An infinite execution can visit the state c repeatedly only if the property e is recurrent and the property $\neg f$ is persistent. Thus, the model checking problem can be reformulated as checking the repeatability of the property $(mode = c)$ for the composed system $C \| M$.

Generalized Büchi Automata

Consider the LTL-formula $\square \lozenge\, e \wedge \square \lozenge \neg e$, which says that the variable e should be 1 infinitely often and should also be 0 infinitely often. A convenient way to capture this requirement is to use the same automaton structure as the one shown in figure 5.5 and to use *two* accepting sets: $F_1 = \{a\}$ and $F_2 = \{b\}$. An execution of such a machine over an input trace is accepting if both of these sets repeat infinitely often. This extension of Büchi automata with such a conjunctive accepting requirement is called *generalized* Büchi automata.

Formally, a generalized Büchi automaton has input variables V, states Q, initial states *Init*, and edges of the form $(m, Guard, m')$, as in the case of a Büchi automaton, and has sets $F_1, F_2, \ldots F_k$ of accepting sets of states. An execution $q_0 \xrightarrow{v_1} q_1 \xrightarrow{v_2} \cdots$ of the automaton over an input trace $v_1 v_2 \cdots$ is said to be accepting if for each j, for infinitely many indices i, the state q_i belongs to the accepting set F_j. In other words, the trace $q_0 q_1 \ldots$ over states corresponding to the execution should satisfy the formula $\wedge_{j=1,\ldots k} \square \lozenge\, (mode \in F_j)$.

It turns out that generalized Büchi automata are no more expressive than Büchi automata. Expressing a requirement on input traces as a generalized Büchi automaton with multiple accepting sets can allow the design of a machine with fewer states, but it is possible to compile it into a Büchi automaton (with a single accepting set) without changing the set of input traces it accepts.

Proposition 5.1 [From Generalized Büchi automata to Büchi automata] *For a given generalized Büchi automaton M over the input variables V, there exists a Büchi automaton M' over the input variables V such that for every trace ρ over V, the automaton M accepts the trace ρ exactly when the automaton M' accepts it.*

Proof. Let M be a generalized Büchi automaton over inputs V with states Q, initial states *Init*, edges E, and accepting sets $F_1, \ldots F_k$. We want to construct a Büchi automaton M' such that visiting one of its accepting states repeatedly ensures that the original automaton M has visited each of the sets F_j repeatedly. For this purpose, the automaton M' maintains the state of M and, additionally, a counter that cycles through the values $1, 2, \ldots k, 0$. Initially the counter is 1. When the state of M is in the set F_1, the counter is incremented to 2. When a state in the accepting set F_2 is encountered, it is incremented to 3. More generally, when the counter is j, the automaton is waiting to visit a state in the accepting set F_j. When such a state is encountered, the counter is incremented to $j + 1$. When the counter j equals k, when a state in the accepting set F_k is encountered, the counter is updated to 0, and in the next transition, it is changed to 1. If along the execution, the counter is 0 repeatedly, then it has cycled through all the values repeatedly, and the execution has visited each of the accepting sets F_j repeatedly. Conversely, if an accepting set F_j repeats infinitely often along an execution, then the counter cannot get "stuck" at the value j, and thus if all the accepting sets repeat infinitely often, then the counter cycles through 0 repeatedly.

Formally, the set of states of M' is the set $Q \times \{0, 1, \ldots k\}$. The initial states of M' are of the form $\langle q, 1 \rangle$ with $q \in Init$. For every edge $(q, Guard, q')$ of M, the automaton M' has the edge $(\langle q, 0 \rangle, Guard, \langle q', 1 \rangle)$; for every $1 \leq c < k$, if $q \in F_c$, then the automaton M' has the edge $(\langle q, c \rangle, Guard, \langle q', c + 1 \rangle)$ or else the edge $(\langle q, c \rangle, Guard, \langle q', c \rangle)$; and if $q \in F_k$, then the automaton M' has the edge $(\langle q, k \rangle, Guard, \langle q', 0 \rangle)$ or else the edge $(\langle q, k \rangle, Guard, \langle q', k \rangle)$. The accepting set for M' is the set of states of the form $\langle q, 0 \rangle$.

To complete the proof, we need to show that an input trace ρ is accepted by the generalized Büchi automaton M exactly when it is accepted by the Büchi automaton M', but this follows in a straightforward manner from the definitions.
∎

Exercise 5.9: For each of the LTL-formulas below, construct a Büchi automaton that accepts exactly those traces that satisfy the formula:

(1) $\Box \Diamond e \vee \Diamond \Box f$;
(2) $\Box \Diamond e \wedge \Box \Diamond f$;
(3) $\Box (e \rightarrow e \, \mathcal{U} \, f)$.

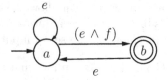

Figure 5.9: Exercise: From a Büchi Automaton to LTL

■

Exercise 5.10: Write an LTL-formula that exactly describes the set of traces that are accepted by the Büchi automaton shown in figure 5.9. Explain your answer. ■

Exercise 5.11*: Given two Büchi automata M_1 and M_2, both over the same set V of input variables, show how to construct a Büchi automaton M over the inputs V such that the automaton M accepts an input trace ρ over V exactly when both the automata M_1 and M_2 accept the trace ρ. ■

Exercise 5.12*: Given a Büchi automaton M with states Q and accepting states F, consider the Büchi automaton M' obtained by toggling the role of accepting states in M: the states, the initial states, and the edges of the automaton M' are the same as the ones in the original automaton M, but its accepting states are $Q \setminus F$ (that is, a state is accepting in M' exactly when it is not accepting in M). Consider the claim "an input trace ρ is accepted by the automaton M' exactly when it is not accepted by the automaton M." Does the claim hold? If your answer is "yes," justify with a proof. If your answer is "no," give a counterexample. In this latter case, does the claim hold if the automaton M is deterministic? ■

5.2.2   From LTL to Büchi Automata*

The construction of Büchi automata for detecting violations of LTL specifications can be automated. An LTL-formula φ can be compiled into a (generalized) Büchi automaton M_φ, which accepts exactly those traces that satisfy the formula φ. States of the desired automaton are *sets of subformulas* of φ. Such an automaton is called a *tableau*. We first illustrate the construction using an example.

Sample Tableau Construction

To illustrate the principles of the tableau construction, let us consider the LTL-formula $\varphi = \Box\,(e \lor f) \land \Diamond\,e$. The states of the tableau are collections of LTL-formulas derived from φ. Each state q is a set of formulas, and we would like to ensure that every formula contained in a state q is satisfied by the input trace along every infinite accepting execution starting in the state q.

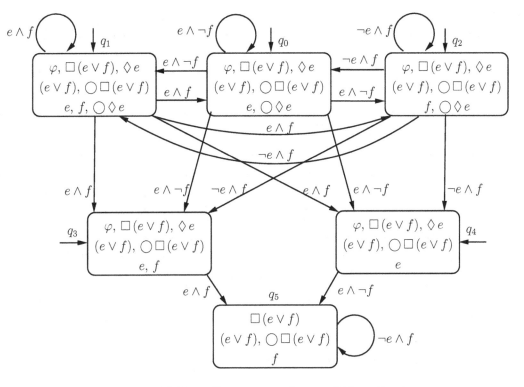

Figure 5.10: Tableau Construction for $\varphi = \Box(e \vee f) \wedge \Diamond e$

An initial state of the tableau is required to contain the given formula φ. From the semantics of conjunction, the formula φ is satisfied when both $\Box(e \vee f)$ and $\Diamond e$ are satisfied, so an initial state must contain both these formulas. From the semantics of the always operator, $\Box(e \vee f)$ is satisfied if both $(e \vee f)$ and the next-formula $\bigcirc\Box(e \vee f)$ are satisfied, and hence we add both of these to the desired initial state. To satisfy the disjunction $(e \vee f)$, a state must contain either e or f or both e and f. The inclusion of the next-formula $\bigcirc\Box(e \vee f)$ does not create any additional requirements on the current state, but the rules for adding transitions between states will ensure its satisfaction. From the semantics of the eventuality operator, $\Diamond e$ is satisfied if at least one of e and the next-formula $\bigcirc\Diamond e$ is satisfied. Combining the resulting cases with different possible ways of satisfying $(e \vee f)$ gives us five initial states q_0, q_1, q_2, q_3, and q_4 as shown in figure 5.10. For example, the state q_0 corresponds to the set $\{\varphi, \Box(e \vee f), \Diamond e, (e \vee f), \bigcirc\Box(e \vee f), e, \bigcirc\Diamond e\}$ of formulas: for an execution starting in a state q_0, we want each formula in this set to be satisfied and a formula such as f, which is not included in this set to be not satisfied.

Whenever a state includes an atomic expression, say e, it means that the input to be processed must satisfy this expression, and thus e should appear as a conjunct in the guard of the edges out of this state. Similarly, when a state does not contain an atomic expression e, the input to be processed must not satisfy e, and thus $\neg e$ should appear as a conjunct in the guard of the edges out of this state. This explains the guards on all the edges of the automaton. In particular, each edge out of the state q_0 has the guard $(e \wedge \neg f)$.

To obtain successors of a state, we examine the next formulas in the state. For every formula of the form $\bigcirc \psi$, $\bigcirc \psi$ should belong to the current state if and only if the successor state contains ψ. Since the state q_0 contains both $\bigcirc \square (e \vee f)$ and $\bigcirc \lozenge e$, its successor is required to contain $\square (e \vee f)$ as well as $\lozenge e$. Such a successor state then must satisfy the conjunction of these two, which is the original formula φ. This means that the successors of the state q_0, and also of q_1 and q_2 by the same logic, are exactly the initial states containing φ. Now let us consider the state q_3, which contains $\bigcirc \square (e \vee f)$ but not $\bigcirc \lozenge e$. Hence, its successor state should contain $\square (e \vee f)$ but not $\lozenge e$ (and hence cannot contain φ). To satisfy the always-formula $\square (e \vee f)$, a state must contain $(e \vee f)$ and $\bigcirc \square (e \vee f)$. Since it does not contain $\lozenge e$, it cannot contain e, and thus the only way to satisfy $(e \vee f)$ is by including f. The state $q_5 = \{\square (e \vee f), (e \vee f), \bigcirc \square (e \vee f), f\}$ is thus the sole successor of the state q_3 and also of states q_4 and q_5 by the same logic.

We would like to ensure that if $p_0 \xrightarrow{v_1} p_1 \xrightarrow{v_2} \cdots$ is an infinite execution through the tableau corresponding to the input trace $\rho = v_1 v_2 \cdots$ starting at some state p_0, then a formula ψ is in p_0 if and only if the input trace ρ satisfies the formula ψ. This is not quite true yet. For instance, an execution can loop forever at the state q_2 provided each input satisfies f: every state contains $\lozenge e$, but no input satisfies the atomic expression e. Intuitively, along this infinite execution, the choice to satisfy $\lozenge e$ is postponed forever. This can be avoided by adding a Büchi acceptance condition that requires that to satisfy $\lozenge e$, one must satisfy e eventually. This is expressed by the Büchi accepting set $F_1 = \{q_0, q_1, q_3, q_4\}$ containing states that either contain e or do not contain $\lozenge e$. For the always-formula $\square (e \vee f)$, if a state does not contain this formula, then we want to make sure that the formula is indeed not satisfied. Note that the negation of an always-formula is an eventually-formula, so the Büchi accepting condition F_2 contains all states that do not contain $(e \vee f)$ or contain $\square (e \vee f)$, and in our example, this turns out to be the set of all states. Thus, an infinite execution is accepting according to both the accepting sets if it does not end up looping at state q_2 forever. Verify that the Büchi automaton, with the states and edges given by the tableau of figure 5.10, accepts an input trace over $\{e, f\}$ exactly when it satisfies the formula φ.

In summary, in a tableau construction, states are subsets of formulas. Each formula stipulates requirements concerning other formulas that must be satisfied by the sequence of inputs along the paths starting in the current state. The edges are defined so as to ensure propagation of the next formulas from one state to its successor. The generalized Büchi accepting requirements ensure eventual fulfillment of eventuality formulas.

Tableau Construction

We proceed to formalize the tableau construction. For the formal construction, let us assume that LTL-formulas are constructed from atomic expressions using the logical connectives of negation, conjunction, and disjunction and the temporal operators next, always, and eventually. Extending the construction to handle the until operator is left as an exercise.

Given an LTL-formula φ, let V_φ be the set of atomic expressions that occur in φ. To evaluate the formula φ, at every step we need to know whether each of the expressions in V_φ is satisfied. Thus, we can treat every atomic expression as a Boolean variable and interpret the formula φ with respect to a trace over the set V_φ of Boolean variables. These variables are the set of input variables for the Büchi automaton M_φ.

Let us first define the set of formulas that are relevant to evaluating the given formula. The *closure* $Sub(\varphi)$ of an LTL-formula φ is defined as:

1. if ψ is a syntactic subexpression occurring in φ, then ψ belongs to $Sub(\varphi)$; and

2. if ψ is a subexpression of the form $\lozenge \psi'$ or $\square \psi'$, then $\bigcirc \psi$ also belongs to $Sub(\varphi)$.

For the illustrative tableau construction in figure 5.10, for $\varphi = \square (e \vee f) \wedge \lozenge e$,

$$Sub(\varphi) \;=\; \{\, e, f, (e \vee f), \lozenge e, \bigcirc \lozenge e, \square (e \vee f), \bigcirc \square (e \vee f), \varphi \,\}.$$

As illustrated in the sample construction, whether the satisfaction of the formula φ depends on the satisfaction of the formulas in $Sub(\varphi)$. Verify that if the formula φ has length k, then the number of formulas in its closure is at most $2k$.

Now, a state of the tableau is a subset of formulas from the closure such that the set of constraints expressed by these formulas are locally consistent. A subset $q \subseteq Sub(\varphi)$ of the closure of φ is *consistent* if the following conditions are satisfied:

- $\neg \psi$ belongs to q exactly when ψ does not belong to q;

- $\psi_1 \wedge \psi_2$ belongs to q exactly when both ψ_1 and ψ_2 belong to q;

- $\psi_1 \vee \psi_2$ belongs to q exactly when either ψ_1 or ψ_2 or both belong to q;

- $\lozenge \psi$ belongs to q exactly when either ψ or $\bigcirc \lozenge \psi$ or both belong to q; and

- $\square \psi$ belongs to q exactly when both ψ and $\bigcirc \square \psi$ belong to q.

In our example of figure 5.10, note that all six states are indeed consistent. These are not all the consistent states, and figure 5.10 shows only those states that are reachable from the initial states. For example, the state $q_6 = \{f, (e \vee f)\}$ is a consistent state but is not reachable.

Given a consistent subset q of $Sub(\varphi)$, let us denote by $Guard_q$ the expression obtained by conjoining each atomic expression e contained in q and the negation of each atomic expression e that does not belong to q. For example, if q contains the atomic expression e but does not contain the atomic expression f, then $Guard_q$ is the expression $e \wedge \neg f$ and captures the constraint on the next input when in state q.

Now we are ready to define the generalized Büchi automaton M_φ, also known as the tableau corresponding to the formula φ:

- the set of input variables is the set V_φ of atomic expressions that appear in φ;

- the set of states is the set of consistent subsets of the closure $Sub(\varphi)$;

- a state $q \subseteq Sub(\varphi)$ is initial exactly when q contains the formula φ;

- for a pair of states q and q', if it is the case that every next formula $\bigcirc \psi$ in $Sub(\varphi)$ belongs to q exactly when ψ belongs to q', then there is an edge $(q, Guard_q, q')$; and

- for each eventually formula $\psi = \Diamond \psi'$ in the closure $Sub(\varphi)$, there is an accepting set F_ψ containing states q such that $\psi' \in q$ or $\psi \notin q$; and for each always formula $\psi = \Box \psi'$ in the closure $Sub(\varphi)$, there is an accepting set F_ψ containing states q such that $\psi' \notin q$ or $\psi \in q$.

The correctness of the construction is established below.

Proposition 5.2 [Correctness of LTL Tableau Construction] *For every LTL-formula φ over the atomic expressions V, a trace ρ over V satisfies φ exactly when it is accepted by the generalized Büchi automaton M_φ.*

Proof. Let φ be an LTL formula, and let $\rho = v_1 v_2 \cdots$ be a trace over the set V_φ of expressions appearing in φ. Suppose $\rho \models \varphi$. For $i \geq 0$, let $q_i \subseteq Sub(\varphi)$ be the set $\{\psi \in Sub(\varphi) \mid (\rho, i+1) \models \psi\}$ of formulas true at the position $i+1$ in the trace ρ. From the definitions, it follows that (1) for all i, the set q_i is consistent; (2) for all i and all formulas $\bigcirc \psi \in Sub(\varphi)$, $\bigcirc \psi \in q_i$ exactly when $\psi \in q_{i+1}$; (3) the set q_0 is an initial state of M_φ; (4) for all i, the input v_{i+1} satisfies the guard $Guard_{q_i}$; (5) for each $\Diamond \psi \in Sub(\varphi)$, if $(\rho, i) \models \Diamond \psi$ for infinitely many positions i, then $(\rho, j) \models \psi$ for infinitely many positions j; and (6) for each $\Box \psi \in Sub(\varphi)$, if $(\rho, i) \not\models \Box \psi$ for infinitely many positions i, then $(\rho, j) \not\models \psi$ for infinitely many positions j. It follows that $q_0 \xrightarrow{v_1} q_1 \xrightarrow{v_2} \cdots$ is an accepting execution in the tableau M_φ, and the automaton M_φ accepts the trace ρ.

Now consider an accepting execution $q_0 \xrightarrow{v_1} q_1 \xrightarrow{v_2} \cdots$ of the automaton M_φ over the input trace $\rho = v_1 v_2 \cdots$. We want to establish that for all $\psi \in Sub(\varphi)$, for all $i \geq 0$, $\psi \in q_i$ if and only if $(\rho, i+1) \models \psi$. The proof is by induction on the structure of ψ and is left as an exercise. It follows that $\rho \models \varphi$. ∎

Note that the number of states of the automaton M_φ is exponential in the size of the formula. This blow-up is unavoidable. The generalized Büchi automaton M_φ can be converted into a Büchi automaton using the construction described in proposition 5.1.

Model Checking

To check whether a component C satisfies an LTL-formula φ, we first negate the formula φ. We build the Büchi automaton M corresponding to the formula $\neg \varphi$ such that an infinite execution of the composed system $C \parallel M$ in which the Büchi states of M are repeatedly encountered corresponds to an infinite execution of the component C that satisfies the negated specification $\neg \varphi$ and, thus, is a counterexample to the model checking problem. This approach of negating the formula *before* applying the tableau construction avoids the need for the computationally demanding task of complementing the tableau and is essential to the practical applications of model checking. We summarize the result below.

Theorem 5.1 [From LTL Model Checking to Repeatability] *Given an LTL-formula φ over the atomic expressions V that refer to the observable variables of a system C, there is an algorithm to construct a nondeterministic Büchi automaton M with the input variables V, with a subset F of accepting states, such that a system C satisfies the specification φ precisely when for the composed system $C\parallel M$, the property "state of M belongs to F" is repeatable.* ∎

Exercise 5.13: Consider the LTL-formula $\varphi = \square \Diamond e \vee \square f$. First compute the closure $Sub(\varphi)$. Then apply the tableau construction to build the generalized Büchi automaton M_φ. It suffices to show only the reachable states. ∎

Exercise 5.14: The formal description of the tableau construction considers formulas where the only temporal operators are next, eventually, and always. Describe how the modifications necessary when the until operator is also allowed. ∎

Exercise 5.15: Consider the LTL-formula $\varphi = (e \, \mathcal{U} \, \bigcirc f) \vee \neg e$. First compute the closure $Sub(\varphi)$. Then apply the tableau construction to build the generalized Büchi automaton M_φ. It suffices to construct only the reachable states. ∎

Exercise 5.16: In section 5.1.4, we mentioned that the following property cannot be specified in LTL: "the expression e is 1 in every even position." Draw a Büchi automaton M with one input variable e that accepts a trace exactly when it satisfies this property. ∎

Exercise 5.17*: Complete the proof of proposition 5.2: for an accepting execution $q_0 \xrightarrow{v_1} q_1 \xrightarrow{v_2} \cdots$ of the automaton M_φ over the input trace $\rho = v_1 v_2 \cdots$, prove that for all $\psi \in Sub(\varphi)$, for all $i \geq 0$, $\psi \in q_i$ if and only if $(\rho, i+1) \models \psi$, by induction on the structure of the formula ψ. ∎

5.2.3   Nested Depth-First Search *

Given a transition system T and a property φ, to check whether φ is repeatable, we search for a state that violates φ, is reachable from some initial state, and is contained in a cycle. The core computational problem here is detecting cycles. As discussed in section 3.3, we assume that the transition system is countably branching and is represented by the functions *FirstInitState*, *NextInitState*, *FirstSuccState*, and *NextSuccState* that can be used to enumerate initial states and successor states of a given state. We want our algorithm to be *on-the-fly*: it should explore states and transitions out of these states only as needed, and it should terminate returning a counterexample as soon as it finds one. Thus, the ideal algorithm should not first examine all the reachable states and then proceed to finding cycles. As a result, the classical algorithms for detecting cycles in a graph that rely on computation of strongly connected components in the graph are not best suited for our application (a strongly connected component in a directed graph is a maximal subset of the vertices such that there is a path between every pair of vertices in this subset). Instead, we will present a cycle-detection algorithm that employs two depth-first search traversals, one nested in the other.

The algorithm explores the reachable states of the transition system in a manner similar to the depth-first search algorithm of figure 3.16 using the stack *Pending* and the set *Reach* to store states already visited. The key difference is the following: while checking the reachability of a property, when a state satisfying the property is encountered, the search terminates; while checking the repeatability of a property, when a state s satisfying the property is encountered, the algorithm initiates another search for a cycle containing the state s. To implement this, suppose every time a state satisfying the property φ is encountered, a brand-new search to check whether the state s is reachable from itself is initiated, and this search uses its own set of visited states. While such a strategy would lead to a correct algorithm, it has time complexity that is quadratic in the number of states, and this can be significantly improved. The optimal algorithm is shown in figure 5.11.

The algorithm involves two nested searches: a primary search performed by the function *DFS* and a secondary (or nested) search performed by the function *NDFS*. The states encountered during the primary search are stored in the set *Reach*, whereas the states visited during the secondary search are stored in the set *NReach*. As in a standard depth-first search, for every reachable state s of T, the function *DFS* is invoked at most once with the state s as its input. Once the primary search originating at a state s terminates, if the state s satisfies the desired repeatable property φ, then it is a potential candidate for the cyclic counterexample. Then a secondary search is initiated by calling the function *NDFS* with the state s as its input. The objective of this secondary search is to find a cycle starting at the state s. When the function *NDFS*(s) is invoked, the stack *Pending* contains an execution starting from an initial state leading to the state s. The secondary search does not modify the stack *Pending*. Thus,

Input: A transition system T and property φ;
Output: If φ is a repeatable property of T return 1, else return 0;

```
set(state) Reach := EmptySet;
set(state) NReach := EmptySet;
stack(state) Pending := EmptyStack;
state s := FirstInitState(T);

while s ≠ null do {
   if Contains(Reach, s) = 0 then
      if DFS(s) = 1 then return 1;
   s := NextInitState(s, T);
   };
return 0.

bool function DFS(state s)
   Insert(s, Reach);
   Push(s, Pending);
   state t := FirstSuccState(s, T);
   while t ≠ null do {
      if Contains(Reach, t) = 0 then
         if DFS(t) = 1 then return 1;
      t := NextSuccState(s, t, T);
      };
   if Satisfies(s, φ) = 1 then
      if Contains(NReach, s) = 0 then
         if NDFS(s) = 1 then return 1;
   Pop(Pending);
   return 0.

bool function NDFS(state s)
   Insert(s, NReach);
   state t := FirstSuccState(s, T);
   while t ≠ null do {
      if Contains(Pending, t) = 1 then return 1;
      if Contains(NReach, s) = 0 then
         if NDFS(t) = 1 then return 1;
      t := NextSuccState(s, t, T);
      };
   return 0.
```

Figure 5.11: Nested Depth-first Search Algorithm for Checking Repeatability

if the secondary search encounters a transition leading to a state belonging to the stack, then it concludes that there is a cycle that contains the state s. This establishes that whenever the algorithm returns 1, the transition system contains a reachable cycle containing a state that satisfies the property φ.

The secondary search uses a separate set *NReach* to keep track of states encountered during the secondary search. However, this set is shared across all calls to *NDFS*: every time the function *NDFS* is called with a state s as its input, the state s is added to this set, and *NDFS* is invoked with a state t as input only if the state t is not in the set *NReach*. Thus, the secondary search explores a reachable state at most once, and the total running time of the secondary search is the same as the primary search. To understand the interplay between the two searches and the argument about the correctness of the search strategy, consider two states s and t that are encountered by the primary search, and suppose both satisfy the property φ. Suppose the secondary search is invoked from the state s first, and it explores all states that are reachable from the state s, adding them to the set *NReach* but without finding a cycle. Later, when the secondary search is invoked from the state t, it will just skip over states that were added to the set *NReach*. What guarantees that this does not cause the algorithm to miss detection of a cycle containing the state t?

To answer this question, let us order the states according to the termination times of the primary search: with each reachable state s, associate a number d_s such that if the call $DFS(s)$ terminates before the call $DFS(t)$, then $d_s < d_t$. Let $s_0, \ldots s_k$ be the ordering of the states that are reachable and satisfy the property φ according to this numbering. Let s_i be the first state in this ordering that belongs to a cycle, and let Q be the set of all states that are reachable from the states s_j with $j < i$.

We first claim that the cycle that contains the state s_i is disjoint from the set Q. If not, there is a state t belonging to the set Q such that both states s_i and t belong to a cycle. This implies that the state s_i is reachable from some state s_j with $j < i$ (since the state t must be reachable from some such state). Thus, the exploration from the state s_j is guaranteed to examine state s_i, but given the ordering of the states, we know that the call $DFS(s_j)$ terminates before the call $DFS(s_i)$. This can happen only if the call $DFS(s_i)$ is pending when $DFS(s_j)$ is invoked. This implies that the state s_j is also reachable from the state s_i, and thus the state s_j is involved in a cycle, a contradiction to the assumption that the state s_i is the first state in the ordering that belongs to a cycle.

When the primary search from the state s_i terminates, the set *NReach* containing the states visited by the secondary search so far equals the set Q. Since the state s_i does not belong to the set Q, the function *NDFS* will be invoked with the state s_i as its input. Since there is a cycle that contains the state s_i and does not involve any of the states already in the set *NReach*, the secondary search is guaranteed to discover this cycle.

To understand how the algorithm works, let us consider the transition system shown in figure 5.12. The initial state is A, and the property is satisfied in

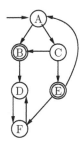

Figure 5.12: Sample Transition System for Illustrating Repeatability Algorithms

states B and E. The execution of the algorithm is illustrated in figure 5.13. The first column lists the calls made to the functions *DFS* and *NDFS*, where the indentation indicates which calls are pending. The subsequent columns list the values of the stack *Pending* (left-most element is at the top of the stack) and the sets *Reach* and *NReach*.

Initially, the function *DFS* is called with the initial state A as input, which invokes *DFS(B)*, which in turn calls *DFS(D)*, which then invokes *DFS(F)*. Since the sole successor of the state F is already in the set *Reach*, the call *DFS(F)* terminates. Thus, $d_F = 1$. Subsequently, the call *DFS(D)* also terminates ($d_D = 2$). At this point, in the execution of *DFS(B)*, the state B has no more successor states, but since the state B satisfies the desired property, the secondary search is initiated for the first time via the call *NDFS(B)*. This in turn calls *NDFS(D)* and then calls *NDFS(F)*. All these calls terminate reporting failure: the stack *Pending* contains the states A and B, and no transitions leading to either of these states are encountered. When the call *DFS(B)* terminates ($d_B = 3$), the set *NReach* contains all the states reachable from the state B, namely, the states B, D, and F. The primary search from the state A now proceeds to call *DFS(C)*, which in turn calls *DFS(E)*. All the successors of the state E are already in the set *Reach*, but since the state E satisfies the property, another secondary search is invoked using the call *NDFS(E)*. At this point, the states B, D, and F are already assumed to be visited for the secondary search, and hence the transition from the state E to F will not be explored for establishing the repeatability of the state E. This is justified by the correctness argument for the algorithm: since *DFS(B)* has already terminated without discovering a cycle containing the state B, a cycle containing the state E cannot involve states reachable from the state B. As it turns out, the state E has a successor state, namely, A, which is in the stack. As a result, the call *NDFS(E)* terminates reporting success, causing all the pending calls to terminate with return value 1.

Note that, just like the depth-first search algorithm of figure 3.16, the nested depth-first search algorithm may detect a cycle before exploring all the reachable

	Pending	Reach	NReach
DFS(A)	[ ]	{ }	{ }
DFS(B)	[A]	{A}	{ }
DFS(D)	[B, A]	{A, B}	{ }
DFS(F)	[D, B, A]	{A, B, D}	{ }
NDFS(B)	[B, A]	{A, B, D, F}	{ }
NDFS(D)	[B, A]	{A, B, D, F}	{B}
NDFS(F)	[B, A]	{A, B, D, F}	{B, D}
DFS(C)	[A]	{A, B, D, F}	{B, D, F}
DFS(E)	[C, A]	{A, B, D, F, C}	{B, D, F}
NDFS(E)	[E, C, A]	{A, B, D, F, C, E}	{B, D, F}

Figure 5.13: Illustrative Execution of the Nested Depth-first Search Algorithm

states. For instance, for the transition system of figure 5.12, if we introduce a transition from the state D to state B, then the call $NDFS(B)$ will discover a cycle, and the algorithm terminates without ever visiting the states C or E. If the number of reachable states of the transition system is finite, then it is guaranteed to terminate with the correct answer. These properties are summarized in the theorem below.

Theorem 5.2 [Nested Depth-first Search for Repeatability Checking] *Given a countably branching transition system T and a property φ, the nested depth-first search algorithm of figure 5.11 has the following guarantees:*

1. *If the algorithm terminates, then the returned value correctly indicates whether the property φ is a repeatable property of T.*

2. *If the number of reachable states of T is finite, then the algorithm terminates, and the number of calls to DFS and to NDFS are bounded by the number of reachable states.*

■

Exercise 5.18: Modify the nested depth-first search algorithm of figure 5.11 so that it outputs a counterexample consisting of an execution leading from an initial state to a state s satisfying φ and a cyclic execution from s back to itself.

■

5.2.4   Symbolic Repeatability Checking

Recall the symbolic breadth-first search algorithm for invariant verification by iterative image computation discussed in section 3.4. We will now develop a symbolic nested search algorithm to check whether the transition system has an infinite execution that repeatedly visits a given property. As before, we assume that a set of states over a set V of typed variables is represented as a region of type **reg**. In the symbolic representation of a transition system with state variables S, the initial states are represented by a region *Init* over the variables S, and the transitions are represented by a region *Trans* over the variables $S \cup S'$. The property φ whose repeatability is to be checked is also represented by a region over the variables S. The data type `reg` for regions supports operations such as `Conj`, `Diff`, and `IsSubset`, as discussed in section 3.4.

Image and Pre-image Computation

The core step of symbolic verification algorithms is image computation. Given a region A over the state variables S, the region that contains all the states that can be reached from the states in A using one transition can be computed using the `Post` operation defined as:

$$\texttt{Post}(A, \textit{Trans}) \;=\; \texttt{Rename}(\texttt{Exists}(\texttt{Conj}(A, \textit{Trans}), S), S', S).$$

The dual operator corresponds to the pre-image computation: given a region A over the state variables, the region that contains all the states from which some state in A can be reached using one transition is called the *pre-image* of the region A. Given a region A, to compute its pre-image, we first rename the unprimed variables to primed variables and then intersect it with the transition region *Trans* over $S \cup S'$ to obtain all the transitions that lead to the states in A. Then we project the result onto the set S of unprimed state variables by existentially quantifying the variables in S'. Thus, the pre-image operator `Pre` is defined as:

$$\texttt{Pre}(A, \textit{Trans}) \;=\; \texttt{Exists}(\texttt{Conj}(\texttt{Rename}(A, S, S'), \textit{Trans}), S').$$

Let us consider an example to illustrate the pre-image computation. Suppose the system has a single variable x of type `real`, and the update is given by the following conditional statement

$$\texttt{if } (1 \leq x \leq 5) \texttt{ then } x := x - 1 \texttt{ else } x := x + 1.$$

Then the transition region is given by the formula

$$[(1 \leq x \leq 5) \wedge (x' = x - 1)] \vee [((1 > x) \vee (x > 5)) \wedge (x' = x + 1)].$$

Consider the region A given by the formula $1 \leq x \leq 2$, and let us apply the sequence of steps needed to compute the pre-image of the region A. First, we rename the variable x to x', and this gives us the formula $1 \leq x' \leq 2$. Then we conjoin this region with the transition formula *Trans*, and this gives the result, which simplifies to:

$$[(1 \leq x \leq 5) \wedge (x' = x - 1) \wedge (1 \leq x' \leq 2)]$$
$$\vee \; [((1 > x) \vee (x > 5)) \wedge (x' = x + 1) \wedge (1 \leq x' \leq 2)].$$

The final step is to apply the operation of existential quantification to eliminate the variable x' leading to the formula

$$[(1 \leq x \leq 5) \wedge (1 \leq x - 1 \leq 2)] \vee [((1 > x) \vee (x > 5)) \wedge (1 \leq x + 1 \leq 2)].$$

Input: A transition system T given by a region *Init* for initial states,
       a region *Trans* for transitions, and a region φ for the property;
Output: If φ is repeatable in T, return 1, else return 0.

```
reg Reach := Empty;
reg New := Init;
while IsEmpty(New) = 0 do {
   Reach := Disj(Reach, New);
   New := Diff(Post(New, Trans), Reach);
   };
reg Recur := Conj(Reach, φ);
while IsEmpty(Recur) = 0 do {
   Reach := Empty;
   New := Pre(Recur, Trans);
   while IsEmpty(New) = 0 do {
      Reach := Disj(Reach, New);
      if IsSubset(Recur, Reach) = 1 then return 1;
      New := Diff(Pre(New, Trans), Reach);
      };
   Recur := Conj(Recur, Reach);
   };
return 0.
```

Figure 5.14: Symbolic Nested Search Algorithm for Checking Repeatability

This formula simplifies to

$$(2 \leq x \leq 3) \ \vee \ (0 \leq x < 1)$$

which precisely describes the desired set of values of x for which executing the conditional assignment causes the resulting value to belong to the interval $[1, 2]$.

Nested Symbolic Search

The symbolic algorithm for checking repeatability shown in figure 5.14 uses both image computation and pre-image computation. The algorithm has two phases: the first phase consists of a single while loop, and the second phase consists of two nested while loops.

The first phase of the algorithm computes the region *Reach* of all states reachable from the region *Init* of initial states by repeatedly applying the image-computation operator Post. This is similar to the algorithm of figure 3.19. Let us illustrate the algorithm using the sample transition system shown in figure 5.12. Figure 5.15 shows the values of the regions *Reach* and *New* at the beginning and, after each iteration, during the first phase of the algorithm.

The second phase attempts to find an infinite execution with repeating occurrences of the property φ. The set of states whose repeated occurrence indicates

	Reach	*New*
Initially	$\{\ \}$	$\{A\}$
After iteration 1	$\{A\}$	$\{B, C\}$
After iteration 2	$\{A, B, C\}$	$\{D, E\}$
After iteration 3	$\{A, B, C, D, E\}$	$\{F\}$
After iteration 4	$\{A, B, C, D, E, F\}$	$\{\ \}$

Figure 5.15: Illustrative Execution of the First Phase of Algorithm 5.14

success is captured by the region *Recur*. This region initially contains all the reachable states that satisfy the property φ and can be computed by intersecting the region *Reach* computed at the end of the first phase and the region representing the property. Let us call this region $Recur_0$. For each of the states s in this set, we want to determine if there exists an execution consisting of one or more transitions starting in the state s and ending in some state in $Recur_0$. To compute this information, the inner loop repeatedly applies the pre-image computation to find those states from which states in $Recur_0$ can be reached in one or more transitions. This computation is similar to the computation of the reachable states: the region *Reach*, initialized to the empty set, contains the states already examined; and the region *New*, initialized to the states from which the current set *Recur* can be reached in one transition, contains the states to be explored. In each iteration of the inner loop, the region *Reach* is updated by adding the unexplored states in the region *New*. The set of states to be newly explored is obtained by computing the pre-image of the current set *New* and removing the already explored states in *Reach* using the set-difference operation. The inner loop terminates when there are no more new states to be examined. At this point, the region *Reach* contains precisely those states that have a path to some state in $Recur_0$. By intersecting this region with *Recur*, we obtain the set $Recur_1$, a subset of $Recur_0$. The outer loop is now repeated again with this revised value of *Recur*.

To illustrate the second phase of the algorithm, let us continue with our example transition system of figure 5.12. Figure 5.16 shows the values of the regions *Recur*, *Reach*, and *New* at each iteration. Initially, the set *Recur* contains the states B and E as potential candidates for the repeating states. One iteration of the outer loop discovers that there is no execution from the state B that can lead back to this set; as a result, *Recur* gets updated to $\{E\}$. During the second iteration of the outer loop, the inner loop computes the set of states from which the state E can be reached. In the third iteration of the inner loop, the state E gets added to the region *Reach*, and this causes the successful termination of the algorithm.

Outer loop	Recur	Inner loop	Reach	New
Initially	$\{B, E\}$			
		Initially	$\{\ \}$	$\{A, C\}$
		After iteration 1	$\{A, C\}$	$\{E\}$
		After iteration 2	$\{A, C, E\}$	$\{\ \}$
After iteration 1	$\{E\}$			
		Initially	$\{\ \}$	$\{C\}$
		After iteration 1	$\{C\}$	$\{A\}$
		After iteration 2	$\{C, A\}$	$\{E\}$
		During iteration 3	$\{C, A, E\}$	

Figure 5.16: Illustrative Execution of the Second Phase of Algorithm 5.14

Correctness

Let $Recur_1, Recur_2, \ldots$ be the successive values assigned to the region variable $Recur$ at the end of the outer while loop. Each such set $Recur_i$ is a subset of the set $Recur_{i-1}$ and contains those states in $Recur_{i-1}$ from which some state in $Recur_{i-1}$ can be reached using an execution with one or more transitions. Hence, each such set $Recur_i$ contains the states s such that the state s is reachable from some initial state, the state s satisfies the property φ, and there is an execution starting from the state s that encounters states satisfying the property φ at least i times.

Suppose the property φ is repeatable for the transition system T. Then there is a state s that is reachable from some initial state, the state s satisfies φ, and there is an infinite execution starting from the state s that encounters states satisfying the property φ infinitely many times. Such a state s will belong to every set $Recur_i$ and thus will never be removed from $Recur$. As a result, if the value of $Recur$ becomes the empty set at any point during the execution of the algorithm, no such state s exists, and the algorithm can terminate claiming that the property φ is not repeatable.

Conversely, suppose for the current nonempty set $Recur_i$, from every state in $Recur_i$, some state in $Recur_i$ can be reached by an execution with one or more transitions. Then in the subsequent iteration of the outer loop, the final value of the set $Reach$ is a superset of $Recur_i$. As the region $Reach$ is computed iteratively by adding more and more states that can reach $Recur_i$ in one or more transitions, when the algorithm finds that $Reach$ is a superset of $Recur$, it terminates reporting repeatability of the property φ. We argue that in this case, indeed there is an infinite execution in which the property φ repeats. Let s_0 be any state in $Recur_i$. Since the set $Recur_i$ is a subset of $Recur_0$, the state s_0 is reachable from an initial state. Hence, it suffices to demonstrate that there exists an infinite execution starting at the state s_0 with the property φ repeating infinitely often. From the state s_0, there is an execution with one or more transitions leading to some state, say s_1, in $Recur_i$. From the state s_1, there is an execution with one or more transitions leading to some state, say s_2, in $Recur_i$. This process can be repeated forever. Concatenating all these finite executions gives us the desired infinite execution with repeating φ.

Complexity Analysis

If the number of reachable states of T is finite, then the termination is guaranteed. Suppose the number of reachable states of T is n, and k of these satisfy the property φ. Then the set $Recur_0$ contains k states, and the number of iterations of the outer loop is at most k (since states are only removed from $Recur$, and the algorithm terminates if the value of $Recur$ does not change). In each iteration of the outer loop, the inner loop can be executed at most n times as it computes the set of states from which $Recur$ is reachable, in an iterative manner. The actual running time of the algorithm depends on how efficiently the various operations on regions are executed, but the number of symbolic operations is quadratic.

The correctness and complexity of the algorithm are summarized below.

Theorem 5.3 [Symbolic Nested Search for Checking Repeatability] *Given a symbolic representation of a transition system T and a property φ, the symbolic nested search algorithm of figure 5.14 has the following guarantees:*

1. *If the algorithm terminates, then the returned value correctly indicates whether the property φ is repeatable for the transition system T.*

2. *If the transition system T has n reachable states, of which k satisfy the property φ, then the algorithm terminates after at most $O(nk)$ operations on regions.*

∎

Exercise 5.19: Consider a transition system with two variables x and y of type **nat**. Suppose the transitions of the system are described by the conditional statement

$$\text{if } (x > y) \text{ then } x := x + 1 \text{ else } y := x.$$

First, describe the transition region as a formula $Trans$ over the variables x, y, x', and y'. Consider the region A given by the formula $1 \leq y \leq 5$. Compute the pre-image of the region A. ∎

Exercise 5.20*: Algorithm of figure 5.14 uses both post-image computation and pre-image computation. Suppose we modify the algorithm by replacing both calls to Pre by Post, that is, in the second phase, replace the assignment $New :=$ Pre($Recur$, $Trans$) by $New :=$ Post($Recur$, $Trans$) and the assignment $New :=$ Diff(Pre(New, $Trans$), $Reach$) by $New :=$ Diff(Post(New, $Trans$), $Reach$). How will this modification impact the correctness of the algorithm? Justify your answer. ∎

5.3   Proving Liveness*

In section 3.2.1, we studied a general proof technique for establishing invariants of transition systems: given a transition system T and a property φ over its state variables, to prove that the property φ is an invariant of the transition system T, we find another property ψ and show that (1) the property ψ is an *inductive* invariant of the transition system T, and (2) the property ψ implies the property φ. This proof technique based on inductive invariants is appealing for the following reasons. First, the method is rooted in the intuitive and informal argument needed to convince oneself about the correctness of the system. Second, the formalization of the rule is mathematically precise, and the rule can be used to produce a machine-checkable proof of the correctness of the system. Third, the rule is general enough so that every invariant property can be established by applying the rule. Now, we focus on identifying proof techniques for proving liveness properties of transition systems.

Consider a transition system T with the state variables S. Liveness properties of such a transition system can be expressed using LTL-formulas over the set S of variables: the transition system T satisfies the LTL-formula φ if every infinite execution of T satisfies φ. The precise details of proof rules for establishing that the transition system satisfies the LTL-formula φ depend on the structure of the formula φ. We focus on the most commonly occurring patterns: eventuality properties and response properties assuming weak fairness.

5.3.1   Eventuality Properties

Let us revisit the leader election protocol of section 2.4.3. We want to establish that every node eventually makes a decision. Since a node announces its decision by updating the output variable *status* when the value of its state variable r equals N, where N is the total number of nodes in the network, we want to show that eventually the value of r becomes N. More precisely, we want to prove that every infinite execution of the transition system corresponding to the protocol satisfies the eventually formula $\Diamond\,(r_n = N)$, where n is an arbitrary node.

To convince yourself that the component of figure 2.35 satisfies the eventuality formula $\Diamond\,(r_n = N)$, observe that the value of the variable r_n is initially 1, and in each round, it is incremented by 1 as long as it is less than N. To formally capture the intuition behind this argument, let us define a function *rank* from the states of the transition system to natural numbers: in a given state s of the transition system, the value of $rank(s)$ is the difference between N and the value that the state s assigns to the variable r_n. The function *rank* captures the *distance* of a state from the desired eventuality goal. To prove that every execution satisfies the eventuality formula $\Diamond\,(r_n = N)$, we show that if a state s does not already satisfy the desired eventuality (that is, the value of the round variable in the state s is not yet N), then executing the protocol for one more step *decreases* the rank; that is, if (s,t) is a transition of the system, then $rank(t) < rank(s)$. Since the rank is a natural number, it cannot decrease forever, implying that a state satisfying the desired eventuality must be encountered in finitely many steps.

The function $rank$ needs to map every state, and not just the states that we informally know to be reachable, to a natural number. In our example, we define the rank $rank(s)$ of a given state s to be $N - s(r_n)$ if $s(r_n) \leq N$ and 0 otherwise. For a state s that assigns, say, the value $N + 1$ to the variable r_n, the rank is 0, and executing a transition in such a state does not decrease the rank. However, we know that such a state is unreachable. More precisely, we show that $0 \leq r_n \leq N$ is an invariant of the system, and this can be established using the proof technique already studied. To show that the execution of one transition decreases the rank, it now suffices to focus on those states that satisfy this invariant. That is, we show that for every state s of the system, assuming that the state s satisfies the invariant $0 \leq r_n \leq N$, if (s, t) is a transition of the system, then either the state t satisfies the eventuality goal or its rank is strictly smaller than the rank of the state s.

This reasoning is summarized in the following proof rule for establishing eventuality properties:

PROOF RULE FOR EVENTUALITY PROPERTIES

To establish that a transition system T satisfies the eventuality formula $\Diamond \varphi$, where φ is a property over the state variables of T, identify a state property ψ and a function $rank$ that maps states of T to **nat** and show that:

1. the property ψ is an invariant of T; and

2. for every state s that satisfies ψ and every transition (s, t) of T, either the state t satisfies φ or $rank(t) < rank(s)$.

To establish that the proof rule is sound, we need to argue that if we establish the two premises (1) and (2) of the rule, then the transition system must satisfy the formula $\Diamond \varphi$. Consider an infinite execution $s_0, s_1, s_2, \ldots$ of the transition system. Clearly, each state s_j appearing in the execution is reachable and by the first premise satisfies the invariant property ψ. To show that some state s_j in the execution must satisfy the desired eventuality φ, assume to the contrary. Then by the second premise, since there is a transition between every pair of adjacent states (s_j, s_{j+1}), we have that $rank(s_{j+1}) < rank(s_j)$ for each $j \geq 0$. However, this is not possible: if $rank(s_0)$ is K, then the rank can decrease at most K times as the rank of each state is a non-negative number and thus cannot decrease at each step of an infinite execution.

Exercise 5.21: Recall the transition system $\text{GCD}(m, n)$ of figure 3.1 capturing the program for computing the greatest common divisor of two numbers m and n. Suppose we want to establish that the program terminates, that is, eventually *mode* equals **stop**. Prove this eventuality property using the proof rule for eventuality formulas by selecting an appropriate invariant and a ranking function. ∎

5.3.2   Conditional Response Properties

A canonical liveness property is the response property "every request φ_1 is eventually followed by the response φ_2," expressed by the LTL-formula $\Box\,(\varphi_1 \rightarrow \Diamond\,\varphi_2)$.

Recall the rule for establishing the eventuality formula $\Diamond\,\varphi$: we find a ranking function that maps states to natural numbers, identify an invariant property ψ, and show that executing a transition in any state that satisfies the invariant causes either the fulfillment of the goal or a decrease in the rank. Let us first consider how to generalize this reasoning to establish the response formula $\Box\,(\varphi_1 \rightarrow \Diamond\,\varphi_2)$. Now we want to show that whenever the property φ_1 holds, executing a sequence of system transitions must result in a state satisfying the goal property φ_2. We again use a ranking function that maps states to natural numbers and show that the rank keeps decreasing until the goal is satisfied. This is achieved by the following proof principle:

PROOF RULE FOR RESPONSE PROPERTIES

To establish that a transition system T satisfies the response formula $\Box\,(\varphi_1 \rightarrow \Diamond\,\varphi_2)$, where φ_1 and φ_2 are state properties, identify a state property ψ and a function $rank$ that maps states of T to **nat** and show that:

1. every state that satisfies the property φ_1 also satisfies the property ψ, and

2. for every state s that satisfies the property ψ and for every transition (s,t) of T, either the state t satisfies the response property φ_2 or the state t satisfies the property ψ and $rank(t) < rank(s)$.

As in the case of the eventuality rule, the property ψ captures the states from which executing a transition causes either the fulfillment of the goal or a decrease in the rank. However, instead of requiring that the property ψ be an invariant of the system, we require that whenever the request property φ_1 holds, the property ψ should hold, and it should continue to hold until the response φ_2 is satisfied.

To illustrate the proof technique for establishing response properties, let us consider a transition system with two variables x and y of type **int**. Suppose initially x equals 1 and y equals 0, and the transitions of the system correspond to executing the following code at each step:

$$\text{if } (x > 0) \text{ then } \{x := x - 1;\ y := y + 1\} \text{ else } x := y.$$

The sequence of values of x is $1, 0, 1, 0, 2, 1, 0, 4, 3, 2, 1, 0, 8, 7, \ldots$ and thus, the program satisfies the recurrence property $\Box\,\Diamond\,(x = 0)$. To prove that the system satisfies this recurrence formula, we can apply the rule for response properties using φ_1 as 1 (that is, always true) and φ_2 as $(x = 0)$. We choose ψ to be the same as φ_1, and thus premise 1 of the rule holds immediately. Consider a state $s = (a, b)$ (that is, x equals a and y equals b). The ranking function *rank* maps such a state s to a if $a > 0$ and to $b+1$ if $a = 0$. To establish premise 2, consider a state $s = (a, b)$. If $a > 0$, then $rank(s) = a$, and letting the system execute one step leads to the state $t = (a - 1, b + 1)$. If $a = 1$, then the state t satisfies the goal $(x = 0)$ (and in this case, $rank(t) = b + 2$, which could be much higher than $rank(s)$), and if $a > 1$, then $a - 1 > 0$ and $rank(t) = a - 1$, which is less than $rank(s)$. However, if $a = 0$, then $rank(s) = b + 1$, and letting the system execute one step leads to the state $t = (b, b)$ with $rank(t) = b$, which is less than $rank(s)$. This means that the proof rule is applicable and allows us to conclude that the system satisfies $\Box \Diamond (x = 0)$.

Establishing Eventual Delivery of Messages for Merge

Let us revisit the example of the asynchronous process Merge of figure 4.3. Suppose we want to establish that if a message v is received on the input channel in_1, it will eventually be delivered on the output channel *out*. This is captured by the response property:

$$\Box \, (in_1 ? v \; \rightarrow \Diamond \; out! v).$$

Here, the request property φ_1 is $in_1 ? v$, and its fulfillment is the response property $\varphi_2 = out! v$. We first choose the strengthening ψ of the request property to be $\texttt{Contains}(x_1, v)$. Note that whenever the process executes the input action $in_1 ? v$, the message v is enqueued in the state variable x_1, and thus, the property $\texttt{Contains}(x_1, v)$ holds.

To define the ranking function, consider the following question: when a message v is in the queue x_1, which quantity do we expect to monotonically decrease until the message v gets removed from the queue? It is the number of messages queued up ahead of the message v. Hence, given a state s, let $rank(s)$ be 0 if the queue $s(x_1)$ does not contain the message v; otherwise, let $rank(s)$ be k if v is the kth message in the queue $s(x_1)$. For instance, if the queue x_1 contained five messages when the message v gets enqueued, then it will be the sixth message in the queue, and the rank will be 6. Whenever a message gets dequeued from the queue x_1 (and transmitted on the output channel), the message v moves up one slot, causing the rank to become 5. This process repeats until the message v is at the front of the queue. In such a state, the rank is 1, and dequeuing a message from the queue x_1 causes the rank to become 0, and in the resulting state, the goal property $out! v$ is satisfied. Now the condition $\texttt{Contains}(x_1, v)$ no longer holds, and the rank stays unchanged. If at a later step the message v is received again on the input channel in_1, it gets enqueued in the queue x_1, and the rank can increase arbitrarily. For instance, if the size of the queue x_1 is 12 upon the arrival of the next instance of the message v on the input channel in_1, then it gets enqueued in the 13th slot, and the rank is 13. It keeps decreasing from 13 down to 0 until the message v is again transferred from the queue x_1 to the output channel.

In order to apply the proof rule for the response property, let us check the premises with the property φ_1 equal to $in_1\,?\,v$, the property φ_2 equal to $out!\,v$, and the property ψ equal to $\text{Contains}(x_1, v)$. Clearly, the premise 1 of the rule holds: a state satisfying the property φ_1 is guaranteed to satisfy the property ψ. Furthermore, in a transition (s, t), where the state s satisfies ψ (that is, the message v is in the queue x_1 in state s), and the state t does not satisfy the goal (that is, the message v is not transmitted from the queue in_1 to the output channel during this transition), we are guaranteed that the state t continues to satisfy the property ψ. Intuitively, the property $\text{Contains}(x_1, v)$ continues to stay true and is preserved in every transition until the goal property holds.

However, the proof of the response formula fails. The second premise of the rule requires that in states satisfying $\text{Contains}(x_1, v)$, executing a transition either satisfies the goal property $out!\,v$ or causes the rank to decrease. That is, when the message v is in the queue x_1, executing a step of the process Merge, either leads to a state with the message v is output on the output channel or causes the message v to shift one slot in the queue x_1. This condition is satisfied only by the execution of the output task A_1^o that dequeues a message from the queue x_1 and transmits it on the channel out. If the transition corresponds to executing any task other than the task A_1^o, then while the condition $\text{Contains}(x_1, v)$ continues to hold, the rank stays the same. If the task A_1^o is never executed, then the rank will stay unchanged, and the message v will never be output. In fact, not all executions of the process Merge satisfy the response formula $\square\,(\,in_1\,?\,v\ \rightarrow\ \lozenge\,out!\,v\,)$. The formula, however, is satisfied by all the executions that satisfy the weak fairness assumption for the task A_1^o. We need to *strengthen* the proof rule for response properties in order to establish such *conditional* response properties.

Proof Rule for Conditional Response

Suppose we want to establish that the process Merge satisfies the response formula $\square\,(\,in_1\,?\,v\ \rightarrow\ \lozenge\,out!\,v\,)$ assuming weak fairness for the output task A_1^o. We use the same choice for ψ, namely, $\text{Contains}(x_1, v)$, and the same ranking function, namely, for a given state s, $rank(s)$ is 0 if the queue $s(x_1)$ does not contain the message v, and $rank(s)$ is k if v is the kth message in the queue $s(x_1)$. Consider a state s that satisfies the property $\text{Contains}(x_1, v)$ and a transition (s, t) of the process such that the transition does not involve sending the message v on the output channel. Instead of insisting that $rank(t) < rank(s)$ as required by the earlier response rule, we now consider two cases. When the transition from the state s to the state t involves the execution of the task A_1^o, we require that $rank(t) < rank(s)$; otherwise, it suffices that the rank does not decrease (that is, $rank(t) \leq rank(s)$), but the task A_1^o should be enabled in the state t. Intuitively, the task A_1^o has the responsibility to decrease the rank and, hence, cause progress toward the fulfillment of the goal. The execution of a task other than the task A_1^o should maintain enabledness of the task A_1^o without increasing the rank. The weak fairness assumption for the task A_1^o ensures that if the task A_1^o stays continuously enabled, it will eventually be executed, and this would decrease the rank.

Establishing the response formula assuming weak fairness for the task A_1^o is the same as establishing the following conditional response formula for all executions:

$$\square\,[\,Guard(A_1^o) \;\to\; \Diamond\,((taken = A_1^o) \;\vee\; \neg\,Guard(A_1^o))\,]$$
$$\to\; \square\,[\,in_1\,?\,v \;\to\; \Diamond\,out\,!\,v\,].$$

The proof rule formalized below shows how to establish the response formula $\square\,[\,\varphi_1 \to \Diamond\,\varphi_2\,]$ under the assumption of the form $\square\,[\,\psi_1 \to \Diamond\,(\psi_2 \vee \neg\psi_1)\,]$. When the property ψ_1 is $Guard(A)$ and the property ψ_2 is $(taken = A)$, the assumption corresponds to the weak fairness assumption for the task A. If we pick the property ψ_1 to be the constant 1, then the assumption simplifies to $\square\,\Diamond\,\psi_2$.

PROOF RULE FOR CONDITIONAL RESPONSE PROPERTIES

To establish that a transition system T satisfies the conditional response formula

$$\square\,[\,\psi_1 \to \Diamond\,(\psi_2 \vee \neg\psi_1)\,] \;\to\; \square\,[\,\varphi_1 \to \Diamond\,\varphi_2\,],$$

where φ_1, φ_2, ψ_1, and ψ_2 are state properties, identify a state property ψ and a function *rank* that maps states of T to **nat** and show that:

1. every state that satisfies the property φ_1 also satisfies the property ψ,

2. every state that satisfies the property ψ also satisfies the property ψ_1, and

3. for every state s that satisfies the property ψ and for every transition (s,t) of T, either the state t satisfies the response φ_2 or the state t satisfies the property ψ and $rank(t) \leq rank(s)$ and if the state t satisfies the property ψ_2 then $rank(t) < rank(s)$.

Let us summarize the proof that the process Merge satisfies the response property $\square\,(\,in_1\,?\,v \to \Diamond\,out\,!\,v\,)$ assuming weak fairness for the output task A_1^o using the notation of this proof rule. For the proof rule above, we have the property φ_1 equal to $in_1\,?\,v$, the property φ_2 equal to $out\,!\,v$, the property ψ_1 equal to $Guard(A_1^o)$, and the property ψ_2 equal to $(taken = A_1^o)$. We choose the property ψ to be Contains(x_1, v) and *rank* to be the function that maps a state s to the position of the message v in the queue x_1 (and 0 if the message is not in the queue). Verify that all three premises as required by the rule indeed are satisfied.

We conclude by sketching a proof that the reasoning behind the proof rule for conditional response is *sound*: if we show all the premises of the rule, then the desired conditional response formula is indeed satisfied by every execution of the transition system. It turns out that this proof rule, coupled with the application of valid temporal patterns such as the chain rule, is *complete*: if the transition system indeed satisfies the conditional response formula, there exist appropriate choices for the strengthening property ψ and the ranking function *rank* for which all the premises of the rule hold.

Theorem 5.4 [Soundness of the Proof Rule for Conditional Response] *The proof rule for establishing that a transition system T satisfies the conditional response formula φ given by $\Box\,[\,\psi_1 \rightarrow \Diamond\,(\psi_2 \vee \neg\,\psi_1)\,] \rightarrow \Box\,[\,\varphi_1 \rightarrow \Diamond\,\varphi_2\,]$ is sound.*

Proof. Let T be a transition system, and consider the LTL-formula φ of the above form. Let ψ be a state property and *rank* be a function that maps states of T to **nat**, such that (1) every state that satisfies the property φ_1 also satisfies the property ψ, (2) every state that satisfies the property ψ also satisfies the property ψ_1, and (3) for every state s that satisfies the property ψ and every transition (s,t) of T such that the state t does not satisfy the property φ_2, the state t satisfies the property ψ, and $rank(t) \leq rank(s)$, and if the state t satisfies the property ψ_2, then $rank(t) < rank(s)$. Under these assumptions, we want to show that every infinite execution of T satisfies the formula φ.

Let $\rho = s_0 s_1 s_2 \cdots$ be an infinite execution of the transition system T. Assume that the execution ρ satisfies the formula $\Box\,[\,\psi_1 \rightarrow \Diamond\,(\psi_2 \vee \neg\,\psi_1)\,]$. We want to show that the execution ρ satisfies the formula $\Box\,[\,\varphi_1 \rightarrow \Diamond\,\varphi_2\,]$. Let i be a position in the execution. Assume that the state s_i satisfies the request property φ_1. We want to establish that there exists a position $j \geq i$ such that the state s_j satisfies the response property φ_2. We will prove this by contradiction. Assume that the property φ_2 is not satisfied in every state s_j for $j \geq i$.

We claim that for every position $j \geq i$, the state s_j satisfies the property ψ. The proof is by induction on j, where $j = i$ is the base case. Since the state s_i satisfies the property φ_1, by premise 1, it also satisfies the property ψ. Now consider an arbitrary state s_j, with $j \geq i$. Assume that the state s_j satisfies the property ψ. There is a transition from the state s_j to the state s_{j+1} since they appear as consecutive states along the execution ρ. By assumption, the state s_{j+1} does not satisfy the property φ_2. By premise 3, we conclude that the state s_{j+1} satisfies the property ψ.

By premise 2, we conclude that every state s_j, for $j \geq i$, also satisfies the property ψ_1. Since the execution ρ satisfies the formula $\Box\,[\,\psi_1 \rightarrow \Diamond\,(\psi_2 \vee \neg\,\psi_1)\,]$, by the semantics of temporal operators, we conclude that there exist infinitely many positions $j_1, j_2, \ldots$ with $i < j_1 < j_2 < \cdots$ such that for each k, the state s_{j_k} satisfies the property ψ_2.

Let $rank(s_i) = m$. We know that at every step $j \geq i$, the state s_j satisfies the property ψ, the state s_{j+1} does not satisfy the property φ_2, and (s_j, s_{j+1}) is a transition of T. By premise 3, at every step $j \geq i$, $rank(s_{j+1}) \leq rank(s_j)$. Since for each k the state s_{j_k} satisfies the property ψ_2, rank must strictly decrease at this step: $rank(s_{j_k}) < rank(s_{j_k-1})$.

To summarize, the rank is m at step i, it stays the same or decreases at every step after i, and it strictly decreases at infinitely many steps $j_1, j_2, \cdots$. This is a contradiction. ∎

Exercise 5.22: Consider an asynchronous process with the state variables x and y, both of type `nat`, initialized to 0. The process consists of two tasks both of which are always enabled. For the task A_1, the update is specified by

$$\text{if } (\,x > 0\,) \text{ then } x := x - 1 \text{ else } x := y$$

and for the task A_2, the update is specified by $y := y + 1$. Weak fairness is assumed for both the tasks. Prove that the process satisfies the recurrence property $\Box \Diamond\, x = 0$ using the proof rule for conditional response properties. ∎

Bibliographic Notes

While the origins of *temporal logic* are rooted in philosophy, the use of linear temporal logic for expressing formal requirements of reactive systems was introduced by Pnueli [Pnu77]. Subsequently, the expressiveness and decision procedures for many variants of temporal logics were studied (see [Eme90] for a survey of the theoretical foundations). The specification language PSL is an industrial standard that is supported by commercial tools for hardware design [EF06] (see also integration of specifications in the hardware description language VERILOG [BKSY12]).

The concept of *model checking* was introduced in [CE81] and [QS82] in the context of checking branching-time temporal properties of finite-state protocols and has received significant attention in both academia and industry (see the textbooks [CGP00] and [BK08] for an introduction and the 2009 ACM Turing Award lecture for an overview of its impact [CES09]).

Automata over infinite traces were introduced by Büchi in the context of decision procedures for monadic second-order logic [Büc62] (see [Tho90] for a survey of results on such automata). The translation from LTL to Büchi automata appears in [VW86], and this work led to the *automata-theoretic* approach to model checking. The *nested* depth-first search algorithm of figure 5.11 was introduced in [CVWY92]. The model checker SPIN [Hol04] includes state-of-the-art implementations of these techniques.

The symbolic nested fixpoint computation of figure 5.14 was introduced in [BCD+92, McM93] and is supported by the model checker NuSMV [CCGR00].

The proof rules for establishing liveness properties of transitions systems were first studied in [MP81] (see also [Lam94]) and are supported by the verification toolkit TLA+ [Lam02].

6

Dynamical Systems

Controllers such as a thermostat for regulating the temperature in a room or a cruise controller for tracking and maintaining the speed of an automobile interact with the physical world via sensors and actuators. The relevant information about the physical environment corresponds to quantities such as temperature, pressure, and speed that evolve continuously obeying laws of physics. As a result, design and analysis of control systems requires construction of models of the physical system. In this chapter, we focus on continuous-time models of dynamical systems. This mathematically rich area is explored in details in a course on control systems. The purpose of this chapter is to give a brief introduction to the core concepts.

6.1 Continuous-Time Models

6.1.1 Continuously Evolving Inputs and Outputs

The typical architecture of a control system is depicted in figure 6.1. The physical world that is to be controlled is modeled by a component called the *plant*. The evolution of the plant can be influenced by the controller using actuators, but it also depends on uncontrollable factors from the environment, usually called *disturbances*. The controller responds to the commands from the user, called *reference inputs*, and can make its decisions based on measurements of the plant provided by the sensors. For example, in a thermostat design, the plant is the room whose temperature is to be controlled. The sensor is a thermometer that can measure the current temperature. The task of the controller is to regulate the temperature so that it stays *close* to the reference temperature set by the occupant on the thermostat. The controller can influence the temperature by adjusting the heat flow from the furnace. The plant model, in this case, needs to capture how the temperature of the room changes as a function of the heat-flow from the furnace and the difference between the room temperature and the outside temperature, which is the uncontrolled disturbance.

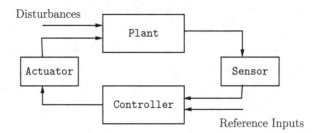

Figure 6.1: Block Diagram of a Control System

Models of dynamical systems can also be conveniently described as components with inputs and outputs that are connected to one another using block diagrams. The underlying model of computation is *synchronous* as in chapter 2 with one essential difference: while the values of the variables of a synchronous reactive component are updated in a sequence of *discrete* logical rounds, the values of the variables of a dynamical system are updated *continuously* with the passage of time. We call such components *continuous-time components*. The variables of a continuous-time component typically range over a compact set such as a closed interval of the set of reals with specified units of measurement. For example, the velocity v of a car can be modeled by a variable that ranges over the real numbers in the interval from 0 to 150 miles per hour. In our specifications of dynamical systems, we assume that every variable has the type **real** without explicitly mentioning the associated interval range.

Signals

The values a variable takes over time can then be described as a function from the time domain to **real**. Such functions from the time domain to the set of reals are called *signals*. Throughout we will assume that the time domain consists of the set of non-negative real numbers and denote this set by **time**. For a variable x, a signal over x is a function that assigns a real value to the variable x for every time t in **time**. We will denote such a function by $\bar{x}$. Thus, $\bar{x}(0)$ denotes the value of the variable x at time 0, and for every time t, $\bar{x}(t)$ denotes its value at time t.

Given a set V of variables, a signal $\overline{V}$ over V assigns values to all the variables in V as a function of time. If the set V contains k variables, then a signal over V is a function from the time domain **time** to k-tuples or vectors over **real**. The number k of variables is called the *dimensionality* of the signal. Alternatively, a k-dimensional signal can be viewed as a k-tuple of single-dimensional signals, one for each of the variables in V.

Since the domain and the range of a signal consists of real numbers or vectors, we can use the standard notion of Euclidean distance over reals, or any other notion of measuring distance that satisfies the classical properties of a metric,

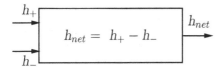

Figure 6.2: Continuous-time Component NetHeat

to measure how far apart two quantities are. For two vectors u and v, let $\|u - v\|$ denote the distance between u and v, and we will denote the distance of a vector u from the origin 0 by $\|u\|$. Now the standard mathematical notions of *continuity* and *differentiability* apply to signals. For example, a signal $\overline{V}$ is *continuous* if for all time values $t \in$ time, for all $\epsilon > 0$, there exists a $\delta > 0$ such that for all time values $t' \in$ time, if $\|t - t'\| < \delta$, then $\|\overline{V}(t) - \overline{V}(t')\| < \epsilon$ holds.

Example: Heat Flow

As a first example, figure 6.2 shows a continuous-time component NetHeat that has two input variables h_+ and h_- and a single output variable h_{net}, which denote the heat inflow, heat outflow, and net heatflow, respectively. The component NetHeat is a mapping from two input signals to an output signal, and its dynamics is expressed by the equation:

$$h_{net} = h_+ - h_-,$$

which says that at every time t, the value of the output h_{net} equals the expression $h_+ - h_-$. Given a signal $\overline{h}_+$ for the input variable h_+ and a signal $\overline{h}_-$ for the input variable h_-, the output signal $\overline{h}_{net}$ for the output variable h_{net} is defined by $\overline{h}_{net}(t) = \overline{h}_+(t) - \overline{h}_-(t)$ for all times t in time. This unique output signal is called the *response* of the component to the input signals $\overline{h}_+$ and $\overline{h}_-$. If the input signals are continuous, then so is the output signal.

Note that the computation of the output value based on the input values is expressed in a *declarative* style using *(algebraic) equations* instead in an *operational* style using *assignments*. Indeed such a declarative description is the norm in the modeling of control systems since the models are primarily used to express relationships between various signals in a mathematically precise manner so they can be subjected to analysis.

Example: Motion of a Car

The component NetHeat is stateless. As an example of a stateful continuous-time component, let us build a model of how the speed of a car changes as a result of the force applied to it by the engine. For the purpose of designing a cruise controller, it typically suffices to make a number of simplifying assumptions. In particular, let us assume that the rotational inertia of the wheels is negligible

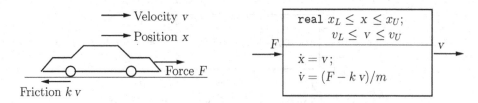

Figure 6.3: Continuous-time Component **Car** Modeling the Car Motion

and that the friction resisting the motion is proportional to the car's speed. The forces acting on the car are shown in figure 6.3. If x denotes the position of the car (measured with respect to an inertial reference) and F denotes the force applied to the car, then using the classical Newton's laws for motion, we can capture the dynamics of the car by the *differential equation*:

$$F - k\dot{x} = m\ddot{x}.$$

Here k is the coefficient of the frictional force, and m denotes the mass of the car. The quantity $\dot{x}$ denotes the first-order time derivative of the signal assigning values to the position variable x and thus captures the velocity of the car. Similarly, $\ddot{x}$ denotes the second-order derivative of this signal, that is, the acceleration of the car.

This equation of motion is modeled by the continuous-time component **Car** of figure 6.3. It uses two state variables: the variable x modeling the position of the car and the variable v modeling the velocity of the car. For every state variable, the component needs to specify its initial value. In our example, the initial value of the position x is specified by the constraint $x_L \leq x \leq x_U$, where x_L and x_U are constants that give lower and upper bounds on the initial position. This declarative specification of the initialization is equivalent to the nondeterministic assignment using the **choose** construct discussed in chapter 2: **real** x := **choose**$\{z \mid x_L \leq z \leq x_U\}$. Similarly, the initial value of the velocity v is given by the constraint $v_L \leq v \leq v_U$, where v_L and v_U are constants that give lower and upper bounds on the initial velocity.

For every state variable, the dynamics is given by specifying the first-order time derivative of the value of the state variable as a function of the input and state variables. The differential equation $\dot{x} = v$ says that the rate of change of the state variable x at each time t equals the value of the state variable v at time t, and the rate of change of the state variable v at each time t equals the value of the expression $(F - kv)/m$ at time t. The two equations together are equivalent to the original equation $F - k\dot{x} = m\ddot{x}$. The output of the car is its velocity. For every output variable, the component specifies the value of the output variable as a function of the input variables and the state variables. In this example, the output simply equals the state variable v.

To illustrate the behaviors of this model, let us choose the initial position to be x_0 and initial velocity to be v_0. Now consider the case when the input force F equals the value kv_0 at all times. Then the position x and the velocity v of the car at all times can be obtained by solving the system of differential equations

$$\dot{x} = v; \quad \dot{v} = k(v_0 - v)/m$$

with the initial condition $\overline{x}(0) = x_0$ and $\overline{v}(0) = v_0$. These equations have a unique solution: the velocity stays constant at the value v_0 at all times, and the distance x increases linearly with time t and is given by the expression $x_0 + t\, v_0$. In other words, given the (constant) input signal $\overline{F}(t) = kv_0$ and the initial state (x_0, v_0), the corresponding signal describing the dynamics of the state/output variable v is given by $\overline{v}(t) = v_0$, and the signal describing the state variable x is given by $\overline{x}(t) = x_0 + t\, v_0$.

Now let us consider the case when the input force F is 0 at all times, the initial position of the car is 0, and the initial velocity is v_0. Then the position x and the velocity v of the car at all times can be obtained by solving the system of differential equations

$$\dot{x} = v; \quad \dot{v} = -kv/m$$

with the initial condition $\overline{x}(0) = 0$ and $\overline{v}(0) = v_0$. Using rules of differential calculus, we can solve these equations, and the resulting signals corresponding to the position and the velocity of the car are given by:

$$\overline{v}(t) = v_0\, e^{-kt/m}; \quad \overline{x}(t) = (mv_0/k)\left[1 - e^{-kt/m}\right].$$

Note that the velocity decreases exponentially converging to 0, while the position increases converging to the value mv_0/k.

As a third scenario, suppose the initial position is 0, and the initial velocity is 0, and we apply a constant force F_0 to the car. Then the position x and the velocity v of the car can be obtained by solving the system of differential equations

$$\dot{x} = v; \quad \dot{v} = (F_0 - kv)/m$$

with the initial condition $\overline{x}(0) = 0$ and $\overline{v}(0) = 0$. If we assume that the mass m of the car is 1000 kg, the coefficient k of friction between the tires and the road is 50, and the input force F_0 is $500N$, then the corresponding signal for the velocity is shown in figure 6.4. The velocity increases converging to the value $F_0/k = 10\, m/s$.

Definition of Continuous-Time Components

In general, the initial values for the state variables are specified by a constraint *Init*, which typically specifies an interval of possible values for each variable. As usual, $[\![Init]\!]$ specifies the set of initial states, that is, the set of states that satisfy the constraint specified by *Init*.

The dynamics of the component is specified by (1) a real-valued expression h_y for every output variable y, and (2) a real-valued expression f_x for every state variable x. Each of these expressions is an expression over the input and state variables. The value of the output variable y at time t is obtained by evaluating the expression h_y using the values of the state and input variables at time t, and the signal for a state variable x should be such that its rate of change at each time t equals the value of the expression f_x evaluated using the values of the state and input variables at time t. Thus, the execution of a continuous-time component is similar to the execution of a deterministic synchronous reactive component, except the notion of a round is now infinitesimal: at every time t, the outputs at time t are determined as a function of the inputs at time t and the state of the component at time t, and then the state is updated using the rate of change specified by the derivative evaluated using the inputs and state at time t.

The definition of a continuous-time component is now summarized below:

CONTINUOUS-TIME COMPONENT

A continuous-time component H has a finite set I of real-valued input variables, a finite set O of real-valued output variables, a finite set S of real-valued state variables, an initialization *Init* specifying a set $[\![Init]\!]$ of initial states, a real-valued expression h_y over $I \cup S$ for every output variable $y \in O$, and a real-valued expression f_x over $I \cup S$ for every state variable $x \in S$. Given a signal $\overline{I}$ over the input variables I, a corresponding execution of the component H is a differentiable signal $\overline{S}$ over the state variables S and a signal $\overline{O}$ over the output variables O such that

1. $\overline{S}(0) \in [\![Init]\!]$;

2. for every output variable y and time t, $\overline{y}(t)$ equals the value of h_y evaluated using the values $\overline{I}(t)$ for the input variables and $\overline{S}(t)$ for the state variables; and

3. for every state variable x and time t, the time derivative $(d/dt)\,\overline{x}(t)$ equals the value of f_x evaluated using the values $\overline{I}(t)$ for the input variables and $\overline{S}(t)$ for the state variables.

A continuous-time component with no inputs is called *closed*.

Existence and Uniqueness of Response Signals

Determining the signals for the state and output variables corresponding to a given initial state s_0 and a given input signal using mathematical analysis amounts to solving an *initial value problem* for a system of *ordinary differential equations*. We need to impose restrictions on the expressions used to define the state and output responses to ensure uniqueness of such solutions since not all differential equations are well behaved. For example, the *conditional* differential

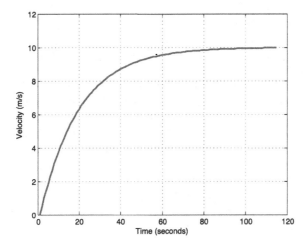

Figure 6.4: Velocity of `Car` in Response to Constant Input Force

equation
$$\dot{x} \;=\; \text{if } (x = 0) \text{ then 1 else 0}$$

states that the derivative of the signal is 1 at time 0 and 0 everywhere else. There is no differentiable signal $\bar{x}$ that can satisfy the given equation for the initial condition $\bar{x}(0) = 0$. The source of this problem lies in the fact that the right-hand side of this differential equation has a discontinuity at $x = 0$. If the right-hand side of a differential equation is a *continuous* function, then a solution is guaranteed to exist.

To illustrate another potential problem in solving differential equations, consider the differential equation $\dot{x} = x^{1/3}$. For the initial value $\bar{x}(0) = 0$, this differential equation has two solutions: the constant signal $\bar{x}_1(t) = 0$ and the signal $\bar{x}_2(t) = (2t/3)^{3/2}$. A classical way to avoid this problem and ensure *uniqueness* of the solution to a differential equation is to require the right-hand side to be *Lipschitz continuous*. Intuitively, Lipschitz continuity means that there is a constant upper bound on how fast a function changes: a function $f : \text{real}^n \mapsto \text{real}^n$ is said to be Lipschitz continuous if there exists a constant K such that for all vectors u and v in real^n, $\|f(u) - f(v)\| \le K \|u - v\|$.

In the example component `Car` of figure 6.3, the right-hand side f_x equals v, which when viewed as a function from `real` to `real` is Lipschitz continuous, and the right-hand side f_v equals $(F - kv)/m$, which when viewed as a function from real^2 to `real` is Lipschitz continuous. The function $x^{1/3}$ (in the right-hand side of the differential equation we just considered) is *not* Lipschitz continuous since its rate of change grows unboundedly as x approaches 0. The quadratic function x^2 is not Lipschitz continuous as its rate of change grows unboundedly as x increases unboundedly. However, if we know that x ranges over a bounded set D, which is the case in typical modeling of dynamical systems, then the domain of the function x^2 is D, and over this domain, it is Lipschitz continuous. A classical result in calculus, known as the Cauchy-Lipschitz Theorem, tells us that for the initial value problem, given by $\dot{S} = f(S)$ and $\overline{S}(0) = s_0$, if f is Lipschitz continuous, then the solution signal $\overline{S}$ exists and is unique.

With this motivation, we can require that, in the definition of a continuous-time component, the expression f_x defining the rate of change of each state variable x is a Lipschitz continuous function over the state and input variables. In this case, for a given initial state, assuming that the input signal is continuous, the state signal is uniquely defined. Since the value of an output variable is a function of the state and input variables, in this case, the output signal is also uniquely defined. If the expression h_y defining an output variable is a continuous function, then this output signal is guaranteed to be continuous. Since the output of one component can be connected as an input to another component in a block diagram and can appear in a right-hand side of a differential equation in the definition of that component, to ensure the desired uniqueness and existence of solutions, we can demand that each output expression h_y is also Lipschitz continuous. Components that meet these restrictions, such as the component Car of figure 6.3, are said to have Lipschitz-continuous dynamics.

LIPSCHITZ-CONTINUOUS DYNAMICS

A continuous-time component H with input variables I, output variables O, and state variables S is said to have *Lipschitz-continuous* dynamics if:

1. for each output variable y, the real-valued expression h_y over $I \cup S$ is a Lipschitz-continuous function; and

2. for each state variable x, the real-valued expression f_x over $I \cup S$ is a Lipschitz-continuous function.

It follows that if a continuous-time component H has Lipschitz-continuous dynamics, then for a given initial state and a given continuous input signal $\overline{I}$, the corresponding response of the component as a signal over the state and the output variables exists and is unique.

Example: Helicopter Spin

As another example, let us consider the classical problem of controlling a helicopter so as to keep it from spinning. A helicopter has six degrees of motion, three for position and three for rotation. In our simplified version, let us assume that the helicopter position is fixed and the helicopter remains vertical. Then the only freedom of motion is the angular rotation around the vertical, that is, Z-axis. This rotation is called the *yaw* (see figure 6.5). The friction of the main rotor at the top of the helicopter causes the yaw to change. The tail rotor then needs to apply a torque to counteract this rotational force to keep the helicopter from spinning. In this setting, the helicopter model has a continuous-time input signal T, denoting the torque around the Y-axis. The moment of inertia of the

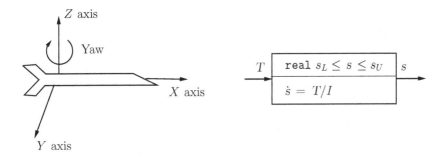

Figure 6.5: Simplified Modeling of Helicopter Motion

helicopter in this simplified setting can be modeled by a single scalar I. The output signal of the model is the angular velocity around the vertical axis and is modeled by the spin $s = \dot{\theta}$, where θ gives the yaw. The equation of motion is then given by

$$\ddot{\theta} = T/I$$

The corresponding continuous-time component is also shown in figure 6.5. There is a single state variable s modeling the spin, a single input variable T corresponding to the torque, and a single output variable, which equals the state s. The initial value of the state is in the interval $[s_L, s_U]$, and its rate of change is given by the expression T/I. Note that this model has Lipschitz-continuous dynamics.

The models expressed as continuous-time components with state variables and differential equations are sometimes called the *state-space representation* of dynamical systems. The dynamics can alternatively be expressed using equations that specify the output signals by integrating over input signals. For instance, for our helicopter model, the value of the output spin at time t equals the sum of its initial value and the integral of the input torque up to time t. This is captured by the *integral equation*

$$\overline{s}(t) = s_0 + (1/I) \int_0^t \overline{T}(\tau)d\tau,$$

where s_0 is the initial state. Note that the internal state is implicit in the integral model. Modeling tools such as SIMULINK allow modeling by both state-space representation and integral equations.

Example: Simple Pendulum

Consider a simple pendulum shown in figure 6.6. The pendulum is a rod with a rotational joint at one end and a mass at the other end. For simplicity, we assume that the friction and the weight of the rod are negligible. A motor is placed at the pivot to provide an external torque to control the pendulum.

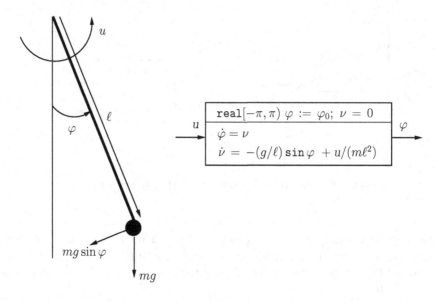

Figure 6.6: Simple Pendulum and Its Model

From Newton's law for rotating objects, the dynamics of the pendulum system is described by the following (nonlinear) differential equation:

$$m\ell^2 \ddot{\varphi} \;=\; u \;-\; mg\ell \sin\varphi$$

where m is the mass, $g = 9.8\,m/s^2$ is the gravitational acceleration, ℓ is the length of the rod, φ is the angle of the pendulum measured counterclockwise from the downward vertical, and u is the external torque applied in the counterclockwise direction around the pivot. Note that the range of values for the angle φ is the interval $[-\pi, \pi)$.

In the continuous-time component modeling the pendulum (see figure 6.6), the second-order differential equation of motion is replaced by a pair of (first-order) differential equations by introducing the state variable ν that captures the angular velocity of the pendulum. The constant φ_0 gives the initial angular position of the pendulum.

To understand the resulting motion, let us assume that the external torque u is set to 0, the length ℓ of the rod is $1m$, and the initial angular displacement is $\pi/4$ radians. Then the motion of the pendulum is described by the equation

$$\ddot{\varphi} \;=\; -9.8\sin\varphi; \quad \overline{\varphi}(0) \;=\; \pi/4.$$

The resulting oscillatory motion is depicted in figure 6.7, which plots the angular displacement as a function of time.

Exercise 6.1: Is the dynamics of the continuous-time component of figure 6.6 modeling the pendulum Lipschitz continuous? ∎

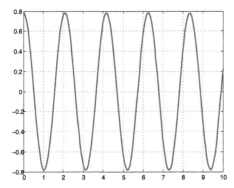

Figure 6.7: Angular Displacement of Simple Pendulum

Exercise 6.2: Consider the model of the simple pendulum from figure 6.6. Suppose the external torque u is set to 0. Analyze the motion of the pendulum if the initial angular displacement is set to $-15\pi/16$ radians, that is, slightly displaced from the vertically upward position. Plot the resulting response signal $\overline{\varphi}$ using MATLAB (use $\ell = 1m$). ∎

6.1.2 Models with Disturbance

Let us revisit the model of the motion of a car. The model described in figure 6.3 assumes that the car is moving in a single dimension on a flat road. Suppose we now want to account for the *grade* of the road: on an up hill, the weight of the car works against the force applied by the engine, and on a down hill, the weight of the car adds to this force (see figure 6.8). The cruise controller needs to adjust the force F to keep the net velocity v in the direction along the road constant. We can model the grade of the road by an additional input, denoted θ, that captures the angle of the road with the horizontal: a positive angle indicates an up hill slope, and a negative angle indicates a down hill slope. The weight of the car equals $m\,g$ in the vertically downward direction, where $g = 9.8\,m/s^2$ is the gravitational acceleration. The modified dynamical system model is also shown in figure 6.8. The forces acting on the car in the direction along the road are: F in the forward direction controlled by the engine, $k\,v$ in the backward direction modeling the friction, and $m\,g\sin\theta$ capturing the gravitational force along the road.

The control design problem for the model of figure 6.8 differs from the corresponding problem for the model of figure 6.3 in a crucial way: the input signal θ modeling the grade of the road is not controlled by the controller and is not known in advance. Such an uncontrolled input is typically called a *disturbance*. The controller should be designed to produce the controlled input signal F so that it works no matter how the input θ varies within a reasonable range of values (for instance, all values in the range $[-\pi/6, +\pi/6]$).

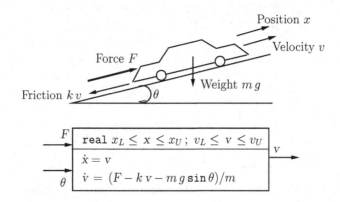

Figure 6.8: Modeling Car Motion on a Graded Road

6.1.3 Composing Components

Continuous-time components can be composed using block diagrams in a way similar to synchronous components. We can define the operations of variable renaming, output hiding, and parallel composition in the same way as in chapter 2. To ensure determinism and well-formed composition, while composing components, we would like to establish absence of cyclic await dependencies. The await dependency of an output variable on input variables is defined the same way as in synchronous components: an output variable y of a continuous-time component H awaits an input variable x if the value of the output y at time t depends on the value of the input x at time t.

For the component NetHeat of figure 6.2, the output h_{net} awaits both the input variables h_+ and h_-. For the components of figures 6.3 and 6.8 modeling the motion of a car and the component of figure 6.5 modeling the motion of a helicopter, the output variable at time t equals the value of one of the state variables at time t and, hence, does not await any of the corresponding input variables.

Since the evolution of each output variable y of a continuous-time component is described by an expression h_y over the state and input variables, the await dependencies can be determined by a simple syntactic check: if the set of input variables that occur in the expression h_y is J, then the output y awaits the input variables in J but does not await the remaining input variables.

Exercise 6.3: Consider a continuous-time component H_1 with an input variable u, a state variable x, and an output variable v, as well as a continuous-time component H_2 with the input variable v, a state variable y, and an output variable w. Assuming that both of these components have Lipschitz-continuous dynamics, prove that the parallel composition $H_1 \| H_2$ of the two components also has Lipschitz-continuous dynamics. ∎

6.1.4 Stability

In earlier chapters, we have explored safety and liveness requirements for synchronous as well as asynchronous systems. Dynamical control systems also have similar requirements. For example, a safety requirement for a cruise controller can demand that the speed of the car should always be below some maximum speed, and a liveness requirement can demand that the difference between the actual speed and the desired speed should *eventually* be close to zero. A cruise controller, however, has a new kind of requirement, namely, that small perturbations in input values, such as the grade of the road, should not cause disproportionately large changes in the speed of the car. This requirement, which is relevant only for continuous-time systems, is called *stability*. We will first define the notion of Lyapunov stability for dynamical systems and then consider the notion of bounded-input-bounded-output stability for continuous-time components.

Equilibria

To define stability, we first need to understand the notion of an equilibrium state of operation of a dynamical system. For this purpose, we assume that the system is closed: if the original system has inputs, then the stability is analyzed by setting the input signal to a fixed value, say 0, at all times.

Consider a closed continuous-time component whose state S is an n-dimensional vector, with the dynamics given by the Lipschitz-continuous differential equation $\dot{S} = f(S)$. A state s_e of the system is said to be an *equilibrium* state if $f(s_e) = 0$. For an equilibrium state s_e, the constant signal $\overline{S}(t) = s_e$ is a solution to the initial value problem $\dot{S} = f(S)$ and $\overline{S}(0) = s_e$, and since the dynamics is Lipschitz-continuous, this is the only solution. Thus, if the initial state of a system with Lipschitz continuous dynamics is an equilibrium state s_e, then the state of the system is guaranteed to stay at this equilibrium state at all times.

As an example, let us revisit the continuous-time component Car of figure 6.3, and let us set the input force F to be the constant signal with value 0. Then the dynamics of the component is described by the equations

$$\dot{x} = v; \quad \dot{v} = -k\,v/m.$$

A state (x_e, v_e) is an equilibrium state of this system exactly when $v_e = 0$. When the velocity v_e equals 0, the signal that sets the position x always equal to x_e and sets the velocity v always equal to 0 satisfies the above differential equations. In contrast, if the initial velocity v_e is nonzero, the velocity will keep changing at an exponential rate. Thus, for the input signal $\overline{F}(t) = 0$, the system has infinitely many equilibria of the form $(x_e, 0)$.

As another example, consider the pendulum model of figure 6.6, and suppose the input torque u is set to the constant value 0. Then the dynamics of the component is described by the equations

$$\dot{\varphi} = \nu; \quad \dot{\nu} = -g\sin\varphi/\ell.$$

Recall that the angular displacement φ ranges over the interval $[-\pi, \pi)$, and in this range, $\sin \varphi$ equals 0 for two values of φ, namely, $\varphi = 0$ and $\varphi = -\pi$. Thus, the system has two equilibria: one where $\varphi = 0$ and $\nu = 0$ (and this corresponds to the pendulum in the vertically downwards position), and one where $\varphi = -\pi$ and $\nu = 0$ (and this corresponds to the pendulum in the vertically inverted position).

Lyapunov Stability

Let us consider a continuous-time component whose state S is an n-dimensional vector, with the dynamics given by the Lipschitz-continuous differential equation $\dot{S} = f(S)$. Consider an equilibrium state s_e of the system. We know that if the initial state of the system is s_e, then the corresponding system evolution is described by the constant signal $\overline{S}_e(t) = s_e$. Now suppose the initial state is perturbed slightly, that is, the initial state is chosen to be s_0 such that the distance $\|s_e - s_0\|$ is small. Consider the signal $\overline{S}_0$ that is the unique response of the system starting from the initial state s_0. If this signal stays *close* to the constant signal $\overline{S}_e$ at all times, then we can conclude that a small perturbation from the equilibrium state causes the system state to stay close to the equilibrium. In such a case, the equilibrium s_e is said to be stable. If, in addition, the state $\overline{S}_0(t)$ *converges* to the equilibrium state s_e as time t advances, then we are guaranteed that after a small perturbation from the equilibrium state, the system state stays close to the equilibrium eventually returning to the equilibrium. When this additional convergence requirement holds, the equilibrium s_e is said to be *asymptotically* stable.

These notions of stability, usually referred to as Lyapunov stability in the literature on dynamical systems, are summarized below.

LYAPUNOV STABILITY

Consider a closed continuous-time component H with n state variables S and dynamics given by the equation $\dot{S} = f(S)$, where $f : \mathtt{real}^n \mapsto \mathtt{real}^n$ is Lipschitz continuous. A state s_e is said to be an equilibrium of the component H if $f(s_e) = 0$. Given an initial state $s_0 \in \mathtt{real}^n$, let $\overline{S}_0 : \mathtt{time} \mapsto \mathtt{real}^n$ be the unique response of the component H from the initial state s_0.

- An equilibrium s_e of H is said to be stable if for every $\epsilon > 0$, there exists a $\delta > 0$ such that for all states s_0, if $\|s_e - s_0\| < \delta$, then $\|\overline{S}_o(t) - s_e\| < \epsilon$ holds at all times $t \geq 0$.

- An equilibrium s_e of H is said to be asymptotically stable if it is stable, and there exists a $\delta > 0$ such that for all states s_0, if $\|s_e - s_0\| < \delta$, then the limit $\lim_{t\to\infty} \overline{S}_0(t)$ exists and equals s_e.

Let us revisit the continuous-time component Car of figure 6.3 again, with the input force F set to the constant value 0. We have already noted that the state with the position $x = x_e$ and the velocity $v = 0$ is an equilibrium for every choice of x_e. Suppose we perturb this equilibrium; that is, consider the behavior of the component starting at the initial state (x_0, v_0) such that the distance $\|(x_e, 0) - (x_0, v_0)\|$ is small. The car will slow down according to the differential equation $\dot{v} = -kv/m$ starting from the initial velocity v_0, with the velocity converging to 0. The position of the car will converge to some value x_f that is a function of the initial position x_0 and the initial velocity v_0: $x_f = x_0 + mv_0/k$. Thus, for the resulting signal, the value $\|(\overline{x}_o(t), \overline{v}_0(t)) - (x_e, 0)\|$ is bounded. Thus, the equilibrium $(x_e, 0)$ is stable. However, it is not asymptotically stable since along this signal, as time advances, the position does not converge to x_e but converges to a different value x_f.

Now, let us consider the pendulum of figure 6.6 with the input signal u set to the constant value 0. We know that the model $\dot{\varphi} = \nu$; $\dot{\nu} = -g\sin\varphi/\ell$ has two equilibria, namely, $(\varphi = 0, \nu = 0)$ and $(\varphi = -\pi, \nu = 0)$. The latter corresponds to the vertically inverted position and is not stable: if we displace the pendulum slightly from this vertically upward position, the angular velocity ν will keep increasing positively, thereby increasing the displacement φ, pushing the pendulum away from the vertical. In contrast, the equilibrium $(\varphi = 0, \nu = 0)$ corresponding to the vertically downward position is stable. For example, if we set the initial angle to a small positive value, the angular velocity will be negative and will cause the angle to decrease back toward the equilibrium value 0 (see figure 6.7 for a representative behavior). The pendulum will oscillate around the equilibrium position, and if the initial perturbation is (φ_0, ν_0), then the value of $\|(\overline{\varphi}_0(t), \overline{\nu}_0(t)) - (0, 0)\|$ stays bounded, with the bound dependent on the values of φ_0 and ν_0. In this model, this equilibrium is not asymptotically stable: for example, if the initial starting position is ϵ with initial angular velocity 0, then the pendulum will keep swinging forever in the arc from ϵ to $-\epsilon$ around the vertical. In reality, of course, such a pendulum asymptotically converges to the vertical coming to a halt due to the damping effects, which are not captured in our model.

Input-Output Stability

The notion of Lyapunov stability is based on the state-space representation of a dynamical system and concerns the behavior of the system when its state is perturbed from the equilibrium state, with the input signal set to 0. An alternative notion of stability views the dynamical system as a *transformer*, mapping input signals to output signals, and demands that a small change to the input signal should cause only a small change to the output signal. This notion of *input-output stability* is formalized next.

A signal $\overline{x}$ assigning values to the real-valued variable x as a function of time is said to be *bounded* if there exists a constant Δ such that $\|\overline{x}(t)\| \leq \Delta$ for all times t. Here are some typical signals analyzed for their boundedness:

- The constant signal defined by $\overline{x}(t) = a$, for a constant value a, is bounded.

- The linearly increasing signal defined by $\overline{x}(t) = a + bt$, for constants a and $b > 0$, is not bounded.

- The exponentially increasing signal defined by $\overline{x}(t) = a + e^{bt}$, for constants a and $b > 0$, is not bounded.

- The exponentially decaying signal defined by $\overline{x}(t) = a + e^{-bt}$, for constants a and $b > 0$, is bounded.

- The step signal defined by $\overline{x}(t) = a$ for $t < t_0$ and $\overline{x}(t) = b$ for $t \geq t_0$, for constants t_0, a, b, is bounded.

- The sinusoidal signal defined by $\overline{x}(t) = a \cos bt$, for constants a and b, is bounded.

A signal over a set V of variables is bounded if the component of the signal corresponding to each variable x in V is bounded.

In a stable system, when started in the initial state $s_0 = 0$, whenever the input signal is bounded, then the output signal produced by the component in response is also bounded. The bound on the output signal can be different from the bound on the input signal. This particular formalization of stability is known as *Bounded Input Bounded Output* (BIBO) stability.

BIBO STABILITY

A continuous-time component H with Lipschitz-continuous dynamics with input variables I and output variables O is *Bounded-Input-Bounded-Output stable* if, for every bounded input signal $\overline{I}$, the output signal $\overline{O}$ produced by the component H starting in the initial state $s_0 = 0$, in response to the input signal $\overline{I}$, is also bounded.

Let us consider the helicopter model from figure 6.5:

$$\dot{s} = T/I$$

and assume that the initial spin is 0. Suppose we apply a constant torque T_0 to the system. Then the rate of change of the spin s is constant and the spin keeps increasing linearly. Thus the system is unstable: the input signal is the bounded constant function $\overline{T}(t) = T_0$, and the corresponding output signal is described by the unbounded function $\overline{s}(t) = (T_0/I)\, t$.

Exercise 6.4: Consider the model of the car moving on a graded road shown in figure 6.8. Suppose the input force F is 0 at all times, and the grade θ of the road is constant at 5 degrees uphill. What are the equilibria for the resulting dynamical system? ∎

Exercise 6.5: Consider a two-dimensional dynamical system whose dynamics is given by

$$\dot{s}_1 = 3\, s_1 + 4\, s_2; \quad \dot{s}_2 = 2\, s_1 + s_2.$$

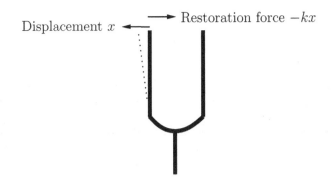

Figure 6.9: Vibrations of a Tuning Fork

Find the equilibria of this system. For each equilibrium, analyze if the equilibrium is asymptotically stable, stable but not asymptotically stable, or unstable. ■

Exercise 6.6: Consider a dynamical system whose dynamics is given by

$$\dot{x} = x^2 - x.$$

Find the equilibria of this system. For each equilibrium, analyze if the equilibrium is asymptotically stable, stable but not asymptotically stable, or unstable. ■

Exercise 6.7*: A tuning fork consists of a metal tine that can be displaced from its vertical position by striking it with a hammer (see figure 6.9). Once displaced, assuming no friction, the fork vibrates forever, generating a musical tone. Let x denote the displacement of the tine in the horizontal direction. The force that pushes the tine back toward the vertical position is proportional to its displacement at any point in time, and thus the equation of motion is given by

$$m\ddot{x} = -k\,x,$$

where k is a constant depending on the properties of the tuning fork. (a) Design a continuous-time component that captures this dynamics by finding a suitable state-space representation. (b) Assuming initial displacement $\bar{x}(0) = x_0$, find the closed-form solution for the response signal $\bar{x}(t)$ (hint: recall the rules of derivatives of basic trigonometric functions). (c) What are the equilibria of this system? For each equilibrium, analyze if the equilibrium is asymptotically stable, stable but not asymptotically stable, or unstable. ■

6.2 Linear Systems

When the dynamics of a continuous-time component is specified using linear expressions, a number of questions regarding the behavior of the system can be answered using mathematical analysis.

6.2.1 Linearity

A *linear* expression over a set of variables is formed using the operations of addition and multiplication by a numerical constant. If the variables are $x_1, x_2, \ldots x_n$, then a linear expression has the form $a_1 x_1 + a_2 x_2 + \cdots + a_n x_n$, where the coefficients $a_1, a_2, \ldots a_n$ are either real, rational, or integer constants. A linear system is a continuous-time component where the expressions used to specify the dynamics of the state and output variables are such linear expressions.

LINEAR COMPONENT

A continuous-time component H with input variables I, output variables O, and state variables S is said to be a *linear* component if:

1. for every output variable y, the expression h_y is a linear expression over the variables $I \cup S$; and

2. for every state variable x, the expression f_x is a linear expression over the variables $I \cup S$.

In our examples, the components of figures 6.2, 6.3, and 6.5 are linear. The model of a car on a graded road in figure 6.8 with the nonlinear term $m\, g \sin \theta$ in the dynamics is not linear. However, notice that the input θ is not controlled and simply represents disturbance or noise that the controller must handle. As a result, we can replace the input θ by another variable d with the meaning that d captures the value of the expression $m\, g \sin \theta$. Now the dynamics becomes $\dot{v} = (F - k\, v - d)/m$ and is linear. The range of values for the input disturbance $\theta \in [-\pi/6, +\pi/6]$ must be replaced by the range $[m\, g \sin(-\pi/6), m\, g \sin(+\pi/6)]$ for the new variable d. For the model of the simple pendulum in figure 6.6, the dynamics is specified by the differential equation $\ddot{\varphi} = -(g/\ell) \sin \varphi + u/(m\ell^2)$, and thus this component is nonlinear.

A linear expression of n variables defines a function from $\mathbf{real}^n$ to $\mathbf{real}$, and this function is Lipschitz continuous (see exercise 6.8). As a result, every linear component has Lipschitz-continuous dynamics.

It is worth noting that an expression such as $ax + b$, with $b > 0$, is *not* considered linear, and thus a single-dimensional system with the dynamics $\dot{x} = ax + b$ is not a linear component. In control theory literature, expressions with constants, that is, expressions of the form $a_1 x_1 + a_2 x_2 + \cdots + a_n x_n + a_{n+1}$, are called *affine* expressions. A particular quantity with affine dynamics is *time* itself as it can be modeled by a variable t with the dynamics $\dot{t} = 1$, and then this variable t can be used in the right-hand sides defining the dynamics of other state variables (for example, the differential equation $\dot{x} = t$ expresses time-dependent evolution of the state variable x). The analysis techniques developed in this chapter focus on linear systems, but some of them can be extended to affine systems also.

Matrix-Based Representation

The conventional form for expressing the dynamics for linear systems uses matrices. Consider a linear component with m input variables $I = \{u_1, \ldots u_m\}$, n state variables $S = \{x_1, \ldots x_n\}$, and k output variables $O = \{y_1, \ldots y_k\}$. In this case, we can view an input as a vector of dimension m, a state as a vector of dimension n, and an output as a vector of dimension k. The dynamics is expressed by four matrices each with real-valued coefficients: a matrix A of dimension $n \times n$, a matrix B of dimension $n \times m$, a matrix C of dimension $k \times n$, and a matrix D of dimension $k \times m$. The dynamics is given by

$$\dot{S} = AS + BI, \ O = CS + DI.$$

That is, for each state variable x_i, the differential equation modeling its rate of change as a function of the state/input variables is given by the linear differential equation:

$$\dot{x}_i = A_{i,1} x_1 + \cdots + A_{i,n} x_n + B_{i,1} u_1 + \cdots + B_{i,m} u_m.$$

For each output variable y_j, its value is defined in terms of the state/input variables by the linear expression:

$$y_j = C_{j,1} x_1 + \cdots + C_{j,n} x_n + D_{j,1} u_1 + \cdots + D_{j,m} u_m.$$

In our example of the car model of figure 6.3, $S = \{x, v\}$, $I = \{F\}$, and $O = \{v\}$. Thus, $n = 2$ and $m = k = 1$. The dynamics can be rewritten as

$$
\begin{aligned}
\dot{x} &= 0\,x + 1\,v + 0\,F \\
\dot{v} &= 0\,x + (-k/m)\,v + (1/m)\,F \\
v &= 0\,x + 1\,v + 0\,F
\end{aligned}
$$

The matrices are then given by:

$$A = \begin{bmatrix} 0 & 1 \\ 0 & -k/m \end{bmatrix}; \ B = \begin{bmatrix} 0 \\ 1/m \end{bmatrix}; \ C = [0 \ 1]; \ D = [0].$$

Linear Response

We have defined linearity based on the state-space representation of the component. Alternatively, linearity can be studied based on the properties of the transformation from the space of input signals to the space of output signals induced by the component.

Consider a continuous-time component H with Lipschitz-continuous dynamics with an input variable x and an output variable y. Let us set the initial state of the system to the origin, that is, the state where x equals 0. Given an input signal $\overline{x}$: time $\mapsto$ real, there is a unique output signal $\overline{y}$: time $\mapsto$ real corresponding to the execution of the component H on the input signal $\overline{x}$

starting at the origin. Thus, a continuous-time component is a function from the set of input signals to the set of output signals. This transformation is guaranteed to be *linear* for a linear component. Linearity of transformations means the following two properties:

- **Scaling:** If the input signal is scaled by a constant factor, then the output signal also gets scaled by the same factor. Given an input signal $\overline{x}$ and a constant α, let $\alpha\,\overline{x}$ be the input signal whose value at each time t is $\alpha\,\overline{x}(t)$. Then for a linear component H, for all input signals $\overline{x}$ and all scaling factors α, if $\overline{y}$ is the output signal corresponding to the input signal $\overline{x}$, then $\alpha\,\overline{y}$ is the output signal corresponding to the input signal $\alpha\,\overline{x}$.

- **Additivity:** If an input signal can be expressed as a sum of two input signals, then the corresponding output signal is also the sum of the output signals corresponding to the component input signals. That is, if the input signals $\overline{x}$, $\overline{x}_1$, and $\overline{x}_2$ are such that $\overline{x}(t) = \overline{x}_1(t) + \overline{x}_2(t)$ for all times t, and if $\overline{y}$, $\overline{y}_1$, and $\overline{y}_2$ are the output signals produced by the component corresponding to the input signals $\overline{x}$, $\overline{x}_1$, and $\overline{x}_2$, respectively, then it must be the case that $\overline{y}(t) = \overline{y}_1(t) + \overline{y}_2(t)$ for all times t.

In general, the component has multiple input and multiple output variables, and we need to consider signals that are mappings from the time domain to the set of real-valued vectors. In this case, linearity is defined by considering the sum of vectors and scaling of vectors. For two signals $\overline{V}$ and $\overline{V}'$ over a set V of variables and constants $\alpha, \beta \in \mathbf{real}$, the signal $\alpha\,\overline{V} + \beta\,\overline{V}'$ is defined to be the signal that assigns, for every time t, the value $\alpha\,\overline{V}(t)(x) + \beta\,\overline{V}'(t)(x)$ to each variable $x \in V$. Linearity of transformations is now captured by the following theorem:

Theorem 6.1 [Linearity of Input-Output Transformation] *Let H be a linear component with input variables I and output variables O. For all input signals $\overline{I}_1$ and $\overline{I}_2$ and constants $\alpha, \beta \in \mathbf{real}$, if the output signals generated by the component H from the initial state 0 in response to the input signals $\overline{I}_1$ and $\overline{I}_2$ are $\overline{O}_1$ and $\overline{O}_2$, respectively, then the output signal generated by the component H from the initial state 0 in response to the input signal $\alpha\,\overline{I}_1 + \beta\,\overline{I}_2$ is $\alpha\,\overline{O}_1 + \beta\,\overline{O}_2$.*

Proof. Suppose the dynamics of a linear component H is given by the equations $\dot{S} = AS + BI$ and $O = CS + DI$, where S is the state vector, I is the input vector, and O is the output vector. Suppose the initial state is 0.

For a given input signal $\overline{I}_1$, suppose the state response signal is $\overline{S}_1$ and the output response signal is $\overline{O}_1$. Then we know that the following conditions must hold: $\overline{S}_1(0) = 0$, and for all times t, $(d/dt)\overline{S}_1(t) = A\,\overline{S}_1(t) + B\,\overline{I}_1(t)$ and $\overline{O}_1(t) = C\,\overline{S}_1(t) + D\,\overline{I}_1(t)$.

Similarly, for another input signal $\overline{I}_1$, suppose the state response signal is $\overline{S}_2$ and the output response signal is $\overline{O}_2$. Then $\overline{S}_2(0) = 0$, and for all times t, $(d/dt)\overline{S}_2(t) = A\,\overline{S}_2(t) + B\,\overline{I}_2(t)$ and $\overline{O}_2(t) = C\,\overline{S}_2(t) + D\,\overline{I}_2(t)$.

Given constants $\alpha, \beta \in$ real, define the signals $\overline{I} = \alpha \overline{I}_1 + \beta \overline{I}_2$ and $\overline{S} = \alpha \overline{S}_1 + \beta \overline{S}_2$ and $\overline{O} = \alpha \overline{O}_1 + \beta \overline{O}_2$. From basic properties of linear arithmetic and differential calculus, it follows that the following must hold: $\overline{S}(0) = 0$, and for all times t, $(d/dt)\overline{S}(t) = A\overline{S}(t) + B\overline{I}(t)$ and $\overline{O}(t) = C\overline{S}(t) + D\overline{I}(t)$. Thus, for the input signal $\overline{I}$, the state response of the component H must be the signal $\overline{S}$ and the output response must be the signal $\overline{O}$. ∎

Exercise 6.8: Prove that a linear expression $e = a_1 x_1 + a_2 x_2 + \cdots + a_n x_n$, viewed as a function from real^n to real, is Lipschitz continuous. ∎

Exercise 6.9: Recall that the component corresponding to the motion of a car on a graded road (see figure 6.8) can be viewed as a component with two inputs, force F and disturbance d, with the dynamics given by the differential equation $\dot{v} = (F - kv - d)/m$. What are the matrices A, B, C, and D for the standard matrix-based representation of the resulting linear system? ∎

Exercise 6.10: Let us revisit the nonlinear model of the pendulum of figure 6.6. A classical approach to designing controllers for nonlinear systems is to linearize the model about an operating point, design a controller for that linearized model, and use it for the original system. For the pendulum example, the operating point of interest is $\varphi = 0$ corresponding to the vertical position of the pendulum. Using the fact that $\sin \varphi \approx \varphi$ for small values of φ, build a corresponding linear component model of the pendulum. ∎

Exercise 6.11: Consider a closed linear component H. Prove that the output response of the system is a linear function of the initial state. That is, suppose that $\overline{O}_0$ is the output response of the system starting from the initial state s_0 and $\overline{O}_1$ is the response signal of the system starting from the initial state s_1, and $\alpha, \beta \in$ real are constants. Then prove that the output response of the system starting from the initial state $\alpha s_0 + \beta s_1$ is the signal $\alpha \overline{O}_0 + \beta \overline{O}_1$. ∎

6.2.2 Solutions of Linear Differential Equations

For linear systems, a number of analysis techniques are available to understand how the output signal is related to the input signal. Let us first consider the linear differential equation $\dot{S} = AS$ and suppose the initial state is given by the vector s_0. To solve this equation, we can construct a sequence of signals $\overline{S}_0, \overline{S}_1, \cdots$ that approximate the desired solution in the following manner. Let $\overline{S}_0$ be the constant signal defined by $\overline{S}_0(t) = s_0$ for all times t. For each $m > 0$, define:

$$\overline{S}_m(t) = s_0 + \int_0^t A\overline{S}_{m-1}(\tau)\, d\tau.$$

We can use mathematical calculations based on solving integrals to find closed forms for these signals:

$$\overline{S}_1(t) = s_0 + \int_0^t A s_0\, d\tau$$

$$= \ s_0 + A\,t\,s_0$$
$$= \ [\mathbf{I} + A\,t]\,s_0.$$

In these calculations, A is an $n \times n$ matrix, s_0 is an $n \times 1$ vector, and t is a scalar. The identity matrix is denoted $\mathbf{I}$ (that is, $\mathbf{I}_{i,j}$ equals 1 if $i = j$ and 0 otherwise). Repeating the calculation for one more step gives:

$$\overline{S}_2(t) \ = \ s_0 + \int_0^t A\,([\mathbf{I} + A\,\tau]\,s_0)\,d\tau$$
$$= \ s_0 + A\,t\,s_0 + A^2\,(t^2/2)\,s_0$$
$$= \ [\mathbf{I} + \sum_{j=1}^{2} A^j\,t^j/j\,!]\,s_0$$

After repeating the pattern, we obtain

$$\overline{S}_m(t) \ = \ [\mathbf{I} + \sum_{j=1}^{m} A^j\,t^j/j\,!]\,s_0.$$

The sequence of functions $\overline{S}_0, \overline{S}_1, \overline{S}_2, \ldots$ converges to the unique solution of the differential equation given by

$$\overline{S}(t) \ = \ [\mathbf{I} + \sum_{j=1}^{\infty} A^j\,t^j/j\,!]\,s_0.$$

Recall that, for a real number a, the quantity e^a is defined as:

$$e^a \ = \ 1 + \sum_{j=1}^{\infty} a^j/j! \ .$$

Similarly, the *matrix exponential* e^A for a matrix A is defined by the equation:

$$e^A \ = \ \mathbf{I} + \sum_{j=1}^{\infty} A^j/j!$$

With this notation, the solution of the differential equation $\dot{S} = AS$, with the initial state s_0, is given by
$$\overline{S}(t) \ = \ e^{At}\,s_0.$$

A similar analysis can be performed for the model of a linear component with inputs. Consider the dynamics $\dot{S} = AS + BI$. Suppose the initial state is given by the vector s_0. Given an input signal $\overline{I}$, the resulting state signal $\overline{S}$ is given by the equation:

$$\overline{S}(t) \ = \ e^{At}\,s_0 + \int_0^t e^{A(t-\tau)}\,B\,\overline{I}(\tau)\,d\tau.$$

The response of the system to a given input signal can be computed using this equation: the output value at time t equals $C\,\overline{S}(t) + D\,\overline{I}(t)$.

The Matrix Exponential

Let us examine the definition of the matrix exponential operation:

$$e^A = \mathbf{I} + \sum_{j=1}^{\infty} A^j/j!$$

For a square $n \times n$ matrix A, observe that each term $A^j/j!$ is a square $n \times n$ matrix, and so is the exponential e^A. If we calculate the matrix e^A, the matrix e^{At} appearing in the solution $\overline{S}(t)$ of the differential equation is easily obtained by multiplying each matrix entry by t.

A number of mathematical tools exist to compute the quantity e^A depending on the structural properties of the matrix A. As an illustrative example, suppose the matrix A is a *diagonal* matrix (that is, each entry $A_{i,j}$, for $i \neq j$, equals 0). Let us denote the ith diagonal entry of the diagonal matrix A by a_i, and the diagonal matrix is denoted $\mathbf{D}(a_1, a_2, \ldots a_n)$. In this case, observe that, for every j, the matrix A^j is also a diagonal matrix whose ith diagonal entry is given by a_i^j. Then e^A is also a diagonal matrix and its ith diagonal entry is the sum $1 + \sum_{j=1}^{\infty} a_i^j/j!$, which equals e^{a_i}. Thus,

$$e^{\mathbf{D}(a_1, a_2, \cdots a_n)} = \mathbf{D}(e^{a_1}, e^{a_2}, \cdots e^{a_n}).$$

As another example, consider the two-dimensional matrix

$$A = \begin{bmatrix} 0 & a \\ 0 & 0 \end{bmatrix}.$$

Observe that for this matrix A, A^2 equals the matrix $\mathbf{0}$ with all entries 0. As a result,

$$e^A = \mathbf{I} + A = \begin{bmatrix} 1 & a \\ 0 & 1 \end{bmatrix}.$$

In general, whenever $A^k = \mathbf{0}$ for some k, only the first k terms in the infinite series defining the matrix exponential e^A are nonzero, and thus one can obtain an explicit matrix representation of e^A.

Eigenvalues and Eigenvectors

Let us consider a standard tool for calculating the matrix exponential using the *similarity* transformation. This transformation is based on computing the eigenvalues and eigenvectors of a matrix.

For an $(n \times n)$-matrix A, if the equation $A x = \lambda x$ holds, for a scalar λ and a non-zero vector x of dimension n, then the value λ is called an *eigenvalue* of the matrix A, and the vector x is called an *eigenvector* of the matrix A corresponding to the eigenvalue λ. An $(n \times n)$-matrix A has at most n distinct eigenvalues, and these correspond to the *characteristic equation* of A given by

$$\det(A - \lambda \mathbf{I}) = 0,$$

where det represents the *determinant* of a matrix. Note that a value λ_i is an eigenvalue of the matrix exactly when the term $(\lambda - \lambda_i)$ is a factor of the characteristic polynomial $\det(A - \lambda I)$.

As an example, consider the two-dimensional matrix

$$A_1 = \begin{bmatrix} 4 & 6 \\ 1 & 3 \end{bmatrix}.$$

The eigenvalues of this matrix A_1 are the solutions of the equation:

$$\det\left(\begin{bmatrix} 4 - \lambda & 6 \\ 1 & 3 - \lambda \end{bmatrix} \right) = 0.$$

Recall that the determinant of a (2×2)-matrix A is given by the expression $A_{1,1} A_{2,2} - A_{1,2} A_{2,1}$. Thus, the desired eigenvalues are the roots of the polynomial

$$(4 - \lambda)(3 - \lambda) - 6 = \lambda^2 - 7\lambda + 6 = (\lambda - 6)(\lambda - 1).$$

Thus, the eigenvalues of the matrix A_1 are $\lambda_1 = 6$ and $\lambda_2 = 1$. To obtain the eigenvector x_1 corresponding to the eigenvalue 6, we need to solve the equation $A_1 x_1 = 6 x_1$. If the entries of the vector x_1 are x_{11} and x_{12}, then we get the system of linear equations:

$$\begin{bmatrix} 4 & 6 \\ 1 & 3 \end{bmatrix} \begin{bmatrix} x_{11} \\ x_{12} \end{bmatrix} = 6 \begin{bmatrix} x_{11} \\ x_{12} \end{bmatrix}.$$

This corresponds to

$$4 x_{11} + 6 x_{12} = 6 x_{11}; \quad x_{11} + 3 x_{12} = 6 x_{12}.$$

These equation are satisfied whenever $x_{11} = 3x_{12}$, and every vector of this form is an eigenvector corresponding to the eigenvalue 6. In particular, let us set $x_1 = [3 \ 1]^T$ (note: x_1 is a column vector with two rows and one column and, thus, is the *transpose* of the row vector $[3 \ 1]$ and denoted $[3 \ 1]^T$). The eigenvector corresponding to the eigenvalue $\lambda_2 = 1$ is obtained by a similar analysis, and in particular, $x_2 = [2 \ -1]^T$ is a corresponding eigenvector. Note that the two vectors x_1 and x_2 are *linearly independent*. This is no coincidence: if the eigenvalues $\lambda_1, \lambda_2, \ldots \lambda_n$ are all distinct and the vectors $x_1, x_2 \ldots x_n$ are eigenvectors corresponding to these eigenvalues, respectively, then these n vectors are guaranteed to be linearly independent.

Note that if the matrix A is a diagonal matrix with diagonal entries a_i, then $A - \lambda I$ is also a diagonal matrix with entries $a_i - \lambda$. The characteristic polynomial of the matrix A then is the product of the terms $(a_i - \lambda)$. Furthermore, for each i, the vector x_i with 1 in the ith entry and 0 s everywhere else satisfies the equation $A x_i = a_i x_i$ and is thus an eigenvector corresponding to the eigenvalue a_i. For example, for the three-dimensional diagonal matrix $\mathbf{D}(1, 2, 1)$, the characteristic polynomial is $(1 - \lambda)^2(2 - \lambda)$. As a result, this three-dimensional matrix has two eigenvalues, namely, 1 and 2, and in this case, the (algebraic) multiplicity of the eigenvalue 1 is 2 (since $(\lambda - 1)^2$ is a factor of the characteristic polynomial). The vectors $x_1 = [1\ 0\ 0]^T$, $x_2 = [0\ 1\ 0]^T$, and $x_3 = [0\ 0\ 1]^T$ are eigenvectors, corresponding to the eigenvalues 1, 2, and 1, respectively. In this case, all these eigenvectors are linearly independent.

In the examples so far, the matrix has n linearly independent eigenvectors with real-valued entries. However, this need not be the case as indicated by the following two examples.

Consider the following two-dimensional (upper triangular) matrix:

$$A_2 = \begin{bmatrix} 1 & 2 \\ 0 & 1 \end{bmatrix}.$$

The characteristic polynomial for this matrix is $(1 - \lambda)^2$. Thus, there is only one eigenvalue, namely, 1. The eigenvectors of this matrix are of the form $[a\ 0]^T$ for an arbitrary constant a. All these eigenvectors are linearly dependent on one another, that is, the two-dimensional matrix A_2 has only one linearly independent eigenvector.

Consider the following two-dimensional matrix:

$$A_3 = \begin{bmatrix} 0 & 1 \\ -1 & 0 \end{bmatrix}.$$

The characteristic polynomial for this matrix is $\lambda^2 + 1$. However, the equation $\lambda^2 + 1 = 0$ has no real-valued solutions. In such a case, we want to interpret matrices as linear transformers over the field of *complex numbers*. With this interpretation, the matrix A_3 has two eigenvalues j and $-j$, both of which are imaginary numbers (note: the imaginary number j is the square-root of -1, and every complex number is of the form $a + bj$, where a, b are real numbers). In this case, the eigenvector corresponding to the eigenvalue j is obtained by solving the equation $A_3\, x = j\, x$ and is of the form $[1\ j]^T$.

If $\lambda_1, \ldots \lambda_p$ are all the (complex) eigenvalues of the matrix A, then

$$\det(A - \lambda\, \mathbf{I}) = (\lambda - \lambda_1)^{n_1} \cdots (\lambda - \lambda_p)^{n_p},$$

where n_j is the algebraic multiplicity of the eigenvalue λ_j and $n_1 + \cdots + n_p$ equals n. Note that if a complex number $a + bj$, with $b \neq 0$ is an eigenvalue of the matrix A, then its conjugate, that is, the complex number $a - bj$, must also be an eigenvalue of the matrix A.

Similarity Transformations

Consider the dynamical system H given by

$$\dot{S} = A\, S; \quad \overline{S}(0) = s_0,$$

where S is the n-dimensional state vector, A is an $(n \times n)$-matrix, and s_0 is the initial state. Suppose P is an *invertible* n-dimensional square matrix with real-valued entries, and P^{-1} is its inverse. Thus, $P^{-1}P = PP^{-1} = I$. Consider the vector S' defined by $S' = P^{-1}S$. This defines a linear transformation of the state. Note that the relation $S = PS'$ also holds. Let us denote the matrix $P^{-1}AP$ by J. When such a relationship holds, the matrices A and J are said to be *similar*.

Now, based on the original dynamical system H with state S, let us specify the dynamical system H' with state S':

$$\dot{S}' = (d/dt)(P^{-1}S) = P^{-1}\dot{S} = P^{-1}AS = P^{-1}APS' = JS'.$$

The initial state of this transformed linear system H' is given by

$$\overline{S}'(0) = P^{-1}\overline{S}(0) = P^{-1}s_0.$$

Such a transformation of the linear system H with state S and dynamics matrix A to obtain another linear system H' with state S' and dynamics matrix J is called a *similarity transformation* (since the matrices A and J are similar). Note that the solution of the system H' is given by:

$$\overline{S}'(t) = e^{Jt}\overline{S}'(0).$$

This implies that the solution of the original system H is given by:

$$\overline{S}(t) = Pe^{Jt}P^{-1}s_0.$$

If the matrix J has properties that make the computation of the matrix exponential e^{Jt} easier, then this can be used to compute the response $\overline{S}(t)$ of the system H.

Suppose the matrix A has n linearly independent eigenvectors $x_1, x_2, \ldots x_n$ with real-valued entries, and let λ_i be the eigenvalue corresponding to the eigenvector x_i. Let us then choose the similarity matrix P to be the matrix whose columns are these n eigenvectors:

$$P = [x_1 \ x_2 \ \cdots \ x_n].$$

Since the columns of P are linearly independent, its rank is n, and it is invertible. Note that the ith column of the matrix P is the eigenvector x_i. From the definition of matrix multiplication, the ith column of the matrix product AP is the vector Ax_i. Since x_i is an eigenvector corresponding to the eigenvalue λ_i, it follows that the ith column of the matrix product AP is the vector $\lambda_i x_i$. Hence, the ith column of the matrix product $P^{-1}AP$ is the vector $P^{-1}\lambda_i x_i$ or $\lambda_i P^{-1}x_i$. Recall that the product $P^{-1}P$ is the identity matrix, and the ith column of the product $P^{-1}P$ equals the vector $P^{-1}x_i$ (since x_i is the ith column of the matrix P). It follows that $J = P^{-1}AP$ is the diagonal matrix $\mathbf{D}(\lambda_1, \lambda_2, \ldots \lambda_n)$. It follows that the matrix exponential e^{Jt} is also a diagonal matrix, and its ith diagonal entry is the scalar $e^{\lambda_i t}$.

The method to compute the response signal of a linear system based on similarity transformation using linearly independent eigenvectors is called *diagonalization* and is summarized in the following theorem:

Theorem 6.2 [Linear System Response by Diagonalization] *Consider an n-dimensional linear system with dynamics given by the differential equation $\dot{S} = A\,S$ with initial state s_0. Suppose the matrix A has n linearly independent real-valued eigenvectors $x_1, x_2, \ldots x_n$ with corresponding eigenvalues $\lambda_1, \lambda_2, \ldots \lambda_n$. Let P be the matrix $[x_1\ x_2\ \cdots\ x_n]$, and let P^{-1} be its inverse. Then the execution of the system is given by the state signal:*

$$\overline{S}(t) \;=\; P\,\mathbf{D}(e^{\lambda_1 t}, e^{\lambda_2 t}, \cdots\ e^{\lambda_n t})\,P^{-1}\,s_0.$$

■

To illustrate this method, let us consider the two-dimensional dynamical system H given by

$$\dot{s}_1 \;=\; 4\,s_1 + 6\,s_2; \quad \dot{s}_2 \;=\; s_1 + 3\,s_2.$$

The matrix for the dynamics is the matrix A_1 that we used in illustrating the computation of eigenvalues and eigenvectors. As noted earlier, $x_1 = [3\ 1]^T$ is an eigenvector corresponding to the eigenvalue 6, and $x_2 = [2\ -1]^T$ is an eigenvector corresponding to the eigenvalue 1. Let us choose the transformation matrix as:

$$P \;=\; [\,x_1\ x_2\,] \;=\; \begin{bmatrix} 3 & 2 \\ 1 & -1 \end{bmatrix}.$$

We now need to compute the inverse P^{-1} of this matrix P. This can be done, for instance, by viewing the entries in the desired matrix P^{-1} as unknowns and setting up a system of simultaneous linear equations given by $P\,P^{-1} = I$:

$$\begin{bmatrix} 3 & 2 \\ 1 & -1 \end{bmatrix} \begin{bmatrix} a & c \\ b & d \end{bmatrix} = \begin{bmatrix} 3a + 2b & 3c + 2d \\ a - b & c - d \end{bmatrix} = \begin{bmatrix} 1 & 0 \\ 0 & 1 \end{bmatrix}.$$

Solving these equations gives:

$$P^{-1} \;=\; \begin{bmatrix} 1/5 & 2/5 \\ 1/5 & -3/5 \end{bmatrix}.$$

Verify that indeed $P\,P^{-1} = P^{-1}P = I$. Furthermore, verify that $P^{-1}A_1P$ is the diagonal matrix $\mathbf{D}(6,1)$. What this means is that if we consider the linear system H' with state variables s_1' and s_2' defined by

$$s_1' \;=\; (s_1 + 2s_2)/5; \quad s_2' \;=\; (s_1 - 3s_2)/5,$$

it has a simpler dynamics given by $\dot{s}_1' = 6s_1'$ and $\dot{s}_2' = s_2'$ and thus is easier to analyze. Putting all the pieces together, starting in the initial state s_0, the state of the system H at time t is described by $P\mathbf{D}(e^{6t}, e^{t})\,P^{-1}s_0$. If the initial state vector s_0 is $[s_{01}\ s_{02}]^T$, then by calculating the matrix products, we get a closed-form solution for the state of the system H at time t:

$$\begin{aligned}
\overline{s}_1(t) &= [(3e^{6t} + 2e^{t})s_{01} + 6(e^{6t} - e^{t})s_{02}]/5 \\
\overline{s}_2(t) &= [(e^{6t} - e^{t})s_{01} + (2e^{6t} + 3e^{t})s_{02}]/5.
\end{aligned}$$

As discussed earlier, the matrix A may not have n independent eigenvectors. In this case, it is possible to choose the similarity transformation matrix P in such a way that the matrix $J = P^{-1}AP$ is in *Jordan canonical form*. This is a special form of matrix that is *almost* diagonal, and it is possible to get an explicit representation of the exponential matrix e^{Jt}.

Exercise 6.12: Consider a single-dimensional linear component with one input with dynamics given by $\dot{s} = a\,s + b\,u$. Suppose we set the input signal to be the constant signal $\overline{u}(t) = c$ for a constant value c. Find a closed-form formula for the system response $\overline{s}(t)$ starting from the initial state s_0 corresponding to this input signal (hint: the integral $\int_0^t e^{-a\tau}\,d\tau$ evaluates to $(1 - e^{-at})/a$). ∎

Exercise 6.13: Consider the two-dimensional dynamical system given by

$$\dot{s}_1 = -s_1 + 2s_2; \quad \dot{s}_2 = s_2.$$

Compute the closed form description of the state signal $\overline{S}(t)$ given the initial state vector s_0 using the method of similarity transformation. ∎

Exercise 6.14: Consider the two-dimensional dynamical system given by

$$\dot{s}_1 = s_2; \quad \dot{s}_2 = -2s_1 - 3s_2.$$

Compute the closed-form description of the state signal $\overline{S}(t)$ given the initial state vector s_0 using the method of similarity transformation. ∎

Exercise 6.15: Consider the three-dimensional dynamical system given by

$$\dot{s}_1 = 3s_1 + 4s_2; \quad \dot{s}_2 = 2s_2; \quad \dot{s}_3 = 4s_1 + 9s_3.$$

Compute the closed-form description of the state signal $\overline{S}(t)$ given the initial state vector s_0 using the method of similarity transformation. Note that the determinant of a 3×3 matrix A is given by the formula

$$a_{11}(a_{22}a_{33} - a_{23}a_{32}) + a_{12}(a_{23}a_{31} - a_{21}a_{33}) + a_{13}(a_{21}a_{32} - a_{22}a_{31}),$$

where a_{ij} denotes the entry of the matrix in ith row and jth column. ∎

6.2.3 Stability

Consider the n-dimensional linear system H given by $\dot{S} = AS$. The response of the system starting from the initial state s_0 is described by the signal $\overline{S}(t) = e^{At}s_0$. Our goal is to develop analytical methods to determine stability of the equilibria of this system.

A state s_e is an equilibrium of the system H if the condition $As_e = 0$ holds. To compute the equilibria of the system H, we can view the n elements of the vector s_e as the unknowns and solve the system of n linear equations corresponding to the condition $As_e = 0$. Observe that the state 0, which assigns 0 to all the state variables, is an equilibrium of the system H. If the matrix A is invertible (that is, if the rank of the matrix A is n), then the equation $As_e = 0$ has a unique solution, and 0 is the only equilibrium. If a nonzero state s_e is an equilibrium of the system H, then we can consider a transformed linear system H' with the state vector S' given by $S' = S - s_e$. The dynamics of this transformed system H' is given $\dot{S}' = AS'$, and at each time t, the equation $\overline{S}'(t) = \overline{S}(t) - s_e$ holds. Note that the state 0 is an equilibrium of the system H'. Thus, the behavior of this transformed system around its equilibrium 0 corresponds exactly to the behavior of the original system around its equilibrium s_e. Thus, if we know how to analyze whether the equilibrium 0 is stable, the same analysis technique can be used to determine whether an arbitrary equilibrium is stable: the stability properties of the equilibrium s_e of the system H are exactly the same as the corresponding properties of the equilibrium 0 of the system H'.

For the remainder of this section, we abbreviate "the state 0 of the linear system H is a stable equilibrium" by "the linear system H is stable" and "the state 0 of the linear system H is an asymptotically stable equilibrium" by "the linear system H is asymptotically stable."

Single-Dimensional System

By definition, the linear system H is stable if, for every $\epsilon > 0$, there exists a $\delta > 0$ such that for all state s_0, if $\|s_0\| < \delta$, then $\|e^{At}s_0\| < \epsilon$ for all times t. The system is asymptotically stable if, in addition, there exists a $\delta > 0$ such that for all states s_0, if $\|s_0\| < \delta$, then the vector $e^{At}s_0$ converges to 0 as t increases. In the latter case, the values of δ for which the condition "for all states s_0, if $\|s_0\| < \delta$, then the signal $e^{At}s_0$ converges to 0" holds is called the *region of attraction*.

First, let us focus on linear systems with dimension 1. Then the dynamics is given by $\dot{x} = ax$. If x_0 denotes the initial state, then the state at time t equals $e^{at}x_0$. Depending on the sign of a, we have three cases:

- If the coefficient a is negative, then the magnitude of the value $e^{at}x_0$ decreases with increasing t and is bounded by the magnitude of the initial value x_0. Thus, the system is stable. Also, observe that no matter what the initial value is, the value of x decays exponentially and will become 0 in the limit. In this case, the system is asymptotically stable, and in fact, the region of attraction includes all the states.

- If the coefficient a is 0, then the value of x stays equal to its initial value x_0. Thus, the magnitude of the state does not change with time, and the system is stable. However, the signal *does not converge* to 0, and thus the system is not asymptotically stable.

- If the coefficient a is positive, then observe that the quantity e^{at} grows exponentially with increasing t. As a result, the magnitude of the value of x increases exponentially and grows in an unbounded manner, and the system is not stable.

Thus, the stability of a one-dimensional linear system depends on the sign of the coefficient of the term capturing dependence of the rate of change on the state: the system is unstable if $a > 0$, stable if $a \le 0$, and asymptotically stable if $a < 0$.

Diagonal State Dynamics Matrix

Now suppose that the matrix A is an n-dimensional diagonal matrix and equals $\mathbf{D}(a_1, a_2, \ldots a_n)$. In this case, we know that the matrix exponential e^{At} is also a diagonal matrix. If the initial state of the system is given by the vector $[s_{01} \ s_{02} \ \cdots \ s_{0n}]^T$, then the state $\overline{S}(t)$ at time t is given by:

$$\overline{S}(t) = [\, e^{a_1 t} s_{01} \ \ e^{a_2 t} s_{02} \ \ \cdots \ \ e^{a_n t} s_{0n} \,]^T.$$

Observe that, for each i, with increasing t, the quantity $e^{a_i t}$ increases exponentially if $a_i > 0$, stays unchanged if $a_i = 0$, and decreases exponentially if $a_i < 0$. The case analysis for a single-dimensional system now generalizes naturally:

- If none of the coefficients a_i s is positive, then for every initial state s_0, for the resulting signal, $\|\overline{S}(t)\| \le \|s_0\|$ holds at all times, and thus the system is stable.

- If all the coefficients a_i s are negative, then in addition to the stability as in the case above, independent of the initial state s_0, for the resulting signal, $\|\overline{S}(t)\|$ converges to the equilibrium 0, and thus the system is asymptotically stable. The region of attraction consists of the entire state space $\mathbf{real}^n$.

- Suppose there exists an index i such that the coefficient a_i is positive. Then if we choose the initial state s_0 such that s_{0i} is positive and $s_{0j} = 0$ for all $j \ne i$, then no matter how small the value s_{0i} is, the ith component of the response signal $\overline{S}(t)$ given by $e^{a_i t} s_{0i}$ will grow unboundedly with increasing t. In this case, the system is unstable.

In summary, when $A = \mathbf{D}(a_1, a_2, \ldots a_n)$, the system is unstable if $a_i > 0$ for some i, stable if $a_i \le 0$ for all i, and asymptotically stable if $a_i < 0$ for all i.

Diagonalizable State Dynamics Matrix

Now recall the similarity transformation that we used to diagonalize a matrix and compute the matrix exponential. Given a linear system H with the dynamics $\dot{S} = A S$, we choose a suitable invertible matrix P and consider the transformed linear system H' with the dynamics $\dot{S}' = J S'$, where $J = P^{-1} A P$. The state vector S' of H' is obtained from the state vector S of H by linear transformation: $S' = P^{-1} S$. Observe that a linear transformation of a signal preserves the properties of being bounded in magnitude and convergence to 0. This implies that the similarity transformation preserves the stability properties of the equilibrium. This is captured by the following proposition:

Proposition 6.1 [Stability Preservation by Similarity Transformation] *Given an n-dimensional linear system H with state vector S and dynamics $\dot{S} = A S$, consider another n-dimensional linear system H' with state vector S' and dynamics $\dot{S}' = J S'$, where $J = P^{-1} A P$ for some invertible matrix P. Then the system H is stable if and only if the system H' is stable, and the system H is asymptotically stable if and only if the system H' is asymptotically stable.* ∎

If we can choose the similarity transformation matrix P so that the matrix J is diagonal, then we can conclude that the system H is unstable if some diagonal entry of J is positive, stable if all entries of J are nonpositive, and asymptotically stable if all diagonal entries of J are negative.

Recall that if all eigenvalues of the matrix A are real valued and distinct, then we can choose the matrix P using the corresponding eigenvectors and the diagonal entries of J correspond to the eigenvalues of A. As a result, if the matrix A has n distinct real-valued eigenvalues $\lambda_1, \lambda_2, \ldots \lambda_n$, then the equilibrium 0 is unstable if $\lambda_i > 0$ for some i, asymptotically stable if $\lambda_i < 0$ for all i, and stable but not asymptotically stable if $\lambda_i \leq 0$ for all i and $\lambda_j = 0$ for some j.

Let us revisit the model of the car in figure 6.3. As noted earlier, the state dynamics matrix for this system is

$$A = \begin{bmatrix} 0 & 1 \\ 0 & -k/m \end{bmatrix}.$$

The eigenvalues of this matrix are the real numbers 0 and $-k/m$. We can immediately conclude that the system is stable but not asymptotically stable, which coincides with the analysis following the definition of Lyapunov stability in section 6.1.4.

In general, the eigenvalues of a matrix can be complex numbers. It is possible to generalize the above case analysis for real-valued eigenvalues to this case. The system is asymptotically stable exactly when all the eigenvalues have negative real parts.

Theorem 6.3 [Stability Test for Linear Dynamics] *The linear system given by $\dot{S} = A S$ is asymptotically stable if and only if every eigenvalue of the matrix A has a negative real part.* ∎

If some eigenvalue of the state-dynamics matrix A has a positive real part, then the system is unstable. If some eigenvalues have negative real parts and the remaining are purely imaginary numbers (that is, with real part equal to 0), then the system is not asymptotically stable, and its stability depends on the structure of the Jordan blocks corresponding to each such imaginary eigenvalue.

BIBO Stability

Now let us turn our attention to components with inputs. Consider the system given by

$$\dot{S} = AS + BI; \ O = CS + DI.$$

To check whether the system is BIBO-stable we set the initial state $s_0 = 0$, and consider the behavior of the system for a bounded input signal $\bar{I}$. We note a sufficient condition for checking BIBO-stability in terms of Lyapunov-stability criterion:

Theorem 6.4 [Reducing BIBO-Stability to Lyapunov Stability] *If the linear dynamical system $\dot{S} = AS$ is asymptotically stable, then the continuous-time component with the dynamics $\dot{S} = AS + BI$ and $O = CS + DI$ is BIBO-stable.* ∎

The proof of this theorem relies on understanding the dynamics of a continuous-time component using *transfer functions* and is beyond the scope of this text-book.

Note that for the helicopter model of figure 6.5, if we set the input torque to 0, dynamics is given by $\dot{s} = 0$. In this dynamics, if the initial spin is s_0, then it will remain constant at the value s_0, and thus the system, while stable, is not asymptotically stable. As noted earlier, if the system is applied a constant torque as input, the spin increases linearly in an unbounded manner, and the system is not BIBO-stable. Thus, stability of the system $\dot{S} = AS$ does not imply BIBO-stability of the continuous-time component with the dynamics $\dot{S} = AS + BI$. In the reverse direction, BIBO-stability of the continuous-time component with the dynamics $\dot{S} = AS + BI$, in itself, does not ensure that Lyapunov-style stability properties of the system $\dot{S} = AS$ and additional properties of the matrices need to be established for this purpose.

Exercise 6.16 *: Prove proposition 6.1. ∎

Exercise 6.17: For each of the systems in exercises 6.13, 6.14, and 6.15, determine whether the system is asymptotically stable, stable but not asymptotically stable, or unstable. ∎

6.3   Designing Controllers

Given a dynamical system model of the plant, the controller is designed to provide the controlled input signals to maintain the output of the system close to the desired output adjusting to changes in the uncontrolled disturbances. Designing controllers in a principled manner is a well-developed discipline. We will review some basic terminology in control design and get familiar with the most commonly used class of controllers in industrial practice.

6.3.1   Open-Loop vs. Feedback Controller

An *open-loop* controller does not use measurements of the state or outputs of the plant to make its decisions. Such a controller relies on the model of the plant to decide on the controlled input for the plant, and its implementation does not require sensors. For example, consider the first model of the car from figure 6.3. Suppose the controller's objective is to maintain a constant velocity (that is, we want $\dot{v} = 0$ at all times). Then if the initial velocity is v_0, the desired value of the input force F equals $k\, v_0$: the controller can simply apply this constant force to the car to maintain the velocity constant at the value v_0. Such a controller is called *open-loop*: when compared to the architecture shown in figure 6.1, the block for sensors and the flow of information from the plant to the controller is missing. Obviously, such a controller would be less expensive to implement than the one that requires sensors to estimate the state of the plant. However, its design heavily relies on the assumption that the behavior of the plant is entirely predictable and accurately captured by the idealized mathematical model. In practice, operation of such a controller is acceptable, provided there is a possibility of manual intervention. If the driver finds the speed of the car unacceptable, she would simply increment or decrement the desired speed triggering a recalculation of the force applied by the open-loop controller.

A *feedback* controller uses sensors to measure the output, and thus indirectly the current state of the plant, to update the values of the controlled input variables. For example, in the revised model of the car in figure 6.8, the model accounts for the change in the grade of the road. Suppose that the controller is applying the correct amount of force to maintain the velocity of the car at the desired cruising speed. A positive change in the grade θ causes the car to slow down, while a negative change in θ causes the car to speed up. The speed of the car, as measured by the sensors, is an input to the controller. It notices the change in speed and adjusts the force to make the velocity again equal to the desired cruising speed. A feedback controller not only can cope with disturbances (such as the grade) whose variation with time is not predictable in advance, but it can work well even when the mathematical model of the plant is only a rough approximation of the real-world dynamics. Implementation of a feedback controller requires sensors, and its performance is related to the accuracy of measurements by these sensors.

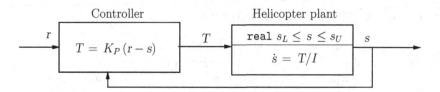

Figure 6.10: Stabilizing Controller for the Helicopter Model

6.3.2   Stabilizing Controller

Stabilizing Helicopter Model

We now describe a simple and typical pattern for designing a controller using the helicopter example of figure 6.5. Recall that the given helicopter model is unstable. The controller is shown in figure 6.10. It takes two input signals: the input variable r represents the *reference signal* that captures the desired spin, and the signal s is the plant output, namely, the (measured) spin of the helicopter. Given two such input variables corresponding to the desired and actual values, we can define the *error signal* e by the equation $e = r - s$. The goal of the controller is to keep the magnitude of the error as small as possible and also ensure that the closed-loop system obtained by composing the helicopter model and the controller is stable. Note that a positive value of e means that the controller should try to increase the actual spin s by applying a positive torque, and a negative value of e means that the controller should try to decrease the actual spin s by applying a negative torque. The controller of figure 6.10 computes the torque by simply scaling the error signal by a positive constant factor K_P. Such a controller is called a *proportional controller*, and the constant K_P is called the *gain* of the controller.

For the closed-loop system consisting of the composition of the controller and the helicopter, the input signal is the reference value r and the output is the spin s. The dynamics of the composite system is given by the equation

$$\dot{s} = K_P (r - s)/I.$$

This is a one-dimensional linear system, and the coefficient capturing the dependence of the rate of change of the state variable s on itself is $-K_P/I$. We know that such a system is asymptotically stable exactly when this coefficient is negative. Thus, the composite system is asymptotically stable as long as the gain K_P is positive.

When the reference input is 0, that is, when the objective of the controller is to keep the helicopter from spinning, the controller applies the torque equal to $-K_P/I$. No matter what the initial spin s_0 is, this causes the spin to decay to 0 exponentially. The higher the value of K_P, the faster the rate of convergence.

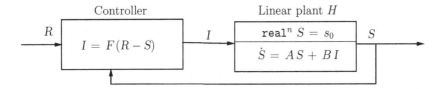

Figure 6.11: State Feedback Controller for Linear Systems

Linear State Feedback

The general architecture of a state feedback controller for a linear system is shown in figure 6.11. The original system is modeled by the linear component H. It has n state variables S and m input variables I, and its dynamics is given by the linear differential equation $\dot{S} = AS + BI$. In this setup, the assumption is that the controller can observe the state fully, that is, the set of output variables of the plant coincides with its state variables. The controller computes its output I based on the state S of the plant and the *reference* input R. The reference input R has the same dimension n as the state vector S.

The controller is a stateless linear component and is defined by the linear transformation

$$I = F(R - S).$$

The transformation matrix F has dimension $m \times n$ and is called the *gain matrix*. The closed-loop system has n-dimensional state S and n-dimensional reference input R, and its dynamics is given by

$$\dot{S} = (A - BF)S + BFR.$$

The control design problem is to choose the $(m \times n)$-matrix F such that the composite system is asymptotically stable. Then by theorem 6.4, we are guaranteed that the system is also BIBO-stable, and thus small variations in the reference inputs do not cause large perturbations in the state signal.

Design of the Gain Matrix

Recall that a linear system with dynamics $\dot{S} = AS$ is asymptotically stable exactly when every eigenvalue of the matrix A has a negative real part (theorem 6.3). Thus, to design a stabilizing controller for figure 6.11, given the matrices A and B, we need to choose the gain matrix F such that every eigenvalue of the matrix $(A - BF)$ has a negative real part.

To illustrate this computation, let us consider the system with two state variables and one input variable given by:

$$\dot{s}_1 = 4\,s_1 + 6\,s_2 + 2\,u; \quad \dot{s}_2 = s_1 + 3\,s_2 + u.$$

The matrix A for this system is the matrix A_1 for which we computed eigenvalues, eigenvectors, and the matrix exponential in section 6.2.2. Recall that the eigenvalues are 6 and 1, and thus the system is not stable. The matrix B is the column vector $[2 \ 1]^T$. The desired gain matrix F is a (1×2)-matrix, and let its entries be f_1 and f_2. We want to select values for these unknowns f_1 and f_2 so that the following matrix has eigenvalues with negative parts:

$$\begin{bmatrix} 4 & 6 \\ 1 & 3 \end{bmatrix} - \begin{bmatrix} 2 \\ 1 \end{bmatrix} [f_1 \ f_2] = \begin{bmatrix} 4 - 2f_1 & 6 - 2f_2 \\ 1 - f_1 & 3 - f_2 \end{bmatrix}.$$

Control design now corresponds to choosing eigenvalues for this matrix and then solving for the unknown f_1 and f_2. The characteristic polynomial for this matrix is

$$\begin{aligned} P(\lambda, f_1, f_2) &= (4 - 2f_1 - \lambda)(3 - f_2 - \lambda) - (6 - 2f_2)(1 - f_1); \\ &= \lambda^2 + (2f_1 + f_2 - 7)\lambda + (6 - 2f_2). \end{aligned}$$

The roots of this characteristic polynomial are λ_1 and λ_2 if the polynomial is of the form $(\lambda - \lambda_1)(\lambda - \lambda_2)$. By matching the coefficients of these quadratic polynomials, we conclude that the eigenvalues of the matrix $(A - BF)$ are λ_1 and λ_2 exactly when the entries of gain matrix F satisfy the following equations:

$$\begin{aligned} 2f_1 + f_2 - 7 &= -\lambda_1 - \lambda_2; \\ 6 - 2f_2 &= \lambda_1 \lambda_2. \end{aligned}$$

If we prefer eigenvalues -1 and -2 (and these would ensure asymptotic stability), then we need to solve

$$2f_1 + f_2 - 7 = 3; \quad 6 - 2f_2 = 2.$$

Solving these equations gives us $f_1 = 4$ and $f_2 = 2$. Thus, the desired matrix F is $[4 \ 2]$, that is, the controller should provide the input signal $u = 4(r_1 - s_1) + 2(r_2 - s_2)$ to the system so that the resulting closed-loop system is asymptotically stable with eigenvalues -1 and -2.

We can also choose the eigenvalues to be complex numbers as long as when we choose a complex number, we also choose its conjugate. For example, if we want the eigenvalues to be $-1 + j$ and $-1 - j$, then we need to solve

$$2f_1 + f_2 - 7 = 2; \quad 6 - 2f_2 = 2.$$

Solving these equations give us $f_1 = 7/2$ and $f_2 = 2$. This means that with the choice of the gain matrix F to be $[7/2 \ 2]$, the resulting closed-loop system is asymptotically stable with eigenvalues $-1 + j$ and $-1 - j$.

Controllability of the Matrix Pair (A, B)

To summarize, to design a stabilizing feedback controller, we choose eigenvalues $\lambda_1, \lambda_2 \ldots \lambda_n$, all with negative real parts, and solve the equation

$$\det[A - BF - \lambda \mathbf{I}] = (\lambda - \lambda_1)(\lambda - \lambda_2) \cdots (\lambda - \lambda_n),$$

with mn unknowns corresponding to the entries of the gain matrix F. This approach to stabilization naturally raises the following questions: (1) when is this equation guaranteed to have solutions? and (2) does the existence of a solution depend on the choice of the eigenvalues? It turns out that when the pair of matrices A and B satisfy a certain property, for every choice of the eigenvalues $\lambda_1, \lambda_2 \dots \lambda_n$, it is possible to choose the entries of the gain matrix F so as to satisfy the above equation.

Given an $(n \times n)$-matrix A and an $(n \times m)$-matrix B, consider the following matrix with n rows and mn columns:

$$\mathbf{C}(A, B) \;=\; [\, B \quad AB \quad A^2B \quad \cdots \quad A^{n-1}B \,].$$

That is, the first m columns of the matrix $\mathbf{C}(A, B)$ are the columns of the $(n \times m)$-matrix B, the next m columns are the columns of the $(n \times m)$-matrix AB, and the last m columns are the columns of the $(n \times m)$-matrix $A^{n-1}B$. This matrix $\mathbf{C}(A, B)$ is called the controllability matrix corresponding to the matrix pair (A, B). The pair (A, B) of matrices is called *controllable* if the rank of the controllability matrix is n, that is, if all the rows of the matrix $\mathbf{C}(A, B)$ are linearly independent. In such a case, the linear system with the dynamics $\dot{S} = AS + BI$ is also called controllable.

The following theorem tells us that controllability is a necessary and sufficient condition for the existence of a gain matrix corresponding to the chosen eigenvalues. If the pair (A, B) of matrices is controllable, then it is possible to obtain the desired gain matrix F for any arbitrary choice of eigenvalues as long as we choose only real numbers or when we choose a complex number, we also choose its conjugate. If the pair (A, B) of matrices is not controllable, then not all eigenvalues can be chosen freely.

Theorem 6.5 [Controllability and Eigenvalue Assignment] *Let A be an $(n \times n)$-matrix and B be an $(n \times m)$-matrix. The following two statements are equivalent:*

- *The $(n \times mn)$-controllability matrix $\mathbf{C}(A, B)$ whose columns are the columns of the matrices $B, AB, \dots A^{n-1}B$, has rank n.*

- *For every choice of (complex) numbers $\lambda_1, \lambda_2, \dots \lambda_n$ such that a complex number appears in this list exactly when its conjugate also appears in the list, there exists a $(m \times n)$-matrix F such that the eigenvalues of the matrix $(A - BF)$ are $\lambda_1, \lambda_2, \dots \lambda_n$.*

∎

Continuing with our example,

$$A_1 \;=\; \begin{bmatrix} 4 & 6 \\ 1 & 3 \end{bmatrix}; \; B \;=\; \begin{bmatrix} 2 \\ 1 \end{bmatrix}; \; \mathbf{C}(A_1, B) \;=\; \begin{bmatrix} 2 & 14 \\ 1 & 5 \end{bmatrix}.$$

In this example, $m = 1$ and $n = 2$. The controllability matrix $\mathbf{C}(A_1, B)$ is a (2×2)-matrix, and its rank is 2. Thus, the pair (A_1, B) is controllable. As we have analyzed already, the eigenvalues of the matrix $(A - BF)$ are λ_1 and λ_2 exactly when the entries of gain matrix F satisfy the following equations: $2f_1 + f_2 = -\lambda_1 - \lambda_2 + 7$ and $2f_2 = 6 - \lambda_1\lambda_2$. This system of linear equations is guaranteed to have a solution for the entries f_1 and f_2 no matter what values we choose for λ_1 and λ_2 (note: if λ_1 is a complex number, then λ_2 must be its conjugate, and this ensures that their sum and product are both real numbers). By theorem 6.5, we know that this is not a coincidence.

For a linear system, controllability entails many other appealing properties. If the linear system is controllable, then for any given initial state and a target state, it is possible to provide an input signal to the system so that the state of the system starting from the initial state becomes equal to the target state in a finite time. More precisely, consider a linear system with dynamics $\dot{S} = AS + BI$ and an initial state s_0, such that the pair (A, B) of matrices is controllable. Then for every state $s \in \mathbf{real}^n$, there exists a time $t^* \in \mathbf{time}$ and an input signal $\bar{I}$ such that for the unique state signal $\bar{S}$ of the system as a response to the input signal $\bar{I}$ starting from the initial state s_0, $\bar{S}(t^*) = s$.

While we have presented a method for designing a controller to ensure the critical property of stability, it is worth mentioning that there are two key aspects of the control design that we are not addressing here, but for which the theory of linear systems provides well-understood tools:

- **Optimality:** there are many choices for the gain matrix F that ensure stability of the resulting closed-loop system. Theory of optimal control, and in particular the technique for designing the *Linear Quadratic Regulator* (LQR) controller, addresses the question of choosing a matrix that satisfies additional criteria (for example, driving certain state variables to 0 as fast as possible).

- **State Estimation:** We have assumed that the input to the controller is the complete state S of the system. What happens when the controller can observe the state of the system only partially via the output vector O? In this case, the controller needs to *estimate* the state of the system based on the observation of the output signal and the theory of observability and state estimation develops techniques for this purpose.

Exercise 6.18 : Consider the linear system with two state variables, one input variable, and the dynamics given by:

$$\dot{s}_1 = s_1/2 + s_2 + u; \quad \dot{s}_2 = s_1 + 2s_2 + u.$$

(1) Show that the system is not stable. (2) Show that the system is controllable. (3) Find the gain matrix F so that the eigenvalues for the resulting closed-loop system are $-1 + j$ and $-1 - j$. ∎

Exercise 6.19 *: Consider the linear system with two state variables, one input variable, and the dynamics given by:

$$\dot{s}_1 = -2s_2 + u; \quad \dot{s}_2 = s_1 - 3s_2 + u.$$

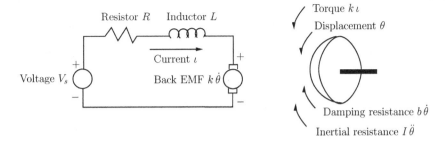

Figure 6.12: Dynamics of a DC Motor

First, show that the matrix-pair (A, B) is not controllable. Then show that it is not possible to choose entries of the gain matrix F so that the matrix $(A - BF)$ has arbitrarily chosen eigenvalues: show that one of the eigenvalues of the matrix $(A - BF)$ is always -1 no matter what the entries in the gain matrix F are (although the second eigenvalue can be set to an arbitrary value by suitably choosing the entries of F). ∎

6.3.3 PID Controllers *

In industrial control systems, the most commonly used design of a controller to correct the discrepancy between the desired reference signal and the measured output signal uses a combination of three terms: a *proportional* term capturing the reaction to the current error, an *integral* term capturing the reaction to the cumulative error, and a *derivative* term capturing the response to the rate of change of error. Such a controller is called a PID controller. Let us illustrate the design of such a controller using a classical example of a control system, namely, a DC motor.

DC Motor

Figure 6.12 shows the design of a DC motor that converts input voltage to rotational motion and is a commonly occurring building block of many electromechanical devices. The electrical circuit is shown on the left and is connected to the rotating shaft shown on the right.

Let us denote the input voltage by V_s, the resistance of the circuit by R, and the inductance of the circuit by L. Let ι denote the electrical current flowing through the circuit and θ denote the angular displacement of the wheel. If k is the electromotive-force (EMF) constant, then the back electromotive-force (EMF) voltage generated by the rotating shaft equals k times its angular velocity. Basic laws of electric circuits tell us that (1) the voltage across a resistor equals the product of the resistance and the electrical current flowing through the circuit, (2) the voltage across an inductor equals the product of the inductance and the

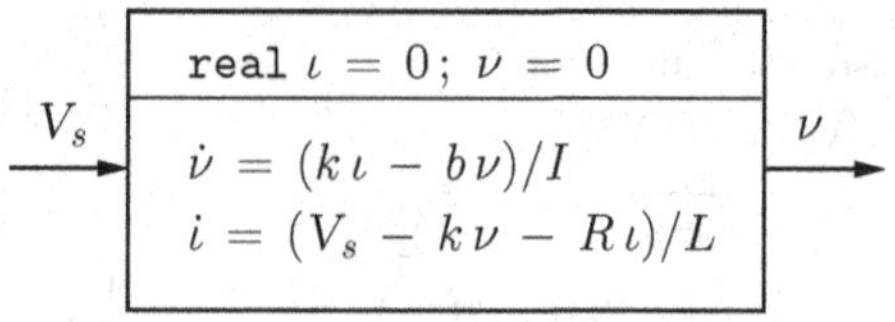

Figure 6.13: Continuous-time Component Modeling DC Motor

rate of change of the electrical current flowing through the circuit, and (3) the sum of voltages around a closed path in an electrical circuit must equal to 0 (Kirchhoff's law). Applying these rules gives us the equation:

$$L\dot{\iota} + R\iota + k\dot{\theta} = V_s.$$

Now let us analyze the motion of the rotating shaft. The torque acting on the shaft equals the EMF constant k times the current flowing through the circuit. If I is the rotational inertia and b is the coefficient of friction corresponding to the damping effects, then Newton's law for rotating bodies gives us:

$$I\ddot{\theta} + b\dot{\theta} = k\iota.$$

The output of the DC motor is the rotational velocity of the shaft. The dynamics then can be captured by the continuous-time component shown in figure 6.13. In this model, ν denotes the rotational velocity, and the angular displacement θ is omitted.

Controller for the DC Motor

The controller for the DC motor of figure 6.13 observes the rotational velocity ν and adjusts the source voltage V_s to achieve the desired response. The canonical task for such a controller is to control the voltage so that the rotational velocity changes from its initial value to a desired speed, say r. We already know that a general technique for designing a *proportional* controller consists of scaling the difference between the reference value and the observed output by a suitably chosen proportional gain constant K_P:

$$V_s = K_p(r - \nu).$$

While we have already analyzed the relationship between the value of the gain constant and the (asymptotic) stability of the resulting closed-loop system, to understand the various requirements other than stability that are of practical interest, let us examine the typical behavior of the DC motor with a proportional controller as shown in figure 6.14. This plot is generated using MATLAB by setting the rotational inertia $I = 0.01$, the damping constant $b = 0.1$, the electromotive force constant $k = 0.01$, the resistance $R = 1$, the inductance $L = 0.5$, the reference input $r = 1$, and the proportional gain constant $K_P = 100$.

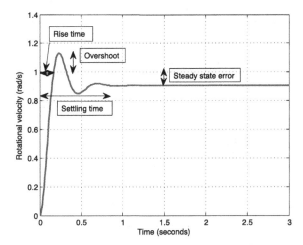

Figure 6.14: Output Response of the DC Motor with a Proportional Controller

The response shown in figure 6.14 is called the *step response* of a continuous-time component: at time t_0 (0 in this example), we want to change the output from some initial value (0 in this example) to a new reference value (1 in this example). The shape of the output response signal shown in figure 6.14 is illustrative of a large number of physical systems. Let us consider intuitively how the output of a system typically changes in response to a change in the reference signal. Let e denote the difference between the reference value and the output of the system. At the initial time instance, the magnitude of this error e is the highest. As e changes, so does the value of the control input supplied by the controller, which impacts the state and the output of the system. The output approaches 1 but overshoots the desired value due to the smoothness of the dynamics of the physical world. The overshoot makes the error negative, causing the controller to change the direction of the derivative of its output. The same phenomenon repeats, causing the output to oscillate for a while before it settles into a steady-state value, which does not change in absence of further external disturbances.

Given such an expected response of the system output, the following metrics capture the performance of the controller:

1. *Overshoot:* The difference between the maximum value of the system output and the desired reference value. For the DC motor response shown in figure 6.14, the maximum rotational velocity is 1.12 rad/s, causing an overshoot of 12%. Ideally, the overshoot should be as small as possible. In particular, a safety requirement can assert that the overshoot should be below some threshold value.

2. *Rise time:* The time difference between the initial time when the reference signal changes and the time at which the output signal crosses the desired

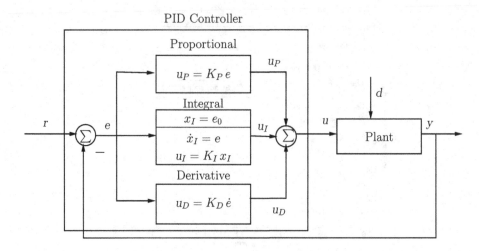

Figure 6.15: A Generic PID Controller

reference value. For the DC motor response shown in figure 6.14, the rise time is 0.15 seconds. Ideally, smaller rise time means better responsiveness of the system, and typically an attempt to reduce the rise time will increase the overshoot.

3. *Steady-state error:* The difference between the steady-state value of the output signal and the value of the reference signal. For the DC motor response shown in figure 6.14, the output stabilizes at 0.9 rad/s, thus leading to 10% steady-state error. Ideally, steady-state error should be 0, but a small error may be acceptable.

4. *Settling time:* The time difference between the initial time when the reference signal changes and the time at which the output signal reaches its steady-state value. For the DC motor response shown in figure 6.14, the settling time is 0.8 seconds. Ideally, settling time should also be small, and the system should reach the desired output value with few oscillations.

For the proportional controller for the DC motor, changing the value of the gain constant K_P affects the rise time and the overshoot, but a purely proportional controller will not get rid of the steady-state error. For optimal performance, we need to incorporate both derivative and integral components.

PID Controller

A generic version of the PID controller is shown in figure 6.15. The controller takes two signals as inputs: the reference signal r and the measured output y of the dynamical system to be controlled. Let the variable e denote the *error signal* capturing the difference between the reference signal r and the measured

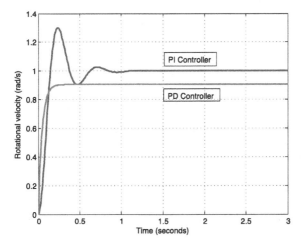

Figure 6.16: Output Response of the DC Motor with PD and PI Controllers

plant output y. Then the controller's output u, which is fed to the plant, is the sum of three terms:

- **Proportional term** $K_P\, e$: The contribution of this term is directly proportional to the current error. The constant K_P is called the *proportional gain*, and the controller scales the error by this factor.

- **Integral term** $K_I \int_0^t e(\tau)\, d\tau$: Note that the integral of the error signal up to time t gives the cumulative error up to time t, and thus the contribution of this term accounts for the cumulative error so far. The constant K_I is called the *integral gain*, and the controller scales the accumulated error by this factor.

- **Derivative term** $K_D\, \dot{e}$: The contribution of this term is correlated to the rate at which the error changes. The constant K_D is called the *derivative gain*, and the controller scales the rate of change of the error by this factor.

Note that the proportional component of the PID controller is stateless. The integral component maintains a state variable x_I that captures the accumulated error, and the rate of change of this state variable equals the error e. The derivative component has a state variable x_D that stores the error e, and the output of the derivative component is the first-order derivative of this state variable. In figure 6.15, using the standard convention, the Sum block is illustrated as a circle labeled with the symbol Σ. Such a component simply outputs the sum of its input signals, where the negative sign on an input signal indicates that the corresponding input should be subtracted. Some of the components may be missing in a specific control design. For instance, a P controller has only the proportional block, and a PI controller has only the proportional and the integral blocks. Both of these are also common in practice.

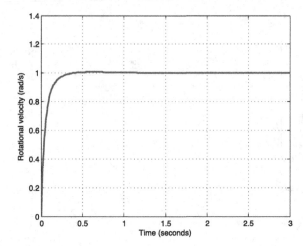

Figure 6.17: Output Response of the DC Motor with a PID Controller

PID Controller for the DC Motor

To understand how the control performance changes with the contributions of various terms, let us revisit the proportional controller for the DC motor (see figure 6.14). Higher values of the proportional gain K_P mean that the rise time will be smaller but with higher overshoot. Too high a value of K_P can cause large oscillations, delaying settling time.

A purely proportional controller has a steady-state error. The effect of the integral term gets rid of the steady-state error. Figure 6.16 shows a PI controller for the DC motor with $K_P = 100$ and $K_I = 200$ (the values of all the other parameters are unchanged). Note that the output now stabilizes close to the desired value 1 (the steady-state error is nonzero but very small), but the overshoot has increased to 30%, which may be unacceptable, and both rise time and settling time have also increased (compared to a purely proportional controller). Higher values of the integral gain K_I lead to better responsiveness but also contribute to higher overshoot.

To reduce the overshoot, we need to use the derivative component. Figure 6.16 also shows response to a PD controller for the DC motor with $K_P = 100$ and $K_D = 10$ (the values of all the other parameters are unchanged). There is no overshoot, but the steady-state error is still large as the output stabilizes at 0.9 rad/s.

Good performance on all metrics is obtained by a judicious use of all three components: figure 6.17 shows a PID controller for the DC motor with $K_P = 100$, $K_D = 10$, and $K_I = 200$. Now the steady-state error is insignificant, there is no overshoot, and the settling time is 0.4 seconds.

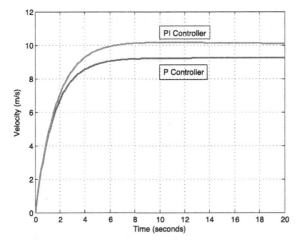

Figure 6.18: Velocity Response with a Cruise Controller

Cruise Controller

Let us revisit the model of a car in figure 6.3. The input to the cruise controller is the car output, namely, the velocity v, and the reference value r, that is, the desired speed set by the driver. The goal of the controller is to supply the input force F so that the velocity becomes equal to the reference value.

We begin the design by first using a proportional controller. The dynamics of the closed loop system is given by:

$$\dot{v} = (F - kv)/m; \quad F = K_P(r - v).$$

If we set the initial velocity v_0 to 0, the mass m to 100 kg, the coefficient of friction k to 50, the desired velocity r to 10 m/s, and the proportional gain K_P to 600, the resulting velocity response is shown in figure 6.18. There is a significant steady-state error: when the driver wants to increase the speed by 10 m/s, it increases by only 9 m/s. Notice that the settling time is about 10 seconds, which seems reasonable. Increasing the value of K_P will reduce the steady-state error but will also significantly decrease the settling time, which is likely to result in an uncomfortably high acceleration for the passengers.

To remove the steady-state error, we can add an integral component. Note that since there is no overshoot, we don't need a derivative component. For the same values of all the other parameters, the velocity response to the PI controller with $K_P = 600$ and $K_I = 40$ is also shown in figure 6.18. The step response in this case seems ideal: the velocity increases by 10 m/s is about 7 seconds, with no overshoot, and stays stable at the desired value with a small steady-state error.

Note that the basic design of the PI cruise controller did not explicitly depend on the model of the car, but the values of the gain constants K_P and K_I

were obtained by running simulations of the model for different choices. For a different model of the car, for instance, for the car on a graded road or for a car with different values of the parameters m and k, suitable values of the gain constants K_P and K_I need to be obtained.

Let us now revisit the design of the synchronous component `CruiseController` discussed in section 2.4.2. The component `ControlSpeed` can indeed be implemented as a PI controller discussed above. The two inputs *speed* and *cruiseSpeed* correspond to the observed velocity and the desired velocity, and its output corresponds to the force. However, we have a semantic mismatch between the discrete and continuous worlds: the components in section 2.4.2 interact with one another in a discrete manner, and variables range over natural numbers; while the models of the car and the PI controller are continuous-time components that process real-valued signals. This gap is usually bridged informally: an actual implementation of the PI controller samples the velocity only at discrete intervals, and the values it operates on are finite-precision numbers with rounding. This discretization can potentially introduce errors, which means that model-based analysis cannot replace extensive testing of the final system integrating all the components.

Exercise 6.20: Recall the linear pendulum model from exercise 6.10. We will use a *Proportional-Derivative* (PD) controller for the linearized model. The feedback control is given as $u = K_P \varphi + K_D \dot\varphi$, where we need to suitably choose the gain parameters K_P and K_D. Write the equations of the closed-loop system, which consists of the composition of the linearized model of the pendulum and the PD controller. For what values of the parameters K_P and K_D is the closed-loop system stable? Suppose the mass m and the length ℓ are such that $m\ell^2 = 1\,(kg.m^2)$ and $mg\ell = 1\,(N.m)$. Simulate the closed-loop system using MATLAB with different values of the gain parameters K_P and K_D. You can choose some suitable initial position and initial angular velocity. Experiment with different parameter values. Plot and discuss your results. ∎

Exercise 6.21*: Figure 6.18 shows the response of the car to a PI controller. Suppose we apply the same controller but now to the model of the car on a graded road as shown in figure 6.8. Consider the input signal $\bar\theta(t) = (\sin(t/5))/3$ (measured in radians) that models a sinusoidal variation in the grade of the road. For this input, plot the velocity of the car in response to the PI controller using the same parameters: the initial velocity v_0 is 0, the mass m is $100\,kg$, the coefficient of friction k is 50, the gravitational acceleration g is $9.8\,m/s^2$, the reference velocity r is $10\,m/s$, the proportional gain K_P is 600, and the integral gain K_I is 40. ∎

6.4   Analysis Techniques *

Given a model of a continuous-time component, which may include the plant model, the feedback controller, and constraints on the initial values and disturbances, we want to analyze the behavior of the system. We first discuss the traditional approach based on *numerical simulation* and then some constraint-based techniques for verifying stability and safety requirements.

6.4.1 Numerical Simulation

Consider a continuous-time component with state variables S and input variables I. The function f gives the rate of change of the state as a function of the state and input variables. Given an input signal $\bar{I}$ that assigns values to inputs as a function of time and an initial state s_0, the evolution of the state of the system, then, can be computed by solving the differential equation $\dot{S} = f(S, I)$ with the initial condition $\bar{S}(0) = s_0$. While for some specific forms of the function f, a closed-form solution for the state response $\bar{S}(t)$ at time t can be computed, a general method for computing this signal is to employ *numerical simulation*. For numerical simulation, the user provides a discretization step parameter Δ, and the simulator attempts to compute the values of the state at times $\Delta, 2\Delta, 3\Delta, \cdots$ that approximate the values of the desired response signal $\bar{S}(t)$ as closely as possible. The simulation algorithm samples the input signal $\bar{I}(t)$ only at times $0, \Delta, 2\Delta, 3\Delta, \cdots$ and the result of the simulation thus depends on the discrete sequence $u_0, u_1, u_2, \ldots$ of input values, where u_i is the value of the input signal $\bar{I}(t)$ at time $t = i\,\Delta$ for each i.

Euler's Method

Euler's method relies on the observation that the rate of change of the desired state signal $\bar{S}$, that is, $d\bar{S}/dt$, at time t, is simply the limit of the quantity $(\bar{S}(t+\Delta) - \bar{S}(t))/\Delta$ as the increment Δ goes to 0. As a result, for small values of Δ, it is natural to approximate the value of $\bar{S}(t+\Delta)$ by the following equation

$$\bar{S}(t + \Delta) = \bar{S}(t) + \Delta\,f(\bar{S}(t), \bar{I}(t)).$$

This approximation assumes that the rate of change of state is constant during the interval $[t, t + \Delta)$ and equals the rate of change at the beginning of the interval. The rate of change at the beginning of the interval is obtained by evaluating the function f using the values of the state and the input at time t. The change in the value of the state is obtained by multiplying this rate by the size Δ of the interval. Thus, given an initial value s_0 for the state and a sequence of values $u_0, u_1, \ldots$ for the input signal, the Euler's method for simulation of the differential equation $\dot{S} = f(S, I)$ computes the following sequence of values: for every $i \geq 0$,

$$s_{i+1} = s_i + \Delta\,f(s_i, u_i).$$

This sequence of values is linearly extrapolated to give the response signal that defines the state $\bar{S}(t)$ at every time $t \in \text{time}$: for every $i \geq 0$ and time value $t \in [i\,\Delta, i\,\Delta + \Delta)$, $\bar{S}(t) = s_i + (t - i\,\Delta)\,f(s_i, u_i)$.

Runge-Kutta Methods

Euler's method estimates the state at the end of an interval by assuming that the rate of change of state stays constant during the interval, and this rate is based only on the state at the beginning of the interval. A better approximation can be obtained if the estimated change in the state at the end of the interval is used to estimate a change in the derivative, and this is used to readjust the state estimate. Runge-Kutta methods comprise a popular class of numerical integration methods based on this idea. In particular, the *second-order Runge-Kutta method* computes the state s_{i+1} by the following calculation:

$$
\begin{aligned}
k_1 &= f(s_i, u_i), \\
k_2 &= f(s_i + \Delta k_1, u_{i+1}), \\
s_{i+1} &= s_i + \Delta (k_1 + k_2)/2.
\end{aligned}
$$

Given the current state s_i and input u_i, the first step computes k_1 to be the current rate of change. However, instead of setting the state s_{i+1} at the end of the interval to be $s_i + \Delta k_1$ as in Euler's method, it uses this estimated state to calculate the estimated rate of change k_2 at the end of the interval (the input value u_{i+1} at the end of the interval is used for this estimate). The third step calculates s_{i+1} assuming that the rate of change is constant during the interval but equals the average of the two values k_1 and k_2.

The higher order Runge-Kutta methods use the same basic idea but use estimates of derivatives at the midpoint as well as the endpoints to compute a weighted average. The most commonly used method in practice is the fourth-order Runge-Kutta method. It computes the state s_{i+1} by the following calculation:

$$
\begin{aligned}
k_1 &= f(s_i, u_i), \\
k_2 &= f(s_i + \Delta k_1/2, (u_i + u_{i+1})/2), \\
k_3 &= f(s_i + \Delta k_2/2, (u_i + u_{i+1})/2), \\
k_4 &= f(s_i + \Delta k_3, u_{i+1}), \\
s_{i+1} &= s_i + \Delta (k_1 + 2k_2 + 2k_3 + k_4)/6.
\end{aligned}
$$

Quality of Approximation

To understand how well these simulation techniques approximate the desired function, let us consider a single-dimensional linear differential equation with no inputs: $\dot{s} = s$ with the initial state $s_0 = 2$. We already know that the solution to this equation is given by the signal $\bar{s}(t) = 2e^t$.

For numerical simulation, let us choose the length Δ of the interval to be 0.1. Over the time interval of $[0, 5]$, the simulation plots resulting from both Euler's method and the second- Runge-Kutta method are shown in figure 6.19. Note that Euler's method introduces significant error that accumulates: the value of

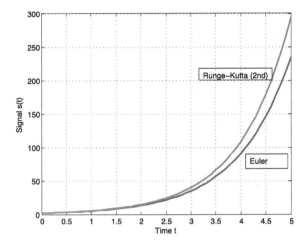

Figure 6.19: Alternative Simulation Algorithms

$\bar{s}(5)$ by Euler's method is 234.78, while the value of $\bar{s}(5)$ by the second-order Runge-Kutta method is 294.54. In this example, switching to the fourth-order Runge-Kutta method improves the accuracy slightly. Since the differences in the plots of second-order and fourth-order methods are not noticeable, figure 6.19 does not show the plot resulting from the fourth-order method, but it turns out that the value of $\bar{s}(5)$ by the fourth-order Runge-Kutta method is 296.83, which coincides with the quantity $2e^5$. Also note that if you were to simulate this differential equation using the standard solvers and plotting routines in MATLAB, the result is identical to the plot given by the fourth-order Runge-Kutta method.

The quality of approximation afforded by numerical simulation can be improved by reducing the step size Δ. Simulation tools automatically tune this parameter to keep the error small. In particular, adaptive techniques are also used to change the value of Δ dynamically, possibly at every step of simulation, to adjust to the current rate of change.

Exercise 6.22: Consider the single-dimensional linear differential equation $\dot{s} = s$. Suppose the initial state is s_0, and let Δ be the length of the interval used for numerical simulation. For Euler's method, find a closed-form formula for the state s_n after n steps of simulation as a function of s_0, Δ, and n. Verify that for $s_0 = 2$, $\Delta = 0.1$, and $n = 50$, s_{50} equals 234.78 (see figure 6.19). ■

Exercise 6.23: Consider the single-dimensional linear differential equation $\dot{s} = s$. Suppose the initial state is s_0, and let Δ be the length of the interval used for numerical simulation. For the second-order Runge-Kutta method of simulation, find a closed-form formula for the state s_n after n steps of simulation as a function of s_0, Δ, and n. Verify that for $s_0 = 2$, $\Delta = 0.1$, and $n = 50$, s_{50} equals 294.54 (see figure 6.19). ■

6.4.2   Barrier Certificates

In chapter 3, we focused on the safety verification problem for (discrete) transition systems: given a transition system and a set of safe states, is it the case that every reachable state of the system is safe? Now let us revisit this problem for continuous-time components.

Safety Verification

Consider a closed continuous-time component H with state variables S, initialization given by $Init$, and dynamics specified by $\dot{S} = f(S)$. For every initial state $s_0 \in [\![Init]\!]$, the system has a unique response signal, and the set of reachable states of the system is the set of states that appear along all these response signals:

$$Reach = \{ \ \overline{S}(t) \ | \ \overline{S}(0) \in [\![Init]\!] \text{ and } t \in \texttt{time} \ \}.$$

Given a state property φ, the safety verification problem is to decide whether every state in the set $Reach$ satisfies the property φ. For example, the property φ can assert that the magnitude of the error e between the reference signal and the system output is less than some constant δ, and a violation of this property indicates an unacceptable overshoot.

Numerical simulation is an effective technique to understand the behavior of a dynamical system starting from a specific initial state. When we know that the initial state belongs to a set, the simulation tool needs to choose different values for the initial state from the given set and run multiple simulations. Such an approach cannot be exhaustive, and thus we need some alternative techniques.

A natural approach would be to develop a *symbolic simulation* algorithm in the style of the symbolic reachability algorithm of section 3.4. However, this turns out to be challenging even for linear components. To illustrate the difficulty, let us consider a two-dimensional linear system whose dynamics is given by:

$$\dot{s}_1 = -7\,s_1 + s_2\,; \quad \dot{s}_2 = 8\,s_1 - 10\,s_2.$$

Suppose the initial set is the rectangle described by the formula

$$Init = 5 \le s_1 \le 6 \ \wedge \ -1 \le s_2 \le 1.$$

Figure 6.20 shows the initial set, and also the system responses when the initial state is chosen to be each of the four corner points. All these system responses converge to the origin, which is no coincidence: verify that both the eigenvalues of the state transition matrix are negative, and thus the system is asymptotically stable (see theorem 6.3).

It is evident from figure 6.20 that the set $Reach$ of reachable states of the system has a complex form: it is *not* convex, and although the initial set is described by linear constraints, linear constraints do not suffice to describe the set $Reach$. As a result, it is not possible to develop symbolic methods to compute the set of reachable states exactly.

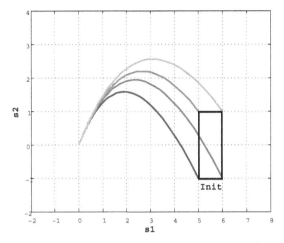

Figure 6.20: Reachable Set of a Continuous-time Component

Inductive Invariants

In chapter 3, we studied the principle of *inductive invariants* to prove safety requirements of (discrete) transition systems. To show that a property φ is an invariant of a transition system T, we find a property ψ such that (1) every state satisfying ψ satisfies φ, (2) the initial states of T satisfy ψ, and (3) if a state s satisfies ψ and (s, t) is a transition of T, then the state t is guaranteed to satisfy ψ. We describe an analogous method for establishing that a given state property is an invariant of a continuous-time system.

Let us revisit the dynamical system of figure 6.20. Suppose we want to establish that the property φ described by $-4 \leq s_2 \leq 4$ is an invariant of the system. In figure 6.21, we want to show that if the initial state belongs to the rectangle, then the execution stays between the two horizontal lines $s_2 = 4$ and $s_2 = -4$.

As in the case of proofs using inductive invariants, the first step is to identify the "strengthening" ψ. For continuous-time components, the desired strengthening is described by a function $\Psi : \mathbf{real}^n \mapsto \mathbf{real}$ mapping states to real numbers, that is, a real-valued expression over the state variables S. The set of states satisfying the equation $\Psi(S) = 0$ is called the *barrier*. The set of states satisfying the formula $\Psi(S) \leq 0$, that is, the set of states on or inside the barrier, is the analog of the inductive invariant. The barrier is chosen to satisfy three obligations. First, we need to establish that every state satisfying $\Psi(S) \leq 0$ also satisfies the desired safety property φ. Equivalently, if a state violates φ, then the value of Ψ must be positive. The second obligation is to show that every initial state s_0 satisfies $\Psi(s_0) \leq 0$. These two obligations imply that the barrier should be chosen so that it separates the initial states from the states violating the desired invariant property: all unsafe states are outside the barrier, and all initial states are inside the barrier.

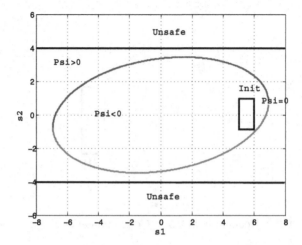

Figure 6.21: Invariant Verification Using a Barrier Certificate

For the example of figure 6.21, the barrier is described by the function

$$\Psi(s_1, s_2) \;=\; 7\,s_1^2 \;-\; 6\,s_1\,s_2 \;+\; 28\,s_2^2 \;-\; 320.$$

The equation $\Psi(s) = 0$ describes the ellipse shown in the figure. Initial states lie inside the ellipse, whereas unsafe states (that is, states above the line $s_2 = 4$ and below the line $s_2 = -4$) are outside the ellipse.

The third obligation in the discrete case corresponds to showing that the inductive invariant is preserved by system transitions. In case of continuous-time components, discrete transitions are replaced by continuous evolution of state over time as described by the differential equation $\dot{S} = f(S)$. A system response $\overline{S}(t)$ demonstrating a violation of the invariant must start in an initial state, which lies inside the barrier, and visit an unsafe state, which lies outside the barrier. The system response is a continuous and differentiable function, and hence it cannot "jump" across the barrier. It suffices to establish that the choice of the barrier with respect to the system dynamics f is such that a crossing of the barrier is impossible.

To understand the required condition, consider the example of figure 6.21. At a given state (s_1, s_2), the value of s_1 changes at the rate $f(s_1) = -7\,s_1 + s_2$, and the value of s_2 changes at the rate $f(s_2) = 8\,s_1 - 10\,s_2$. Thus, the state (s_1, s_2) flows, or evolves, along the vector $(-7\,s_1 + s_2, 8\,s_1 - 10\,s_2)$ given by the dynamics f. A key observation is that the value of Ψ always decreases along these flow directions. In particular, if the state lies on the barrier (that is, the boundary of the ellipse), this vector field points inward. This implies that if the state of the system at time τ lies on the barrier, then the state of the system at time $\tau + dt$ must lie in the interior. It is clear that system executions cannot cross the barrier from inside to outside.

Lie Derivatives

The appropriate mathematical concept to formalize the requirement that the value of the function Ψ decreases along the flow direction is called the *Lie derivative*. The Lie derivative of a function described by the expression Ψ with respect to the vector field described by the dynamics f is denoted $\mathcal{L}_f \Psi$ and gives the rate of change of the value of the expression Ψ as a function of the state variables S as the state evolves according to the dynamics $\dot{S} = f(S)$. In other words, the Lie derivative $\mathcal{L}_f \Psi$ is the *directional derivative* of the function Ψ along the vector field f.

The Lie derivative of the function Ψ can be computed from *partial derivatives* of the function Ψ with respect to each coordinate of the state vector:

$$\mathcal{L}_f \Psi(S) = \sum_{s \in S} (\partial \Psi / \partial s) f(s).$$

In our example, the state vector has dimension 2:

$$
\begin{aligned}
\mathcal{L}_f \Psi(s_1, s_2) &= (\partial \Psi / \partial s_1) f(s_1) + (\partial \Psi / \partial s_2) f(s_2), \\
&= (14 s_1 - 6 s_2)(-7 s_1 + s_2) + (-6 s_1 + 56 s_2)(8 s_1 - 10 s_2), \\
&= -146 s_1^2 - 566 s_2^2 + 564 s_1 s_2.
\end{aligned}
$$

The third obligation for Ψ to be a barrier certificate is that the value of the function $\mathcal{L}_f \Psi$ should always be negative at all points belonging to the barrier. This ensures that for every state s on the barrier, $\Psi(s)$ decreases as the state s evolves according to the dynamics f.

Going back to our example, verify that the expression $-146 s_1^2 - 566 s_2^2 + 564 s_1 s_2$ always has a negative value not just for the states on the barrier.

A technical requirement for the argument outlined above is that the function described by the expression Ψ should be a *smooth* function. A smooth function is a function for which all derivatives, higher order derivatives as well as partial derivatives, exist. This ensures that the Lie derivative is well defined. In the example of figure 6.21, the function Ψ is a quadratic function and is smooth. Every polynomial function is a smooth function. A discontinuous function is not smooth, and this means that the barrier separating the initial states and the bad states cannot be a rectangle (or a polyhedron).

The proof technique for establishing invariants using barrier certificates is summarized below.

Theorem 6.6 [Safety Verification using Barrier Certificate] *Let H be a closed continuous-time component with state variables S, initialization Init, and dynamics given by $\dot{S} = f(S)$. To show that a state property φ is invariant in* *all reachable states of the system H, it suffices to find a smooth real-valued expression Ψ over the state variables, called a* barrier certificate, *such that (1) if Init(S) holds, then $\Psi(S) \leq 0$; (2) if $\varphi(S)$ does not hold, then $\Psi(S) > 0$; and (3) the Lie derivative of Ψ with respect to the vector field f is negative for all states belonging to the barrier: if $\Psi(S) = 0$ then $(\mathcal{L}_f \Psi)(S) < 0$.*

Proof. Let H be a closed continuous-time component with state variables S, initialization *Init*, and dynamics given by $\dot{S} = f(S)$. Let the real-valued expression Ψ over the state variables be a barrier certificate. That is, the value Ψ is non-positive in all initial states, the value of Ψ is positive in all states that violate φ, and the Lie derivative of Ψ with respect to the vector field f is negative for all states for which Ψ is 0.

We want to prove that every reachable state of the system H satisfies the property φ. The proof is by contradiction. Assume that there is a reachable state that does not satisfy φ. Then there exists a system response signal $\overline{S}$ and a time $t^* \in \texttt{time}$ such that $\overline{S}(0)$ is an initial state and $\overline{S}(t^*)$ does not satisfy the property φ. Let us define the function $\overline{\Psi} : \texttt{time} \mapsto \texttt{real}$ such that $\overline{\Psi}(t) = \Psi(\overline{S}(t))$. The function $\overline{\Psi}$ is a continuous function as it is a composition of two continuous functions.

Since the value of Ψ in every initial state is non-positive, we know that $\overline{\Psi}(0) \leq 0$. Since the value of Ψ in every unsafe state is positive, we know that $\overline{\Psi}(t^*) > 0$. From continuity of $\overline{\Psi}$, it follows that there exists a time τ such that $\overline{\Psi}(\tau) = 0$ and $\overline{\Psi}(t) > 0$ for all times t such that $\tau < t \leq t^*$. By definition, $(d/dt)\overline{\Psi}(\tau)$ equals the Lie derivative $(\mathcal{L}_f \Psi)$ evaluated at the state $s_\tau = \overline{S}(\tau)$. Since $\overline{\Psi}(\tau) = 0$, we know that $\Psi(s_\tau) = 0$, and by assumption the value of the Lie derivative $\mathcal{L}_f \Psi$ at the state s_τ on the barrier, must be negative.

For the continuous function $\overline{\Psi}$, the conditions (1) $\overline{\Psi}(\tau) = 0$, (2) $\overline{\Psi}(t) > 0$ for all times t such that $\tau < t \leq t^*$, and (3) $(d/dt)\overline{\Psi}(\tau) < 0$, cannot hold all at once, and this leads to contradiction. ∎

Recipe for Choosing the Barrier Certificate

Theorem 6.6 gives us a general method for establishing safety of linear as well as non-linear continuous-time systems. The key to use this method effectively is to choose the function Ψ so that it satisfies all the necessary assumptions. We next outline a recipe for choosing this function for linear systems. Similar techniques are also used to establish stability of continuous-time systems and are known as *Lyapunov methods* in the literature.

Given an n-dimensional continuous-time system with dynamics $\dot{S} = AS$, we choose a *symmetric* n-dimensional matrix P and a constant k, let the function Ψ be defined as:

$$\Psi(S) = S^T P S + k.$$

Recall that a matrix is called symmetric if it equals its own transpose (that is, lower triangular entries equal upper triangular entries). The function Ψ

defined above is a quadratic expression over the state variables and is a smooth function. In this case, the dynamics of the state is defined by the matrix A. The Lie derivative of the expression Ψ with respect to this dynamics is given by:

$$
\begin{aligned}
(\mathcal{L}_A \Psi)(S) &= (d/dt)\,(S^T P S + k), \\
&= \dot{S}^T P S + S^T P \dot{S}, \\
&= S^T A^T P S + S^T P A S, \\
&= S^T (A^T P + P A)\, S.
\end{aligned}
$$

Let us define the matrix P' to be $A^T P + P A$. Observe that the matrix P' is also a symmetric matrix. The Lie derivative of Ψ is given by the quadratic expression $S^T P' S$. The third requirement for Ψ being a barrier certificate now can be related to a well-understood form of matrices from linear algebra: if the matrix P' is *negative definite*, then for all states S the quantity $S^T P' S$ is guaranteed to be negative. If all the eigenvalues of the matrix P' are negative, then it is guaranteed to satisfy this condition. Thus, given the dynamics matrix A, the coefficients of the desired symmetric matrix P are chosen so that the resulting matrix P' has negative eigenvalues. Once we fix the matrix P, we need to choose the coefficient k in the definition of the barrier certificate Ψ so that it separates the initial states from the unsafe states.

Let us revisit the two-dimensional continuous-time system of figure 6.20. The dynamics matrix is given by

$$
A = \begin{bmatrix} -7 & 1 \\ 8 & -10 \end{bmatrix}.
$$

Let us choose the symmetric matrix P to be

$$
P = \begin{bmatrix} 7 & -3 \\ -3 & 28 \end{bmatrix}.
$$

Observe that

$$
S^T P S = \begin{bmatrix} s_1 & s_2 \end{bmatrix} \begin{bmatrix} 7 & -3 \\ -3 & 28 \end{bmatrix} \begin{bmatrix} s_1 \\ s_2 \end{bmatrix} = 7 s_1^2 - 6 s_1 s_2 + 28 s_2^2.
$$

The coefficient k is chosen to be -320 so that the expression $S^T P S + k$ is negative in the initial rectangle and is positive in the unsafe states. The Lie derivative is of the form $S^T P' S$, where the symmetric matrix P' is given by:

$$
\begin{bmatrix} -7 & 8 \\ 1 & -10 \end{bmatrix} \begin{bmatrix} 7 & -3 \\ -3 & 28 \end{bmatrix} + \begin{bmatrix} 7 & -3 \\ -3 & 28 \end{bmatrix} \begin{bmatrix} -7 & 1 \\ 8 & -10 \end{bmatrix} = \begin{bmatrix} -146 & 282 \\ 282 & -566 \end{bmatrix}.
$$

Observe that the matrix P' is indeed symmetric. If we compute its eigenvalues, we get -5.2984 and -11.7016. This implies that the matrix P' is negative definite. This means that the expression $S^T P' S$, which equals $-146 s_1^2 - 566 s_2^2 + 564 s_1 s_2$, is always negative.

Exercise 6.24: For establishing safety of the system shown in figure 6.21, suppose the barrier function is $7\,s_1^2 - 6\,s_1\,s_2 + 28\,s_2^2 + k$, where k is a constant. What are the possible choices of k for which the resulting function is a barrier certificate? ■

Exercise 6.25 *: Suppose we replace the third condition for an expression Ψ to be a barrier certificate by the weaker condition "if $\Psi(S) = 0$ then $(\mathcal{L}_f\,\Psi)(S) \leq 0$" (that is, instead of requiring the Lie derivative to be negative for states on the barrier, it is required to be only non-positive). With this revised definition, from the existence of a barrier certificate, can we conclude that the property φ is an invariant? Prove or disprove your answer. ■

Exercise 6.26 *: This project concerns implementing a tool for symbolic reachability analysis of *one-dimensional* dynamical systems in MATLAB.

A closed interval of the set of real numbers is denoted $[a, b]$, where $a, b \in$ real and $a \leq b$ and corresponds the set of all real numbers between a and b. A state of a one-dimensional dynamical system is a real number. A set of such states can be naturally represented as a *union* of closed intervals. For example, $[0, 1] \cup [4, 5]$ is the set $\{x \in$ real $\mid 0 \leq x \leq 1 \ \vee \ 4 \leq x \leq 5\}$. For the purpose of this project, let us fix the data type for regions to be such unions of closed intervals, each such region is of the form $A = \bigcup_i [a_i, b_i]$.

Part (a): Implement a programming library for computing with regions. The first step is to choose a data structure to represent regions. Your representation should ensure that the intervals are mutually disjoint. Make sure that the empty set is also represented correctly. Explain succinctly and rigorously your implementation of the following operations: (1) union of regions (the operation Disj), (2) difference of regions (the operation Diff), (3) inclusion test (the function IsSubset(A, B) returns 1 exactly when the region A is a subset of the region B), (4) test for emptiness (the operation IsEmpty), (5) sum of regions (the function Sum(A, B) returns the representation for the set $\{x + y \mid x \in A, y \in B\}$), (6) product of a region and a scalar (the function Product(A, α) returns the representation of the set $\{\alpha x \mid x \in A\}$), and (7) square of a region (the operation Square(A) returns the representation of the set $\{x^2 \mid x \in A\}$).

Part (b): Consider a discretized representation of the dynamical system of the form $x_{k+1} = f(x_k, u_k)$, $k \geq 0$, where $x_k \in$ real is the state, u_k is the control input that belongs to the region U, and f expresses the dynamics of the system. The set of initial states is *Init*. Let $Reach_k$ be the set of reachable states at step k, that is, $Reach_k$ is the set of all states x such that there exists an execution of the system starting from some state $x_0 \in$ *Init* under some control inputs $u_0, u_1, ..., u_{k-1}$ from the region U and ending at the state $x_k = x$. The set *Reach* of all reachable states is $Reach = \bigcup_{k \geq 0} Reach_k$. Consider the breadth-first search algorithm to compute successive values of the regions $Reach_k$ up to a maximum number of iterations N. The algorithm should terminate as soon as there is no new state that can be added to *Reach* or when it iterates for N steps.

Implement the algorithm using the library you developed in part (a), provided that the sets *Init* and U are also regions, that is, unions of closed intervals, and the dynamics f can be described using the operations considered in part (a). Experiment your implementation using the following examples. In each case, report how the algorithm terminates and at which step, Output *Reach* (in its minimal form as a union of disjoint closed intervals).

(1) $x_{k+1} = -0.95x_k + u_k$, *Init* $= [1, 2]$, $U = [-0.1, 0.1]$, $N = 100$;
(2) $x_{k+1} = -0.96x_k + u_k$, *Init* $= [1, 2]$, $U = [-0.1, 0.1]$, $N = 100$;
(3) $x_{k+1} = -0.95x_k + u_k$, *Init* $= [1, 2]$, $U = [-0.2, 0.2]$, $N = 100$;
(4) $x_{k+1} = 0.5x_k^2 + u_k$, *Init* $= [1.8, 1.89]$, $U = [0, 0.1]$, $N = 40$;
(5) $x_{k+1} = 0.5x_k^2 + u_k$, *Init* $= [1.8, 1.9]$, $U = [0, 0.1]$, $N = 40$.

■

Bibliographic Notes

The core topics discussed in this chapter, namely, models of dynamical systems using differential equations, stability, linear systems, and analysis techniques using concepts from linear algebra, are studied in control theory for decades and appear in numerous textbooks. Our presentation is based on the textbook by Antsaklis and Michel [AM06], and some examples are also borrowed from [FPE02], [LV02], and [LS11].

For a detailed introduction to practical control design using PID controllers, see [AH95] (see also www.mathworks.com for tutorials on design of PID controllers using MATLAB).

The use of barrier certificates for verification of safety properties was introduced in [PJ04], and our discussion of this topic is based on the presentation in [Tab09] (see also [Pla10] and [TT09] for discussion on soundness, completeness, and variations of this proof technique).

The numerical calculations and the plots in this chapter are all obtained using the toolkit MATLAB (see [SH92] for an introduction to control design using MATLAB, and the tutorials at www.mathworks.com to learn how to use MATLAB).

7

Timed Model

In the synchronous model of computation, all components execute in lock-step, and the production of outputs by a component is synchronized with the reception of inputs. In the asynchronous model of computation, all processes execute at independent speeds, and there is an unspecified delay between the reception of inputs and the production of outputs by a process. Now we turn our attention to a *timed* model of computation where processes are not tightly synchronized to execute in a sequence of rounds but rely on the global physical time to achieve a loose form of synchronization. The timed model allows us to express phenomena such as "execute the task corresponding to sensing of temperature every 5ms," "the delay between the reception of an input value and the corresponding output response is between 2ms to 4ms," and "if an acknowledgment is not received within 4ms, resend."

7.1 Timed Processes

The formal model of computation for timed processes is a variation of the model of *asynchronous processes* from chapter 4. We will first illustrate the model with examples.

7.1.1 Timing-Based Light Switch

Consider a light switch that uses a single push-button, along with a built-in timer, to control a light bulb with two intensity levels. The switch is initially off. When it is pressed once, it turns on the light at a dim intensity, and if it is pressed twice in rapid succession, then the light is turned on at a bright intensity. Here, *rapid* means that the duration between the successive press events is less than one second. If the delay between the successive press events is more than one second, the second press event is interpreted as a command to switch the light off.

The system is modeled by the timed process `LightSwitch` shown in figure 7.1. It

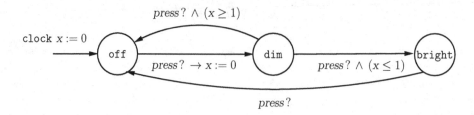

Figure 7.1: A Timed Model of a Light Switch

has an input channel *press* on which it receives events corresponding to pressing of the switch. The dynamics is illustrated using the extended-state machine notation. In this example, the mode can be either `off`, `dim`, or `bright`. The process uses a state variable x whose type is `clock`. A timed process has input, internal, and output actions just like an asynchronous process: an input action receives an input value on an input channel, an output action produces an output value on an output channel, and an internal action updates only the state variables and does not involve any input or output channels. During each such action, the clock variables are tested and updated in the same way as other state variables. The distinguishing new feature of the model is that a timed process also has *timed actions* that capture elapse of time. During a timed action of duration δ, which may be any positive real number, the value of each clock variable is incremented by the amount δ. A state variable, such as the variable *mode* for the process `LightSwitch`, of a type other than `clock` is called a *discrete* variable. A discrete state variable stays unchanged during a timed action.

For the process `LightSwitch`, initially the mode is `off`, the clock variable x is 0, and the process is waiting for the input event *press*. Waiting for a time period of δ_1 is modeled by a timed action of duration δ_1. After such an action, the mode is still `off`, but the value of the clock variable x equals δ_1. When the input event *press* is received, the process updates the mode to `dim` and resets the clock variable x to 0. As the process waits in the mode `dim`, the value of the clock variable x captures the time elapsed since the time instance when the mode-switch from `off` to `dim` occurred. Waiting in the mode `dim` for a total duration of length δ_2 corresponds to a timed action of duration δ_2, and the value of the clock variable x after such a timed action equals δ_2. When the subsequent input event `press` occurs, the value of the clock variable x is used to decide if the process updates the mode to `bright` or to `off`, and this is captured by the conjuncts $(x \leq 1)$ and $(x \geq 1)$ in the guard conditions of the two mode-switches out of the mode `dim`. Note that if the value of x is exactly 1 (that is, the duration between the two successive *press* events is one second), both these mode-switches are enabled, and thus the model behaves nondeterministically switching either to the mode `off` or to the mode `bright`. When the process is in the mode `bright`, it switches back to the initial mode `off` whenever it

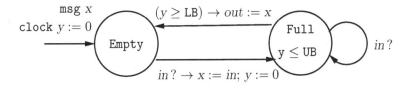

Figure 7.2: A Timed Buffer with a Bounded Delay

receives the next input event.

One possible execution of the process is shown below, where each state is specified by listing the mode and the value of the clock variable x:

$$(\texttt{off}, 0) \xrightarrow{2.3} (\texttt{off}, 2.3) \xrightarrow{press?} (\texttt{dim}, 0) \xrightarrow{0.2} (\texttt{dim}, 0.2) \xrightarrow{0.5}$$
$$(\texttt{dim}, 0.7) \xrightarrow{press?} (\texttt{bright}, 0.7) \xrightarrow{3.0} (\texttt{bright}, 3.7) \xrightarrow{press?} (\texttt{off}, 3.7).$$

Note that during an execution, two timed actions may follow one another. In
such a case, the effect of a timed action of duration δ_1 immediately followed by
a timed action of duration δ_2 is identical to a single timed action of duration
$\delta_1 + \delta_2$.

7.1.2 Buffer with a Bounded Delay

As a second example, let us consider a timed buffer of capacity 1 with an input
channel *in* and an output channel *out*, both of type msg. Whenever an input
value v is supplied to the buffer, it is stored in the internal discrete state variable
x. Now the buffer becomes full, and it simply ignores (or loses) further inputs
until it gets a chance to output the stored value on the output channel. The
timing assumption is that the delay between the reception of an input value and
the transmission of the corresponding output value is at least LB and at most UB
time units, where the constants LB and UB capture the lower and upper bound,
respectively, on the delay. This form of lower and upper bounds on delays is a
commonly occurring pattern for timed systems.

The timed process TimedBuf shown in figure 7.2 captures the desired timed
behavior using one clock variable y. The mode of the state machine indicates
whether the buffer is empty or full. Initially the mode is Empty. When the input
is received on the channel *in*, the message value is stored in the state variable
x, the mode is updated to Full, and the clock y is set to 0. As time elapses
while the mode of the process equals Full, the value of the clock y captures
the duration of time the process has been waiting in this mode. Input events
received in the mode Full do not change the buffer state, and this is modeled
by the mode-switch corresponding to the self-loop on the mode Full. The
mode-switch corresponding to the output actions is guarded with the condition

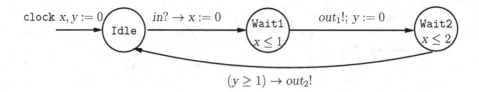

Figure 7.3: A Timed Process with Two Clocks

($y \geq$ LB) and thus captures the assumption that the buffer can issue the message on its output channel only after the lower bound LB on the delay.

The assumption concerning the *upper bound*, namely, that the process is guaranteed to produce the output within UB time units of receiving an input, is captured by the annotation $y \leq$ UB associated with the mode Full. If the mode is Full and the value of the clock y equals δ, then a timed action of duration δ' is allowed only if the constraint $y \leq$ UB holds throughout the transition as the value of the clock y keeps increasing with time, that is, only if the condition $\delta + \delta' \leq$ UB holds. The process and its environment synchronize on the passage of time during a timed action. The process wants to issue the output before the clock y reaches UB and thus is willing to let time elapse only up to a certain limit. The condition $y \leq$ UB associated with the mode Full is called a *clock invariant*. The clock invariant associated with a mode and the guard condition associated with the switches out of that mode together ensure lower and upper bounds on delays.

In general, each mode has an associated clock invariant, which is an expression over the clock variables, along with possibly other discrete state variables, of the process. When the clock invariant associated with a mode is the constant 1, it means that there is no upper bound on how long the process can wait in this mode. In such a case, we omit the annotation, as is the case for the mode Empty of the process TimedBuf and also for all the modes of the process LightSwitch of figure 7.1.

7.1.3   Multiple Clocks

As an example of a timed process that uses two clocks, consider the process shown in figure 7.3 with an input channel *in* and the output channels out_1 and out_2. If an input event on the channel *in* happens at time t, then the process responds by producing an output event on the channel out_1 at time t_1 followed by an output event on the channel out_2 at time t_2 such that (1) the delay $t_1 - t$ is at most 1, (2) the delay $t_2 - t$ is at most 2, and (3) the delay $t_2 - t_1$ is at least 1. Thus, the output event on the channel out_2 is constrained to occur within an interval that depends on the timing of the preceding input event on the channel *in* as well as the output event on the channel out_1. The process does not accept inputs on channel *in* until it has issued both output events.

$$\begin{array}{|l|}
\hline
\texttt{nat } x_1 := 0; \; x_2 := 0 \\
\texttt{clock } y_1 := 0; \; y_2 := 0 \\
\hline
CI: \; (y_1 \leq 2) \wedge (y_2 \leq 2) \\
A_1: \; (y_1 \geq 1) \rightarrow \{x_1 := x_1 + 1; \; y_1 := 0\} \\
A_2: \; (y_2 \geq 1) \rightarrow \{x_2 := x_2 + 1; \; y_2 := 0\} \\
\hline
\end{array}$$

Figure 7.4: Timed Process `TimedInc` with Parallel Increments

The desired constraints are expressed using two clock variables x and y. Initially the mode is `Idle`, and both clock variables equal 0. Waiting in the mode `Idle` for a duration of time δ is modeled by a timed action of duration δ, which updates both clock variables to the value δ. When an input event occurs, the process sets the clock x to 0, the clock y remains δ, and the mode is updated to `Wait1`. Waiting in the mode `Wait1` for a duration of time δ' increments the clock x to δ' and the clock y to $\delta + \delta'$. Such a timed action is allowed only as long as the value of x does not exceed 1 due to the clock-invariant $x \leq 1$ associated with the mode `Wait1`. The output event out_1 can occur at any time instance before the value of the clock x exceeds 1. At this point, the process switches to the mode `Wait2`, the clock y is reset to 0, and the value of the clock x indicates the time spent in the mode `Wait1`. When the mode of the process equals `Wait2`, both clocks increase with time, with the value of the clock x capturing the time elapsed since the occurrence of the input event and the value of the clock y capturing the time elapsed since the occurrence of the output event on the channel out_1. The clock-invariant $x \leq 2$ associated with the mode `Wait2` and the guard condition $(y \geq 1)$ associated with the mode-switch from the mode `Wait2` to the mode `Idle` capture the desired timing constraints on the occurrence of the output event on the channel out_2. Below is a sample execution of the process, where each state is specified by listing the mode, the value of the clock variable x, and the value of the clock variable y:

$$(\texttt{Idle}, 0, 0) \xrightarrow{5.7} ,(\texttt{Idle}, 5.7, 5.7) \xrightarrow{in?} (\texttt{Wait1}, 0, 5.7) \xrightarrow{0.6} (\texttt{Wait1}, 0.6, 6.3) \xrightarrow{out_1!}$$
$$(\texttt{Wait2}, 0.6, 0) \xrightarrow{0.5} (\texttt{Wait2}, 1.1, 0.5) \xrightarrow{0.8} (\texttt{Wait2}, 1.9, 1.3) \xrightarrow{out_2!} (\texttt{Idle}, 1.9, 1.3).$$

Note that if we restrict the process to use only one clock, then the desired timing constraints cannot be expressed accurately.

As another example of a timed process, consider the process `TimedInc` shown in figure 7.4, which is a modified version of the asynchronous process `AsyncInc` of figure 4.2. The process `TimedInc` has two discrete state variables x_1 and x_2, both of which are initialized to 0 and are incremented by the internal tasks A_1 and A_2, respectively. However, unlike the asynchronous process `AsyncInc`, the order in which these two tasks execute is no longer completely unconstrained.

The time delay between the successive executions of the task A_1 is at least one and at most two time units. This constraint is specified using the clock variable y_1: initially its value is 0, the task A_1 can be executed only when the guard $(y_1 \geq 1)$ is satisfied, its execution resets the clock y_1 to 0, and the clock-invariant has a conjunct $(y_1 \leq 2)$, which ensures that the task A_1 must get executed before more than two time units elapse since its last execution. The clock y_2 is used in a similar manner to ensure that the time delay between the successive executions of the task A_2 is at least one and at most two time units.

Although the two tasks A_1 and A_2 do not have any variables in common, the fact that the two clocks y_1 and y_2 need to increase by the same amount of duration during a timed action constrains the relative frequencies at which the two tasks execute. In particular, consider an execution of the process in which the task A_1 executes twice without executing the task A_2. Such an execution then has the form below (the state lists the variables x_1, y_1, x_2, and y_2 in that order):

$$(0,0,0,0) \xrightarrow{\delta_1} (0,\delta_1,0,\delta_1) \xrightarrow{A_1} (1,0,0,\delta_1) \xrightarrow{\delta_2} (1,\delta_2,0,\delta_1+\delta_2) \xrightarrow{A_1} (2,0,0,\delta_1+\delta_2)$$

Based on the guard of the task A_1, we know that $\delta_1 \geq 1$ and $\delta_2 \geq 1$, and based on the clock-invariant, we can conclude that $\delta_1 + \delta_2 \leq 2$. These constraints can be satisfied only if $\delta_1 = \delta_2 = 1$, and thus in the state after executing the task A_1 twice, the clock y_2 must be 2. It means that in this state, time cannot elapse, and the only possible action is the execution of the task A_2. Thus, the variable x_2 must be incremented at least once before the variable x_1 is incremented thrice.

7.1.4   Formal Model

Recall the definition of an asynchronous process from section 4.1: an asynchronous process P has:

1. a finite set I of typed input channels defining the set of inputs of the form $x\,?\,v$ with $x \in I$ and a value v for x;

2. a finite set O of typed output channels defining the set of outputs of the form $y\,!\,v$ with $y \in O$ and a value v for y;

3. a finite set S of typed state variables defining the set Q_S of states;

4. an initialization $Init$ defining the set $[\![Init]\!] \subseteq Q_S$ of initial states;

5. for each input channel x, a set $\mathcal{A}_x$ of input tasks, each described by a guard condition over S and an update from the read-set $S \cup \{x\}$ to the write-set S defining a set of input actions of the form $s \xrightarrow{x\,?\,v} t$;

6. for each output channel y, a set $\mathcal{A}_y$ of output tasks, each described by a guard condition over S and an update from the read-set S to the write-set $S \cup \{y\}$ defining a set of output actions of the form $s \xrightarrow{y\,!\,v} t$; and

7. a set $\mathcal{A}$ of internal tasks, each described by a guard condition over S and an update from the read-set S to the write-set S defining a set of internal actions of the form $s \xrightarrow{\varepsilon} t$.

We can define the formal model for timed processes as an extension of the above definition of asynchronous processes. The notions of input, output, and state variables, and input, output, and internal tasks and actions stay unchanged. The additional notion is that of a clock invariant, which is a Boolean expression over state variables, and is used to define timed actions of duration δ. Given a state s, which is a valuation of all the state variables, and a positive real number δ, let $s + \delta$ denote the state that assigns the value $s(x) + \delta$ to every clock variable x and the value $s(y)$ to every discrete state variable y. The state resulting from the timed action of duration δ starting in a state s is the state $s + \delta$. Such a timed action is allowed only if the Boolean condition specified by the clock invariant holds in all the states encountered during this interval. The definition of a timed process is now summarized below.

TIMED PROCESS

A *timed process* TP consists of:

- an asynchronous process P, where some of its state variables can be of type `clock`; and

- a *clock invariant* CI, which is a Boolean expression over the state variables S.

Inputs, outputs, states, initial states, internal actions, input actions, and output actions of the timed process TP are the same as that of the asynchronous process P. Given a state s and a real-valued time $\delta > 0$, $s \xrightarrow{\delta} s + \delta$ is a *timed action* of TP if the state $s + t$ satisfies the expression CI for all values $0 \le t \le \delta$.

For the timed process `TimedBuf` of figure 7.2, the various components are listed below:

- it has a single input channel *in* of type `msg`;

- it has a single output channel *out* of type `msg`;

- it has a (discrete) state variables *mode* of enumerated type {`Empty`, `Full`} and x of type `msg`, and a variable y of type `clock`;

- the initial value of the clock variable y is 0, the initial value of the mode variable *mode* is `Empty`, and the initial value of the variable x is unconstrained;

- there is a single input task for processing the input channel *in*, its guard condition is 1 (that is, the task is always enabled), and the update corresponding to the extended-state machine of figure 7.2 can be equivalently written as:

if $(mode = \texttt{Empty})$ then { $mode := \texttt{Full}$; $x := in$};

- there is a single output task for producing outputs on the channel *out*, has the guard condition $(mode = \texttt{Full}) \wedge y \geq \texttt{LB}$, and the corresponding update code is $out := x$;

- it has no internal tasks;

- the clock invariant *CI* is given by the expression

$$(mode = \texttt{Full}) \rightarrow (y \leq \texttt{UB}).$$

Note that the clock invariant puts no constraints when the mode is $\texttt{Empty}$. The translation from the extended-state machine notation to the formal definition of timed processes can be automated.

The definition of a timed action requires that starting in a state s, a timed action of duration δ is possible if the state $s + t$ satisfies the clock invariant at every time t during the interval $[0, \delta]$. Typically the expressions used in clock invariants are *convex* functions of values of the clock variables, so it suffices to check that the starting state s and the final state $s + \delta$ both satisfy the clock invariant.

As in the case of asynchronous processes, the operational semantics of a timed process can be captured by defining its executions. An execution starts in an initial state and proceeds by executing an input action, an output action, an internal action, or a timed action at every step. Note that input, output, and internal actions are interleaved as in an asynchronous process. However, during a timed action, the clocks belonging to different processes all increase together, reflecting the passage of the same global time, and thus a timed action is executed synchronously. This is why sometimes this model is called a *partially synchronous model*.

Exercise 7.1: Consider a timed process with an input event x and two output events y and z. Whenever the process receives an input event on the channel x, it issues output events on the channels y and z such that (1) the time delay between x? and y! is between two and four units, (2) the time delay between x? and z! is between three and five units, and (3) while the process is waiting to issue its outputs, any additional input events are ignored. Design a timed state machine that exactly models this description. ∎

Exercise 7.2: Consider a timed process with two input events x and y and an output event z. Initially, the process is waiting to receive an input event x?. If this event occurs at time t, then the process waits to receive an input on the channel y. If the event y? occurs before time $t + 2$ or does not occur before time $t + 5$, then the process simply returns to the initial state, and if the event y? is received at some time t' between times $t + 2$ and $t + 5$, then the process issues an output event on z at some time between times $t' + 1$ and $t + 6$ and returns to the initial state. Unexpected input events (e.g., the event y in the initial mode) are ignored. Design a timed state machine that exactly models this description. ■

Exercise 7.3: Consider an asynchronous OR gate with Boolean input variables x and y and Boolean output variable z. Assume that initially all the variables have value 0. The event x? denotes toggling of the input wire x, and similarly the event y? denotes the toggling of the input wire y. The gate can change its output by issuing the event z!. The desired timing behavior is specified by the following rules:

1. When an input variable changes at time t_1, if this change warrants a change in the output (according to the standard logic of OR gate), then the output should be issued at time t_2 such that the delay $t_2 - t_1$ is between two and four time units (unless the inputs change again during the interval from t_1 to t_2; if so, see the rules below).

2. While a change in the output is pending in the interval $[t_1, t_2]$, if one of the input variables changes again, but this change is consistent with the output change about to happen at time t_2, then the output should change as scheduled.

3. While a change in the output is pending in the interval $[t_1, t_2]$, if one of the input variables changes at time t in a manner so as to make the change in output inconsistent with the revised inputs, then the behavior depends on the relative difference $t - t_1$: if this difference is less than 1, then the pending output change is canceled; if it is more than 1, then the output change will occur as scheduled at time t_2, and at that time, another output event is scheduled with a delay of two to four time units.

Based on this description, design a timed process (as an extended-state machine with one clock variable) that models the OR gate. The process should be input-enabled: it should allow input events to happen at all times. ■

7.1.5 Timed Process Composition

Timed processes can be composed together using block diagrams. Operations such as input-output variable renaming and output hiding are defined in the usual manner. Let us consider the operation of composing timed processes. To compose two timed processes, we first compose the corresponding asynchronous processes using the composition operation for asynchronous processes as described in section 4.1. The clock invariant for the composed process is simply the *conjunction* of the clock invariants of the component processes.

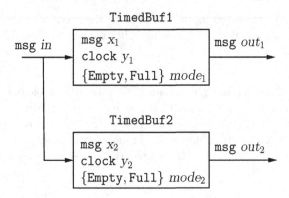

Figure 7.5: Composition of Two Instances of the Process `TimedBuf`

TIMED PROCESS COMPOSITION

For two timed processes $TP_1 = (P_1, CI_1)$ and $TP_2 = (P_2, CI_2)$, such that the output channels of the two processes are disjoint, the *parallel composition* $TP_1 \,|\, TP_2$ is the timed process whose asynchronous process is $P_1 \,|\, P_2$ and whose clock invariant is $CI_1 \wedge CI_2$.

Thus, the internal, input, and output actions of the composite process are obtained from the corresponding actions of the component processes using the asynchronous composition. The conjunction of the clock invariants means that a timed action of duration δ is possible in the composite process only if it is acceptable for each component process to wait for a duration of δ. For states s_1 and s_2 of the processes TP_1 and TP_2, respectively, and a time duration $\delta > 0$, $(s_1, s_2) \xrightarrow{\delta} (s_1 + \delta, s_2 + \delta)$ is a timed action of the composite process $TP_1 \,|\, TP_2$ exactly when $s_1 \xrightarrow{\delta} s_1 + \delta$ is a timed action of TP_1 and $s_2 \xrightarrow{\delta} s_2 + \delta$ is a timed action of TP_2.

Product of Timed State Machines

To understand how the composition works, let us describe the composition of two instances of the timed process `TimedBuf` connected in parallel with a common input channel. Figure 7.5 shows two instances with their variables renamed appropriately. The process `TimedBuf1` responds to the input event on the channel *in* by producing an output event on the channel out_1 after a delay of at least LB_1 and at most UB_1 time units, and the process `TimedBuf2` responds to the input event on the channel *in* by producing an output event on the channel out_2 after a delay of at least LB_2 and at most UB_2 time units. Instead of compiling the extended-state machine of figure 7.2 for each process into a description consisting of tasks and then computing the tasks for the composite process using the composition operation, let us construct the extended-state

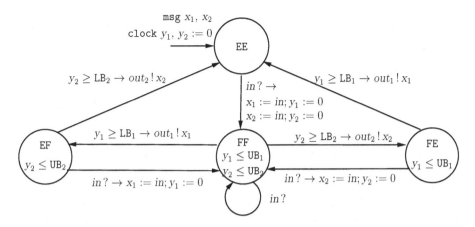

Figure 7.6: State Machine for Composition of Two `TimedBuf` Processes

machine capturing the behavior of the composition by taking a *product* of the state machines for the component processes.

The behavior of the parallel composition of the two processes is captured by the extended-state machine shown in figure 7.6. Since each component has two possible modes, the composite process has four modes. The initial mode is EE indicating that both component processes start in the mode `Empty`. When an input on the channel *in* is processed, the mode changes to FF (that is, the variable $mode_1$ is `Full` and the variable $mode_2$ is `Full`). The variables x_1 and y_1 are updated according to the input action of the first process, and the variables x_2 and y_2 are updated according to the input action of the second process.

The mode FF corresponds to the case when each process is in the mode `Full`. The clock-invariant of this mode is the conjunction $(y_1 \leq UB_1 \wedge y_2 \leq UB_2)$. Thus, the composite process can wait in this mode only as long as the clock y_1 does not exceed UB_1 and the clock y_2 does not exceed UB_2. This conjunctive constraint reflects the synchronization of the two component processes on timed actions. The mode can change in two ways depending on which component process produces the output first. If the second component issues its output on the channel out_2, the mode changes to FE (that is, the variable $mode_1$ is `Full` and the variable $mode_2$ is `Empty`). This switch is guarded by the condition $(y_2 \geq LB_2)$, corresponding to the guard of the output action of the second component, and the variables of the first component stay unchanged during this switch. The clock-invariant in the mode FE is $(y_1 \leq UB_1)$ since the second component does not impose any constraints on how long the process can wait in this mode. In the mode FE, if the first component produces its output, then the mode changes to EE, and if an input event is received, then the mode switches back to FF.

Note that the values of the parameters that capture lower and upper bounds

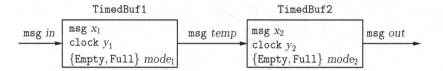

Figure 7.7: Composition of Two Instances of `TimedBuf` Processes in Series

on delays determine the possible executions of this composite process. For example, if the upper bound UB_1 is strictly smaller than the lower bound LB_2, then in response to an initial input event, the first component is guaranteed to produce its output *before* the second component produces its output. That is, a mode-switch from the mode EE to the mode FF is guaranteed to be followed by the mode-switch to the mode EF since the guard condition $(y_2 \geq LB_2)$ cannot be satisfied before the clock invariant $(y_1 \leq UB_1)$ gets violated. Similarly, if the upper bound UB_2 is strictly smaller than the lower bound LB_1, then a mode-switch from the mode EE to the mode FF is guaranteed to be followed by the mode-switch to the mode FE corresponding to the output by the second component. If the intervals $[LB_1, UB_1]$ and $[LB_2, UB_2]$ overlap, then following a mode-switch from the mode EE to the mode FF, both scenarios, the first process producing its output before the second, and vice versa, are feasible. The goal of the timing analysis, to be discussed in section 7.3, is to discover which event sequences are consistent with the timing constraints.

Exercise 7.4: Figure 7.6 shows the *product* extended-state machine that captures the behavior of the composition of two instances of the timed process `TimedBuf` shown in figure 7.5. Now consider the composition of two instances of the timed process `TimedBuf` connected in series as shown in figure 7.7. Draw the extended-state machine with four modes and two clocks that captures the behavior of this composite process. ∎

Exercise 7.5: For the timed process `TimedInc` of figure 7.4, argue that both the properties $(x_1 \leq 2x_2 + 2)$ and $(x_2 \leq 2x_1 + 2)$ are invariants of the system. ∎

7.1.6 Modeling Imperfect Clocks *

In our model of timed processes, the value of a clock variable increases to accurately capture the amount of time elapsed. Now consider a timed process P that has a clock variable x that can measure time only imperfectly. The error is specified using a per-unit drift, say 0.01. This means that if the value of the clock x increases by 1, then the actual time elapsed may be any value in the interval $[0.99, 1.01]$. In general, if the drift is ϵ and the process P resets the clock x to 0 at time t and finds the constraint $LB \leq x \leq UB$ to be satisfied at a later time instance t', then it can conclude that the elapsed time $t' - t$ is at least $LB(1 - \epsilon)$ and at most $UB(1 + \epsilon)$.

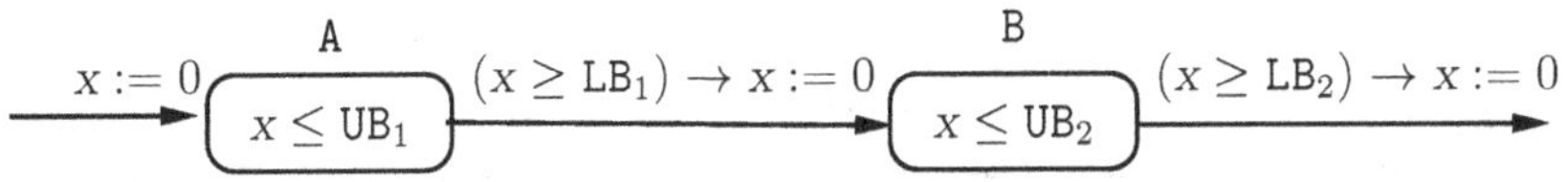

$$x := 0 \quad \xrightarrow{\text{A}} \quad \boxed{x \leq \text{UB}_1} \quad \xrightarrow{(x \geq \text{LB}_1) \rightarrow x := 0} \quad \boxed{x \leq \text{UB}_2}^{\text{B}} \quad \xrightarrow{(x \geq \text{LB}_2) \rightarrow x := 0}$$

Timed process P with an imperfect clock x with drift ϵ

$$y := 0 \quad \xrightarrow{\text{A}} \quad \boxed{y \leq \text{UB}_1^{+\epsilon}} \quad \xrightarrow{(y \geq \text{LB}_1^{-\epsilon}) \rightarrow y := 0} \quad \boxed{y \leq \text{UB}_2^{+\epsilon}}^{\text{B}} \quad \xrightarrow{(y \geq \text{LB}_2^{-\epsilon}) \rightarrow y := 0}$$

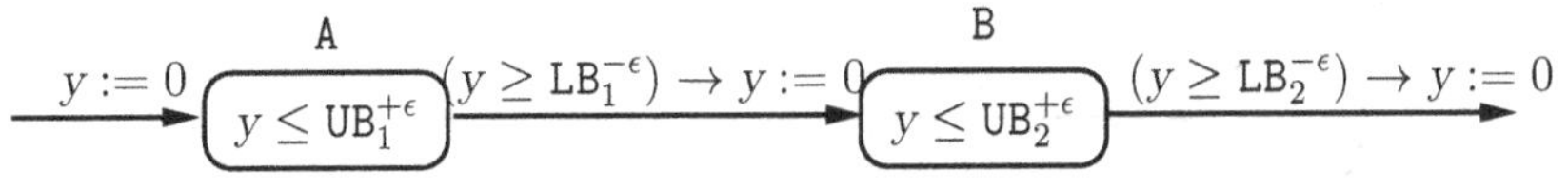

Equivalent timed process P' with (perfect) clock y

Figure 7.8: Simulating an Imperfect Clock with a Drift by a Perfect Clock

Although our basic model does not allow modeling of such imperfect clocks explicitly, we can capture the resulting errors by changing the timing constraints. As an example, consider the timed process P shown in figure 7.8 that measures time using an imperfect clock x with drift ϵ. The clock-invariants associated with the modes and the guards on the mode-switches imply that the process spends between LB_1 and UB_1 time units in the mode A and between LB_2 and UB_2 time units in the mode B, as measured according to its imperfect clock x. We can capture the same behavior by the timed process P' that uses a perfect clock y. The clock-invariants associated with the modes and the guards of the mode-switches are modified by scaling upper bounds upward by a factor of ϵ and scaling lower bounds downward by a factor of ϵ. In figure 7.8, we abbreviate $\text{LB}(1 - \epsilon)$ by $\text{LB}^{-\epsilon}$ and $\text{UB}(1 + \epsilon)$ by $\text{UB}^{+\epsilon}$. The timing constraints of the process P' imply that the process spends between $\text{LB}_1(1-\epsilon)$ and $\text{UB}_1(1+\epsilon)$ time units in the mode A and between $\text{LB}_2(1 - \epsilon)$ and $\text{UB}_2(1 + \epsilon)$ time units in the mode B. As a result, the possible interactions of the process P with an imperfect clock and that of the process P' with a perfect clock with other processes are the same.

Exercise 7.6: We have argued that, in figure 7.8, the timing behavior of the timed process P with an imperfect clock x with drift ϵ and the timed process P' with a perfect clock y with modified clock-invariants and guards are equivalent. Suppose in figure 7.8 we remove the resetting of the clock to 0 on the mode-switch from the mode A to the mode B. That is, consider the processes P and P' with the updates $x := 0$ and $y := 0$, respectively, on the mode-switch from the mode A to the mode B omitted. Are the resulting processes still equivalent in terms of the relative timings of when the mode-switches can occur? ∎

7.2 Timing-Based Protocols

In this section, we illustrate the formal design of timed systems using three case studies: achieving distributed coordination by relying on timing assumptions,

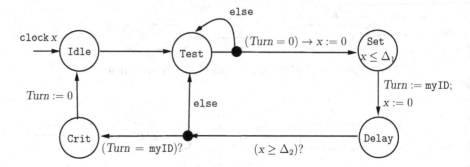

Figure 7.9: Timing-based Mutual Exclusion

achieving reliable transmission in presence of imperfect timing measurements, and design of a pacemaker to deliver timely pulse to the heart to maintain its rhythmic pulsation.

7.2.1   Timing-Based Distributed Coordination

In chapter 4, we studied how to solve problems that require coordination among distributed processes in the asynchronous model of computation, that is, without making any assumptions about the relative speeds of the participating processes. Now we turn our attention to solving such problems in the timed model of computation, where timing assumptions about the relative speeds of concurrent processes can be relied on to design solutions. In chapter 4, we established that there is no solution to the consensus problem if we restrict shared variables to atomic registers. However, if we assume that delays between successive steps of a process are bounded, then the knowledge of these bounds can be used to solve consensus using only atomic registers. Below we describe a timing-based solution to the classical coordination problem of *mutual exclusion*, and the same ideas can be used to solve consensus.

Recall the mutual exclusion problem aimed at allowing asynchronous processes to access a critical shared resource in a safe manner. The allocation of the resource is not governed by a central coordinator, but processes need to coordinate among themselves using atomic registers. The safety requirement is mutual exclusion: no two processes should be in the critical section simultaneously, and the liveness requirement is deadlock freedom: if some process wants to enter the critical section, then some process should be allowed to enter the critical section.

The timing-based solution, known as the Fischer's protocol, uses a single shared register *Turn*, and each process, with identifier myID, executes the timed state machine shown in figure 7.9. Initially, the register *Turn* is 0 and the mode is Idle. There is no clock invariant associated with the initial mode Idle, and thus the process may spend an arbitrary amount of time in this mode. When the process wants to enter the critical section, it switches to the mode Test. Then it reads the shared register *Turn*: if *Turn* equals 0, then the process proceeds to the mode Set, or else returns to the mode Test in order to read the shared register again.

In the mode Set, the next step of the process is to update the value of the shared register *Turn* to its own identifier. Observe that if a process P' tests *Turn* *after* a process P has set *Turn*, then the process P' will have to wait in the mode Test. However, if two processes test *Turn* before either has set it, then the protocol needs to resolve contention between them to allow only one to proceed to the critical section. The protocol assumes that writing to the shared register *Turn* takes at most Δ_1 time units. This is specified using the clock variable x and the invariant $(x \leq \Delta_1)$ associated with the mode Set. After updating *Turn*, a process waits in the mode Delay for at least Δ_2 time units; this delay is ensured by the guard $(x \geq \Delta_2)$ on the mode-switch out of Delay. When a process leaves the mode Delay, it reads the shared register *Turn* again. At this time, assuming $\Delta_2 > \Delta_1$, we can conclude that all the processes that tested *Turn* to be zero, and thus proceeded to set the register, have finished their updates. If a process finds *Turn* unchanged, that is, equal to its own identifier, it proceeds to the critical section, but if it finds that the value has been overwritten by some other process, then it retries by switching back to the mode Test. Once in the critical section, that is, in the mode Crit, a process can spend an arbitrary amount of time, and upon exit, it resets *Turn* to 0 and returns to the initial mode.

To prove mutual exclusion, consider the sequence of events in a typical execution depicted in figure 7.10. Let t_1 be the time when the process P leaves the mode Test and enters the mode Set, that is, the time when the process reads the register *Turn* and finds it to be 0. Let t_2 be the time when it enters the mode Delay, that is, the time when it sets the register *Turn* to its own identifier. Let t_3 be the time when the process P leaves the mode Delay and reads the value of the register *Turn* again. The timing constraints ensure that the condition $t_3 - t_2 \geq \Delta_2$ holds. It is possible that some other process P' attempting to enter the critical section also reads the shared register *Turn* to be 0 sometime during the time interval from t_1 to t_2, say at time t_1' (see figure 7.10). We don't need to worry about processes that had not entered the mode Set before time t_2 since the value of the register *Turn* is guaranteed to be non-zero at all times from t_2 onward, and thus each such process would read a non-zero value from the register *Turn*. Due to the upper bound on time a process spends in the mode Set, the process P' must execute the mode-switch from the mode Set to the mode Delay sometime during the interval $[t_1', t_2']$, where $t_2' = t_1' + \Delta_1$. Now it is easy to see that if $\Delta_2 > \Delta_1$, then $t_2' < t_3$, meaning that any competing process that switched from the mode Test to the mode Set during the interval $[t_1, t_2]$ would have left the mode Set by time t_3. In our illustrative scenario of figure 7.10, the process P modifies the register *Turn* to its identifier at time t_2, and the process P' modifies the register *Turn* to its own identifier at some time after time t_1' and before time t_2'. If this write by the process P' occurs before

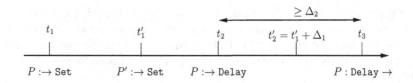

Figure 7.10: Illustrating a Timed Execution of the Mutual-exclusion Protocol

time t_2, then this update is overwritten by that of the process P, and if it occurs after time t_2, then it overwrites the update by the process P. In either case, the test ($Turn = \texttt{myID}$) cannot succeed for both, and it is not possible for both to proceed to the critical section.

In general, among all the competing processes that succeed in entering the mode $\texttt{Set}$, if the process P' is the last process to set the register $Turn$ to its identifier, then every process will find the register $Turn$ to be equal to the identifier of the process P' when it checks the register $Turn$ when it leaves the mode $\texttt{Delay}$. Such a process P' will be the unique winner and will proceed to its critical section. When it leaves the critical section, it sets the register $Turn$ back to 0 so that processes waiting in the mode $\texttt{Test}$ can compete again to enter the critical section.

From the discussion above, for $\Delta_2 > \Delta_1$, it follows that the protocol satisfies both mutual exclusion and deadlock freedom. More precisely, consider the timed process obtained by composing arbitrarily many instantiations of the process P of figure 7.9, each with its own value of the identifier $\texttt{myID}$ and the process modeling the atomic register $Turn$. For the composite process, the mutual exclusion property

$$\varphi_{me}: \ \neg\,(P.mode = \texttt{Crit} \ \wedge \ P'.mode = \texttt{Crit})$$

is an invariant, for every pair P and P' of processes. The composite process also satisfies the deadlock freedom property: if $P.mode = \texttt{Test}$ for a process P, then eventually $P'.mode = \texttt{Crit}$ for some process P'.

Exercise 7.7: For the timing-based mutual exclusion protocol of figure 7.9, consider the starvation-freedom requirement "if a process P enters the mode $\texttt{Test}$, then it will eventually enter the mode $\texttt{Crit}$." Does the system satisfy the starvation-freedom requirement? If not, show a counter-example. ∎

Exercise 7.8: Describe a protocol for solving the consensus problem described in section 4.3.3 using atomic registers and timing assumptions. State the timing assumptions explicitly. Describe the protocol in the state machine notation (using the mutual exclusion protocol of figure 7.9 as a guide). Argue why the protocol meets all three requirements of the consensus problem. ∎

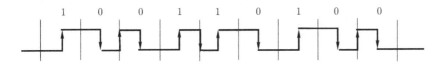

Figure 7.11: Manchester Encoding of the Bit Sequence 100110100

7.2.2   Audio Control Protocol *

We now consider a timing-based protocol for transferring a sequence of bits from the sender to the receiver using imperfect clocks. The encoding used in this protocol is called the Manchester encoding, and this protocol is based on an audio control protocol used by Philips Inc.

Problem Description

A stream of bits, that is, values of Boolean type, is communicated using high- and low-voltage settings on a communication bus. Time is divided into slots of fixed length, and in each slot, a single bit is communicated by changing the voltage in the middle of the slot. The value 0 is encoded as a falling edge from high voltage to low voltage, and the value 1 is encoded as a rising edge from low voltage to high voltage. If the bits to be sent in consecutive slots are the same, then there must be an intermediate change in the voltage, and this happens at the end of a slot. The voltage pulse corresponding to the encoding for the bit sequence 100110100 is shown in figure 7.11.

The clocks of the sender and the receiver are imperfect and have a specified drift. The receiver does not know when the first time slot begins, but both the sender and the receiver know the agreed-on width of the slots. The sender and the receiver synchronize the beginning of the transmission by requiring low voltage when no information is exchanged and by agreeing that each message begins with the bit 1. The receiver does not know the length of the message in advance but can infer the end of the current message when it detects that no information has been communicated during a slot. A challenge for the protocol designer is the constraint that the receiver cannot reliably detect a falling edge. Thus, all decoding must be inferred based purely on the relative timings of the rising edges. As a result, the receiver cannot resolve the ambiguity between messages ending in 10 and in 1. This is because even when the message ends with 1, the sender sets the voltage to low to ensure that the voltage is low when no information is being transmitted. The delay between the last falling edge and the preceding rising edge is a full time slot for messages ending with 10 and is only a half time slot for messages ending with 1. Since the receiver cannot detect a falling edge, it cannot differentiate between these two cases. To resolve this ambiguity, it is assumed that each message ends in 00.

For the design and analysis of the desired protocol, let us assume that the length of the time slot is four time units and the drift of the clocks of the sender and

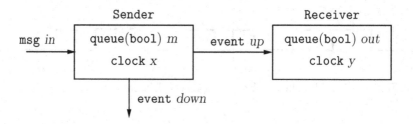

Figure 7.12: Block Diagram for the Audio Control Protocol

the receiver is ϵ per time unit. What this means is that if the sender keeps the voltage high for four time units according to its internal clock, then the actual elapsed time may be anywhere between $4 - 4\epsilon$ and $4 + 4\epsilon$ time units. Similarly, if the receiver finds the duration between two consecutive rising edges to be less than three time units, then it can only assume that the corresponding actual delay is less than $3 + 3\epsilon$ time units.

The correctness requirement is that every message is correctly decoded by the receiver, assuming that the drift in the clock values is bounded by a given ϵ per time unit. More generally, once we design a protocol, we would like to determine the largest value of the drift rate ϵ for which the protocol works correctly.

The Sender Process

The block diagram for the system composed of the sender and the receiver is shown in figure 7.12. The sender process receives the message to be transferred on the input channel *in*. A single message is a sequence of Boolean values. The sender process uses an internal queue m to store the sequence of bits to be transferred. We model the rising and falling edges of the voltage as output events *up* and *down*, respectively. The clock variable x is used to specify the timing constraints. The state machine for the sender process is shown in figure 7.13.

The process starts in the mode A, where it is waiting to receive the input. When the input is received on the channel *in*, it is stored in the queue m. The first bit is immediately dequeued for transfer, and it is assumed to be 1 (it is easy to modify the state machine for the sender so that it checks if the first bit is 1 and enters an error state if the check fails). The process sets its clock variable x to 0 and switches to the mode B. After waiting for two time units, it issues the event *up* to change the voltage to high in the middle of the slot. The process removes the next bit to be transmitted from the queue m. If this bit is 1, then it switches to the mode C, where it waits for two time units, changes the voltage to low by issuing the event *down*, and then returns to the mode B in order to transmit 1 in the following time slot. If the bit is 0, then it switches to the mode D. In mode D, it waits for four time units till the middle of the next time slot, and then changes the voltage to low. It then examines the next bit from the

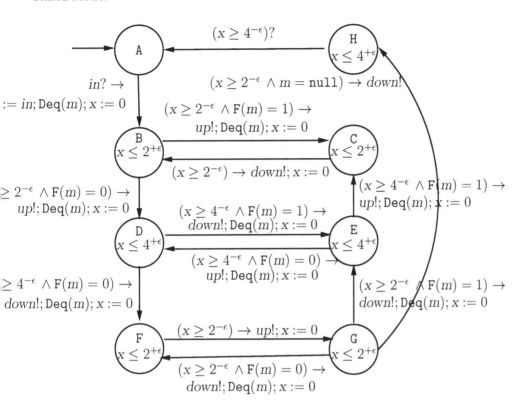

Figure 7.13: The Sender Process in the Audio Control Protocol

message queue. If this bit is 1 (different from the last bit processed), it switches to the mode E, where it waits for four time units before issuing the event *up*. If this bit is 0 (same as the last bit processed), then it waits in the mode F for two time units, raises the voltage, waits in the mode G for two time units, and then lowers it again to send the 0 bit. Each time the next bit is removed from the message queue and decisions are made based on whether the next bit is the same or different compared to the bit most recently sent. The last two bits of the message are guaranteed to be 00. The process will be in the mode G when the message ends, and when the queue m is empty, it returns to the idle location A after waiting in the mode H for a duration of a time slot without changing the output.

The detailed state machine for the sender appears in figure 7.13. In the description $F(m)$ stands for the first element of the queue m, and the action $Deq(m)$ removes the first element from the queue m. Note that all timing constraints are modified to reflect the possible errors in the measurement of time, as explained in section 7.1.6. For example, to specify that the process waits in the mode H for four time units we need to associate the clock-invariant $(x \leq 4)$ with the

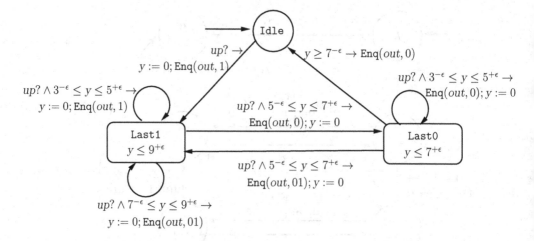

Figure 7.14: The Receiver Process in the Audio Control Protocol

mode H, and associate the guard $(x \geq 4)$ with the mode-switch from the mode H to the mode A. To capture the errors in time measurement, the clock-invariant $(x \leq 4)$ is changed to the constraint $(x \leq 4 + 4\epsilon)$, and the guard $(x \geq 4)$ is changed to the constraint $(x \geq 4 - 4\epsilon)$. We use $4^{-\epsilon}$ as an abbreviation for $4 - 4\epsilon$ and $4^{+\epsilon}$ for $4 + 4\epsilon$.

The Receiver Process

The receiver process is shown in figure 7.14. The receiver uses a clock variable y and an output buffer *out* to store the decoded message. It starts in the mode Idle. When it receives the first *up* event, it initiates the message to be 1. The mode Last1 corresponds to the case that the last decoded bit is 1; analogously, the mode Last0 corresponds to the case that the last decoded bit is 0. The clock y is used to measure the duration between successive *up* events. In the mode Last1, if the next bit is 1, then the exact duration until the next event is expected to be 4. Since the receiver simply needs to distinguish among various cases, if the duration is any time between 3 and 5, then it considers the next bit to be 1. The delay is measured using the receiver's imperfect clock. The check $(3^{-\epsilon} \leq y \leq 5^{+\epsilon})$ is an abbreviation for the check $(3 - 3\epsilon \leq y \leq 5 + 5\epsilon)$, and this accounts for the fact that the receiver's clock has a potential drift of ϵ per time unit compared to the physical elapsed time. In the mode Last1, if the next rising edge is detected after a delay in the interval $[5, 7]$, then the bit 0 is sent, and if the next rising edge is detected after a delay in the interval $[7, 9]$, then the bits 0 and 1 were transmitted (no rising edge is required for a 0 sandwiched between two 1s). In the mode Last0, similar logic is applied by partitioning the expected delays until the next *up* event into different categories: a delay between 3 and 5 means the next bit is 0, a delay between 5 and 7 means that

Time	Event	x	Sender	Queue m	y	Receiver	Queue out
0			B	00110100		Idle	null
2.07	up	2.07	D	0110100		Last1	1
5.97	down	3.9	F	110100	3.9	Last1	1
7.97	up	2	G	110100	5.9	Last0	10
9.92	down	1.95	E	10100	1.95	Last0	10
14.08	up	4.16	C	0100	6.11	Last1	1001
16.1	down	2.02	B	0100	2.02	Last1	1001
18	up	1.9	D	100	3.92	Last1	10011
22.05	down	4.05	E	00	4.05	Last1	10011
25.91	up	3.86	D	0	7.91	Last1	1001101
30.01	down	4.1	F	null	4.1	Last1	1001101
32.11	up	2.1	G	null	6.2	Last0	10011010
34.16	down	2.05	H	null	2.05	Last0	10011010
38.29		4.13	A	null	6.18	Last0	10011010
39.39		1.1	A	null	7.28	Idle	100110100

Figure 7.15: An Execution of the Audio Control Protocol

bits 0 and 1 are sent, and if no event is detected for seven time units, then the receiver concludes that the transmission has ended (recall that the message ends with 0) and returns to the mode Idle.

Example Execution

Figure 7.15 shows a possible execution of the protocol when the message string 100110100 is supplied to the sender at time 0, where the error rate ϵ equals 0.05. Then at time 0, the sender switches to the mode B, setting its variable m to 00110100 and its clock x to 0. Each row in the table shows the time at which the transition occurs, the event issued by the sender during this transition, the value of the clock variable x of the sender at the time of the transition (before it gets updated), the mode of the sender after the transition, the value of the internal message queue m after the transition, the value of the clock variable y of the receiver at the time of the transition (before it gets updated), the mode of the receiver after the transition, and the value of the output message queue *out* after the transition. Note that at the end of this execution, both processes have returned to their respective initial modes, and the value of the output queue equals the original input message 100110100.

Analysis

The parallel composition of the timed processes for the sender and the receiver can be analyzed to check whether the protocol works correctly, that is, whether the message received on the channel *in* by the sender equals the final value of the buffer *out*. This requirement can be captured by a safety monitor. We would also like to find out what is the maximum value of the error rate ϵ for which the protocol works correctly. It turns out the industrial design by Philips allowed an error of 5% (that is, $\epsilon = 1/20$), and the protocol meets the correctness requirement for this error rate. A formal analysis using model checking tools such as HyTech and Uppaal established that the protocol is resilient for errors upto $\epsilon = 1/15$.

Exercise 7.9: Demonstrate that the audio control protocol cannot tolerate the error rate $\epsilon = 0.25$ by showing an incorrect execution corresponding to the input string 100110100. ∎

7.2.3   Dual Chamber Implantable Pacemaker

The design and implementation of software for medical devices is challenging due to their rapidly increasing functionality and the tight coupling of computation, control, and communication. The safety-critical nature of such devices make them an ideal domain for exploring applications of formal modeling and analysis. In this section, we use a dual chamber implantable pacemaker to illustrate the modeling of control software and specification of correctness requirements for such devices. We begin with an overview of the basic functionality of a pacemaker.

Pacemaker Basics

The human heart is an excellent example of a naturally occurring timed system. It spontaneously generates electrical impulses that organize the sequence of muscle contractions during each heart beat. The underlying timing pattern of these impulses is key to the proper functioning of the heart. The implantable cardiac pacemaker is a rhythm management device that monitors these patterns and corrects them via external means when needed.

Controlled by the nervous system, a specialized tissue, called the *SinoAtrial node*, at the top of the right atrium periodically generates electrical pulses. These pulses cause both atria to contract, forcing blood into the ventricles. The electrical conduction gets delayed at the *AtrioVentricular node*, allowing the ventricles to fill fully, but then spreads rapidly across the ventricular muscles, resulting in their coordinated contraction that pumps the blood out of the heart.

A common heart disease, called *bradycardia*, is due to failures in either impulse generation or impulse propagation and results in slow heart rate, leading to insufficient pumping of blood. Bradycardia can be treated by an implantable pacemaker that monitors the heart rate and delivers timely external electrical pulses to maintain an appropriate heart rate as well as atrio-ventricular coordination. Such a pacemaker usually has two leads fixed on the wall of the right atrium and the right ventricle. Activation of local tissue is sensed by these leads, and these sensing events act as inputs to the pacemaker. If these sensed events

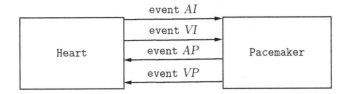

Figure 7.16: Interaction between the Heart and the Pacemaker

do not occur in a timely manner, then the pacemaker responds by producing pacing events that trigger electrical stimuli to the heart.

A modern pacemaker responds to a variety of heart conditions and can operate in different modes. We focus on a mode called DDD: the first character describes the pacing locations and D means that the pacemaker is pacing both the atrium and the ventricle, the second character describes the sensing locations and D means that both chambers are being sensed, and the third character specifies how the pacemaker software responds to sensing and D means that sensing can both activate or inhibit further pacing. While models corresponding to other commonly used modes (for instance, the VDI mode in which the pacemaker paces only the ventricle, senses both chambers, and sensing causes inhibition of pacing) are similar, the decision logic for switching from one mode to another causes additional complexity for the pacemaker software and is not reflected in our model for the DDD mode.

Design Overview

Figure 7.16 shows the top-level block diagram for communication between the two timed processes: the plant process **Heart** and the controller process **Pacemaker**. The events AI and VI are inputs to the pacemaker, while its outputs correspond to the events AP and VP. The event AI is sensed by the lead placed in the wall of the right atrium and corresponds to electrical potential exceeding a specific threshold value. The event VI is analogous and denotes electrical activation in the right ventricle. The pacing events AP and VP induce contractions of the muscles of the atrium and the ventricle, respectively. Note that all communication variables are modeled as events and thus have no associated data values. The behavior of the pacemaker depends on the timing delays between these events.

The design of the pacemaker is composed of four processes as shown in figure 7.17. The event VI denotes raw sensory input indicating electrical activity in the ventricle. The timed process **FilterV** outputs the event VS: the sequence of VS events corresponds to a filtered version of the sequence of VI events, and these events are used for deciding when to produce the pacing events. Similarly, the timed process **FilterA** outputs the event AS, a filtered version of the raw sensory event AI. The processes **PaceA** and **PaceV** produce the pacing events AP

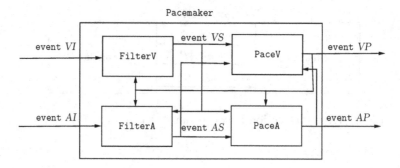

Figure 7.17: Block Diagram for the Pacemaker Subcomponents

and *VP*, respectively, based on the timing pattern of the filtered sense events *AS* and *VS*. The inputs to each of these four subprocesses are as shown in figure 7.17, and we proceed to explain their designs.

Event Sensing

The timed process `FilterV` of figure 7.18 filters the sequence of *VI* events to remove noise. After each ventricular event, there is a blanking period, called *Ventricular Refractory Period* (VRP), during which additional ventricular events are not considered to be new activity and, hence, ignored. The programmable parameter δ_1 in the description of the process `FilterV` corresponds to VRP, and a typical value of this parameter is 100 *ms*.

The process `FilterV` uses a clock variable x to measure the timing delay since the last ventricular event. When the process receives an input event *VI*, if the clock x has not exceeded δ_1, then it means that the VRP has not elapsed, and no output is produced. If the clock x exceeds δ_1, then the process wants to immediately respond by producing the event *VS* and by resetting the clock x to mark the beginning of a new VRP. Note that the underlying formal model is that of asynchronous processes, and thus the output transition producing the event *VS* has to be decoupled from the input transition that receives the event *VI*. To ensure that this output transition immediately follows the input transition without a delay, on receiving the input, the process sets the clock x to 0 and switches to the mode B with the associated clock-invariant ($x \leq 0$). This ensures that this mode B is transient, and the system cannot spend a non-zero amount of time while the process `FilterV` is in this mode. Note that the clock x is reset to 0 also when the ventricular pacing event *VP* is received, and this ensures that an input event *VI* within VRP of such a pacing event will be ignored.

The process `FilterA` is responsible for filtering the sequence of sensed atrial events. Its functioning is defined by the following rule: an input event *AI* should be viewed as the filtered event *AS* if it is not within the *Post Ventricular Atrial*

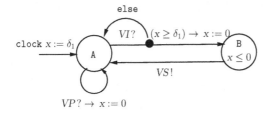

Process **FilterV**

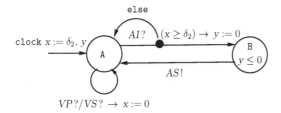

Process **FilterA**

Figure 7.18: Timed Processes for Filtering Sense Events

Refractory Period (PVARP) since the last ventricular event. The parameter δ_2 in the description of the process **FilterA** corresponds to this PVARP, and its typical value is 100 *ms*.

The clock x of the process **FilterA** measures the delay since the last ventricular event and is reset to 0 whenever either of the ventricular input events *VS* or *VP* is received. When the event *AI* occurs, if the value of x is below the threshold δ_2, then it does not result in any output event. Otherwise the process responds by producing the output event *AS* without any delay. To ensure that no time elapses between the reception of *AI* and production of *AS*, it sets a clock variable y to 0 when the input is received and switches to the transient mode B that has the associated clock-invariant $(y \leq 0)$ and a single outgoing transition that produces the desired output.

Timing of Pacing Events

The timed processes **PaceA** and **PaceV** shown in figure 7.19 implement the basic functionality of the pacemaker to keep the heart rate above a basic minimum.

The function of the process **PaceV** is to ensure that a ventricular event occurs within a maximum delay of *Atrio-Ventricular Interval* (AVI) since the last atrial event. The clock x is set to 0 when the event *AS* or *AP* occurs, and the process switches to the mode **Pending**. Note that the mode-switch $AS?/AP? \rightarrow x := 0$ is a short-hand for two mode-switches one triggered by the condition $AS?$ and one triggered by the condition $AP?$. The parameter δ_3 corresponds to AVI, and

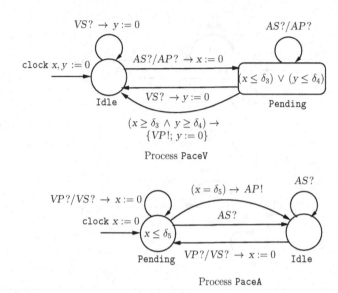

Figure 7.19: Timed Processes for Generating Pacing Events

its typical value is 150 ms. If the ventricular sensing event VS does not occur before the clock x exceeds this threshold δ_3 while waiting in the mode **Pending**, then the process should issue the pacing event VP. However, in order to prevent the pacemaker from pacing the ventricle too fast, the pacing event is issued only when at least *Upper Rate Interval* (URI) time has elapsed since the most recent ventricular event. To capture this constraint, the process uses another clock y that is reset to 0 on every ventricular event. The parameter δ_4 corresponds to URI that enforces a lower bound on the times between consecutive ventricular events, and its typical value is 400 ms. The condition to issue the pacing event VP is the conjunction $(x \geq \delta_3) \wedge (y \geq \delta_4)$. The disjunctive clock-invariant associated with the mode **Pending** is the negation of this guard and ensures that when both time limits expire, the pacemaker responds by pacing.

The process **PaceA** encodes the logic to generate the atrial pacing event AP. In the specification of a pacemaker, *Lower Rate Interval* (LRI) refers to the longest allowed interval between two ventricular events. In the initial mode **Pending**, if the process does not receive a ventricular event or the atrial sensing event AS within δ_5 time units since the previous ventricular event, then it issues the atrial pacing event AP. The value of the parameter δ_5 is chosen to be the difference LRI − AVI (with the assumption that the process **PaceV** paces the ventricle after a delay of AVI time units with respect to an atrial event). With an atrial event, the process **PaceA** switches to the mode **Idle** and switches back on the subsequent ventricular event. A typical value of LRI is 1000 ms.

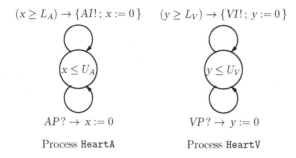

$$(x \geq L_A) \rightarrow \{AI! \, ; \, x := 0 \} \qquad (y \geq L_V) \rightarrow \{VI! \, ; \, y := 0 \}$$

$$x \leq U_A \qquad\qquad\qquad y \leq U_V$$

$$AP? \rightarrow x := 0 \qquad\qquad\qquad VP? \rightarrow y := 0$$

Process HeartA   Process HeartV

Figure 7.20: A Nondeterministic Model of the Heart

Heart Modeling

To analyze the functioning of the pacemaker, we need a model of the heart that generates the sensory events AI and VI. Figure 7.20 shows such an abstract model. The timed process HeartA generates the atrial sensing event AI after a nondeterministically chosen delay in the interval $[L_A, U_A]$ since the previous atrial event. The timed process HeartV is symmetric and generates the ventricular event VI with a nondeterministic delay in the interval $[L_V, U_V]$. The process Heart is the parallel composition of the timed processes HeartA and HeartV.

For establishing basic safety requirements of the pacemaker design, the simplified models in figure 7.20 suffice. However, this modeling does not capture the corelation in the timings of the atrial and ventricular events. A more faithful model can be constructed as a composition of processes, some capturing nodes in the heart tissue and some corresponding to the conduction pathways connecting these nodes.

Illustrative Execution

Let us illustrate the behavior of the pacemaker using a sample execution shown in figure 7.21. At time t_1, the process HeartA outputs the atrial event AI. The pacemaker component FilterA considers this to be a new atrial event and generates the sensing event AS without any delay. Given the absence of a ventricular sensing event, the process PaceV generates the pacing event VP at time t_2, which is the maximum of δ_4 and $t_1 + \delta_3$ (the parameters δ_3 and δ_4 correspond to the periods AVI and URI, respectively).

Following the ventricular event at time t_2, the subsequent atrial pulsation is generated by the heart at time t_3. However, this event is ignored by the process FilterA since $t_3 < t_2 + \delta_2$, where the parameter δ_2 is set to PVARP.

Meanwhile, the process PaceA expects the subsequent atrial sensing within a period of δ_5 corresponding to the difference LRI − AVI after the ventricular

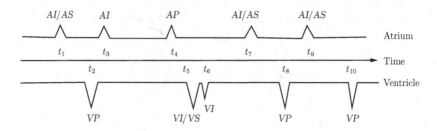

Figure 7.21: An Illustrative Execution of the Pacemaker

event at time t_2. Since such an event does not occur in this sample execution, it responds by generating the event AP at time $t_4 = t_2 + \delta_5$.

The heart generates the following ventricular event VI at time t_5, which gets mapped to the event VS by the filtering process FilterV without any delay. The subsequent ventricular puslation is generated by the heart at time t_6, and since $t_6 < t_5 + \delta_1$, where the parameter δ_1 corresponds to VRP, the event VI at time t_6 is ignored by the pacemaker. The next atrial event is generated by the heart at time t_7, and this event is translated to the atrial sensing event AS without any delay (assuming t_7 exceeds $t_5 + \delta_2$).

The process PaceV expects the subsequent ventricular event within δ_3 time units of the atrial event at time t_7. At time $t_8 = t_7 + \delta_3$, it turns out that $t_8 > t_5 + \delta_4$, implying that sufficient time (URI) has elapsed since the most recent VS, and hence the process PaceV generates the pacing event VP.

The following atrial event AI occurs at time t_9 and is translated to AS without any delay. Subsequently, the process PaceV expects a ventricular sensing before time $t_9 + \delta_3$. However, in this case, $t_9 + \delta_3 < t_8 + \delta_4$, implying that insufficient time has elapsed since it generated the most recent pacing event VP. As a result, it keeps waiting, and only at time $t_{10} = t_8 + \delta_4$ (which is greater than $t_9 + \delta_3$) it paces the heart again by generating the event VP

Requirements

The most basic functionality of a pacemaker is to treat bradycardia by maintaining the ventricular rate above the threshold of LRI (Lower Rate Interval). Thus, the closed-loop system consisting of the parallel composition of Heart and Pacemaker (see figure 7.16) should satisfy the following safety requirement: the delay between two successive ventricular events should not exceed LRI. This property cannot be directly captured as an invariant of the system, but we can use the timed process LRIMonitor of figure 7.22 as a safety monitor. The monitor observes the events VS and VP and enters the error state E if more than LRI time units have elapsed since the last occurrence of a ventricular event. Verifying safety of the pacemaker now corresponds to checking

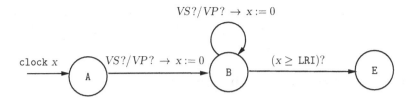

Figure 7.22: Safety Monitor for the Pacemaker

whether the property (LRIMonitor.*mode* $\neq$ E) is an invariant of the system Heart | Pacemaker | LRIMonitor.

For the pacemaker model we have discussed, along with the specified values of the programmable delay parameters and for the nondeterministic heart model of figure 7.20, irrespective of the values of the parameters L_A, U_A, L_V, and U_V, it is the case that the pacemaker is safe (that is, the monitor LRIMonitor cannot enter the error mode). This can be established by a manual proof by identifying a suitable inductive invariant or by using a model checker such as UPPAAL that implements the algorithmic reachability analysis discussed in section 7.3.

There are a number of other requirements that a pacemaker design is expected to satisfy. For example, a pacemaker should pace the ventricles beyond a maximum rate specified as the upper rate limit. A number of timing patterns of atrial and ventricular events are considered undesirable, and the pacemaker is expected to take a corrective action in response. While the modeling, specification, and analysis techniques we have discussed are suitable for the formalization of such requirements, we omit such detailed modeling from this case study.

Exercise 7.10: In the modeling of the timed process PaceV (see figure 7.19), we have used a *disjunctive* clock-invariant expression for the mode Pending. Construct an alternative model where the mode Pending is split into two modes, one with the clock-invariant $(x \leq \delta_3)$ and one withe the clock-invariant $(y \leq \delta_4)$, such that the resulting process has identical input/output behavior as the process PaceV. ∎

7.3   Timed Automata

Given a timed process TP, which may be expressed as a parallel composition of a number of timed processes, including a safety monitor, and a property φ over the state variables, the goal of the reachability analysis is to check whether φ is an invariant of the process TP and, if not, produce a counter-example. A key obstacle in adapting the on-the-fly enumerative depth-first search algorithm for invariant verification (as studied in section 3.3) is the fact that the state of a timed process includes the values for the real-valued clock variables. As a result, we must develop symbolic or constraint-based techniques to handle sets of values for clocks. In this section, we present analysis techniques that are applicable for the most commonly occurring pattern in which clocks are used in the modeling of timed systems.

7.3.1   Model of Timed Automata

In chapter 3, we established that verification problems admit algorithmic solutions provided the system being analyzed is a finite-state system. A timed process is typically not a finite-state system due to the presence of clock variables. However, algorithmic analysis is possible if we restrict the way the clock variables are used. For such an analysis, the use of the clock variables is restricted in the following way. First, the only value assigned to a clock variable is 0. Second, the only tests involving the clock variables are of the form $x \le k$ and $x \ge k$, for some integer constant k, that is, the only atomic expressions used in the clock-invariants and updates compare a clock variable to a constant. All the examples we have considered so far in this chapter obey these restrictions. Such updates and tests are adequate to express lower and upper bounds on the delays between events.

Timed processes with such restricted usage of clock variables are called *timed automata*. In particular, the timed process `TimedBuf` (see figure 7.2) is a timed automaton: the clock y is reset to 0 on the switch from the mode `Empty` to the mode `Full` and the test $(y \le UB)$ in the clock-invariant of the mode `Full` and the test $(y \ge LB)$ in the guard of the mode-switch from the mode `Full` to the mode `Empty`, compare the clock variable with the constants corresponding to the upper and lower bounds on the delay, respectively.

Timed Automaton

A timed process *TP* is said to be a *timed automaton* if for every clock variable x of the process:

1. the only assignment to the variable x occurring in the update description of any of the tasks of the process *TP* is of the form $x := 0$; and

2. each atomic expression involving the variable x occurring either in the clock invariant of the process *TP*, or in the guards or update descriptions of any of the tasks of the process *TP*, is of the form $x \le k$ or $x \ge k$, where k is an integer constant.

Note that a test of the form $(x = k)$, for an integer constant k, can be defined since it is equivalent to the conjunction $(x \le k) \wedge (x \ge k)$, and so can be a test of the form $(x < k)x$, which is equivalent to $\neg (x \ge k)$.

Since the analysis treats clock variables in a special way, it is convenient to consider a state of a timed automaton *TP* as a pair (s, ν), where the *discrete-state* s assigns values to all the discrete variables, and the *clock-valuation* ν assigns values to the clock variables. Note that if a timed automaton has n

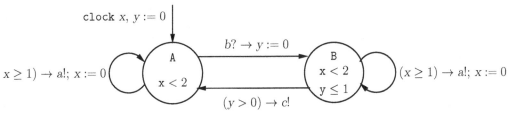

Figure 7.23: Example Timed Automaton for Illustrating Region Equivalence

clock variables, then a clock-valuation is an element of $\mathtt{time}^n$, where $\mathtt{time}$ is the set of non-negative real numbers.

7.3.2   Region Equivalence *

Example Partition

To explain the analysis technique for timed automata, consider the automaton in figure 7.23 with modes A and B, clocks x and y, an event input channel b, and event output channels a and c. The constraints involving the clock x imply that the process issues the event a periodically such that the delay between two consecutive such events is greater than or equal to 1 and strictly less than 2. The constraints involving the clock y imply that whenever the process receives an input event b, it issues an output event c with a delay strictly greater than 0 and less than or equal to 1.

Although this timed automaton has only two discrete states, its state-space is infinite since there are infinitely many clock-valuations. The basic idea of the analysis algorithm is to cluster these clock-valuations into finitely many equivalence classes so that equivalent states behave *similarly*. This equivalence, called the *region equivalence*, is depicted in figure 7.24 for our example automaton. In this example, a clock-valuation is a point in the first quadrant of the two-dimensional x/y plane. This space is split into finitely many clusters by drawing vertical, horizontal, and diagonal lines.

Intuitively, whenever we find a line such that the clock-valuations that lie on the line and that lie in the halves on either side of the line lead to different behaviors, we need to split the space by drawing such a line. To limit the number of partitions, we should draw such lines only when needed. The guards and invariants of the automaton compare clock variables with integer constants, and this motivates drawing horizontal and vertical lines. For example, in the mode A, if the clock variable x is less than 1, then the event a cannot be issued immediately; if the clock variable x is 1 or more, then the event a can be issued immediately; in the mode B, if the clock y is 0, then the event c is not possible without waiting a non-zero amount of time; and if the clock y equals 1, then the event c must be issued immediately. In this example, the largest constant that

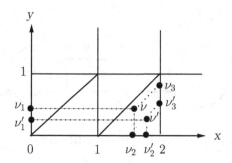

Figure 7.24: Clock Regions

the clock x is ever compared to is 2, and the largest constant that the clock y is ever compared to is 1. This suggests that the actual value of the clock x is not relevant for determining the set of possible future behaviors once it exceeds 2; similarly, once the value of the clock y exceeds 1, it need not be tracked accurately. As a result, to obtain the desired partitioning, we draw the vertical lines $x = 0$, $x = 1$, and $x = 2$ and the horizontal lines $y = 0$ and $y = 1$.

In a given mode, the effect of a timed action of duration δ is adding δ to the values of both clocks. Thus, during a timed action, the clock-valuation evolves along the *diagonal* direction. In our example, in the mode B, consider a clock-valuation where the variable x is between 0 and 1 and the variable y is between 0 and 1. This information is adequate to determine which guards are satisfied and, thus, which mode-switches can be executed immediately. If the constraint $(x > y)$ holds, then as time elapses, the value of the clock x reaches 2 before the clock y reaches 1, at which instance the output event a is enabled. However, if the value of x is less than that of y, then as time elapses, the value of the clock y reaches 1 before the clock x reaches 1, and this implies that the event b is guaranteed to be issued before the event a can be issued. To account for such effects, we split the partition further by drawing the two diagonal lines.

The resulting partition is shown in figure 7.24. Two clock-valuations are considered *region-equivalent* if they belong to the same partition, and each partition consisting of all equivalent clock-valuations is called a *clock-region*. In our example, there are 28 clock-regions: 6 corner points such as $(0, 1)$ and $(2, 0)$; 14 open line segments such as the segment $(0 < y < 1)$ on the line $x = 0$ and the segment $(0 < y < 1)$ on the diagonal line $x = (y + 1)$; and 8 open regions such as the triangular region $(0 < x < y < 1)$ and the unbounded region specified by the constraints $(1 < x < 2)$ and $(y > 1)$.

Two states are region-equivalent if their corresponding discrete states are identical and their clock-valuations are region-equivalent. To understand why region-equivalent states behave similarly, consider two clock-valuations ν and ν' belonging to the open region specified by the constraints $1 < x < 2$ and $0 < y < 1$

and $(x - y) > 1$ shown in figure 7.24. Consider an atomic clock constraint that compares the clock x with 0, 1, or 2 and compares the clock y with 0 or 1. Such a clock constraint is satisfied by the clock-valuation ν if and only if it is satisfied by the clock-valuation ν'. Thus, if a mode-switch is enabled in the state (A, ν), then it is also enabled in the region-equivalent state (A, ν'), and if a mode-switch is enabled in the state (B, ν), then it is also enabled in the region-equivalent state (B, ν'). If during such a mode-switch the clock x gets reset, then the resulting clock-valuations after the mode-switch are ν_1 and ν'_1 (see figure 7.24), which are again region-equivalent; and if the clock y gets reset, then the resulting clock-valuations after the mode-switch are ν_2 and ν'_2, which are again region-equivalent. Starting in the clock-valuation ν, as time elapses, the mode stays the same, and both the clocks increase along the diagonal line. The clock-valuation during such a transition remains region-equivalent to ν until the value of the clock x reaches 2, leading to the clock-valuation ν_3. The effect of a timed action starting from clock-valuation ν' is similar: the clock-valuation evolves along the diagonal line leading to the clock-valuation ν'_3, which is region-equivalent to the clock-valuation ν_3. Note that the duration of the timed action leading the clock-valuation ν to ν_3 is different from the duration of the timed action leading the clock-valuation ν' to ν'_3, but the order in which different clock-regions are encountered as time elapses is identical for two equivalent clock-valuations.

In summary, whatever action the process can take from a state can be matched by taking a corresponding action starting in a region-equivalent state, resulting in states that are equivalent, from which the same argument can be applied again. Thus, any execution starting in a state can be matched by a corresponding execution starting in a region-equivalent state, such that the two executions have matching sequence of input/output/internal/timed actions, with the only difference being in the exact durations used for timed actions.

As an example, consider the following execution of the timed automaton of figure 7.23:

$$(A, 0, 0) \xrightarrow{0.6} (A, 0.6, 0.6) \xrightarrow{b?} (B, 0.6, 0) \xrightarrow{0.5} (B, 1.1, 0.5) \xrightarrow{C!} (A, 1.1, 0.5)$$
$$\xrightarrow{0.2} (A, 1.3, 0.7) \xrightarrow{a!} (A, 0, 0.7) \xrightarrow{1.25} (A, 1.25, 1.95) \xrightarrow{0.61} (A, 1.86, 2.56).$$

Now suppose at the first step the duration of the timed action is changed to 0.1, resulting in the state $(A, 0.1, 0.1)$ that is region-equivalent to the state $(A, 0.6, 0.6)$. Below is another execution whose first step is the timed action of duration 0.1 such that, at every step of the execution, the state remains region-equivalent to the corresponding state of the execution above:

$$(A, 0, 0) \xrightarrow{0.1} (A, 0.1, 0.1) \xrightarrow{b?} (B, 0.1, 0) \xrightarrow{0.91} (B, 1.01, 0.91) \xrightarrow{C!} (A, 1.01, 0.91)$$
$$\xrightarrow{0.05} (A, 1.06, 0.96) \xrightarrow{a!} (A, 0, 0.96) \xrightarrow{1.25} (A, 1.25, 2.21) \xrightarrow{0.61} (A, 1.86, 2.82).$$

Region Equivalence

Now let us define the notion of region equivalence formally for the general case. Consider a timed automaton TP. For two clock-valuations ν and ν' to be considered equivalent, the following conditions must hold. Consider a clock variable x. The clock-valuations ν and ν' must agree on whether the clock x is 0, the clock x is between 0 and 1, the clock x is 1, the clock x is between 1 and 2, and such relationships with respect to the lines parallel to the x-axis. If k_x is the largest constant that the clock x is compared with in the atomic constraints that appear in a guard, update description, or the clock-invariant of TP, then once the value of the clock x exceeds k_x, its actual value does not matter. This condition can be summarized by requiring that the clock-valuations ν and ν' agree on all constraints of the form $(x = d)$ and $(x < d)$ for every integer d between 0 and k_x. The second condition accounts for constraints on the differences of clock values. Consider two clocks x and y such that both of them are assigned values that do not exceed their respective thresholds k_x and k_y. Then the values assigned by the clock-valuations ν and ν' must agree on the relationship with respect to the diagonal lines. This can be formalized by requiring that the ordering of the fractional parts of x and y must be identical according to the clock-valuations ν and ν'. For example, in figure 7.24, in each square, for the clock-valuations on the diagonal line, the fractional parts of the clocks x and y are equal; for the clock-valuations in the lower triangle, the fractional part of the clock x exceeds that of the clock y; and for the clock-valuations in the upper triangle, the fractional part of the clock y exceeds that of the clock x.

Two states are region-equivalent if they assign the same values to all the discrete variables and, thus, have identical discrete states, and their clock-valuations are region equivalent. This definition of the region equivalence is captured below.

REGION EQUIVALENCE

Given a timed automaton TP, for each clock variable x, let k_x be the largest integer constant that the variable x is compared with in the atomic constraints appearing in the guards, update descriptions, and the clock-invariant of the automaton TP. Two clock-valuations ν and ν' of the timed automaton TP are said to be *region-equivalent* if the following conditions hold:

1. for every clock variable x and for every integer $0 \leq d \leq k_x$, $\nu(x) = d$ if and only if $\nu'(x) = d$ and $\nu(x) < d$ if and only if $\nu'(x) < d$; and

2. for every pair of clock variables x and y such that $\nu(x) \leq k_x$ and $\nu(y) \leq k_y$, the fractional part of $\nu(x)$ is less than or equal to the fractional part of $\nu(y)$ if and only if the fractional part of $\nu'(x)$ is less than or equal to the fractional part of $\nu'(y)$.

Two states $s = (t, \nu)$ and $s' = (t', \nu')$ of the timed automaton TP are said to be region-equivalent if (1) the discrete states t and t' are the same, and (2) the clock-valuations ν and ν' are region-equivalent.

When a process has three clocks, the desired partitioning is obtained by creating a three-dimensional grid using axis-parallel planes up to a certain relevant constant on each axis. If we consider a cube given by $(0 < x < 1)$ and $(0 < y < 1)$ and $(0 < z < 1)$, then it is further split by diagonal planes into multiple cells. Examples of such clock regions include $(x < y < z)$, $(x = y < z)$, $(x = y = z)$, and $(y < x = z)$. Each such clock region can be described by giving the relative ordering of the fractional parts of the three clocks.

As illustrated in figure 7.24, if two states s and t are region-equivalent, then the transition from one state can be matched by a transition from the other such that the resulting states continue to be region-equivalent. This is formalized by the following theorem:

Theorem 7.1 [Region Equivalence] *Consider a timed automaton TP and two states s and t of the automaton TP such that s and t are region-equivalent. Then (1) if $s \xrightarrow{\alpha} s'$ is an input, or output, or internal action of TP, then there exists a state t' such that $t \xrightarrow{\alpha} t'$ holds and states s' and t' are region-equivalent; and (2) for every real-valued time duration $\delta > 0$ such that $s \xrightarrow{\delta} s + \delta$ is a timed action of TP, there exists a duration $\delta' > 0$ such that $t \xrightarrow{\delta'} t + \delta'$ is a timed action of TP and the states $s + \delta$ and $t + \delta'$ are region-equivalent.*

Proof. Let *TP* be a timed automaton. Consider two states s and t that are region-equivalent. We want to prove that every input/output/internal/timed action from the state s can be matched by a corresponding action from the state t such that the target states continue to be region-equivalent.

Observe that if two states are region-equivalent, then every expression that appears in the guard or update code of the automaton has the same value in both states.

Consider an internal action obtained by executing an internal task A with the guard condition *Guard* and update code *Update* from the state s. Since $s(Guard) = t(Guard)$, the task A is enabled in the state t also. Now consider the execution of the update code in the two region-equivalent states s and t. At every step of the execution, a conditional expression evaluates to the same value in both states. Executing an assignment of the form $y := e$, for a discrete variable y, preserves the region-equivalence. If a statement involves a nondeterministic choice, then it can be resolved exactly the same way in both executions. Furthermore, executing an assignment to a clock variable of the form $x := 0$ preserves the region-equivalence: it is easy to establish that whenever two clock-valuations ν and ν' are region-equivalent, so are the clock-valuations $\nu[x \mapsto 0]$ and $\nu'[x \mapsto 0]$. It follows that if the execution of the update code *Update* starting from the states s and t results in the states s' and t', respectively, then it must be the case that states s' and t' are region-equivalent.

The case of input and output actions is similar.

Consider a timed action $s \xrightarrow{\delta} s'$ where $s' = s + \delta$ for $\delta > 0$. Suppose the choice of the duration δ is such that for every $0 \le \epsilon \le \delta$, the state $s + \epsilon$ is region-equivalent to either the starting state s or the end-state s' (that is, the time duration is short enough so that multiple regions are not encountered along the way). We want to find a duration δ' so that the states $s + \delta$ and $t + \delta'$ are region-equivalent.

Let us say that a clock x is *integral* in the state s if $s(x)$ is an integer value not greater than the threshold k_x. Similarly, a clock x is *fractional* in the state s if $s(x)$ is not an integer (and thus has a nonzero fractional part) and does not exceed the threshold k_x.

Suppose in the state s there is some clock, say x, that is integral. Then for any $\epsilon > 0$, in the state $s + \epsilon$, the value of the clock x is no longer an integer, and such a state is not region-equivalent to s and, by assumption, must be region-equivalent to s'. Among all the clocks that are fractional in the state s, let y be the clock whose fractional part is the highest in the state s, and let ϵ_s be this fractional value (if there are multiple choices for y with equal fractional values, then we can choose any of them, and if there is no fractional clock in the state s, then this has to be handled as a separate simpler case). Then starting in the state s, if we let $1 - \epsilon_s$ time elapse, then the clock y will have an integer value, and the resulting state will no longer be region-equivalent to s'. Thus it must be the case that $0 < \delta < \epsilon_s$. Since the state t is region-equivalent to the state s, it agrees with the state s in terms of which clocks have integer values, which have fractional values, and the relative ordering of these fractional values. In particular, the clock x has an integer value in the state t also, and for any small $\epsilon > 0$, the state $t + \epsilon$ is no longer region-equivalent to t. Furthermore, among the fractional clocks in the state t, the clock y has the highest fractional value, and let this fraction be ϵ_t. Then verify that for every $0 < \delta' < \epsilon_t$, the state $t + \delta'$ is region-equivalent to the state $s + \delta$, and thus the desired duration can be any value in the interval $(0, \epsilon_t)$.

Now suppose there is no integral clock in the state s, but there is a fractional clock. Then let x be the clock with the highest fractional part, denoted ϵ_s. Then starting in the state s, as time elapses, the state stays region-equivalent to the state s for any duration less than $1 - \epsilon_s$, and at time $1 - \epsilon_s$, the clock x becomes an integer, triggering a change in the region. In this case, the duration δ must be equal to $1 - \epsilon_s$. Now the state t exhibits a similar behavior. If ϵ_t denotes the fractional part of the clock x in the state t, then we can choose $\delta' = 1 - \epsilon_t$ and verify that the state $t + \delta'$ is region-equivalent to the state s'.

When the state s has no integral or fractional clocks, the value of each clock x already exceeds the threshold k_x. In this case, for all values of δ and δ', the states s, $s + \delta$, t, and $t + \delta'$ are all region-equivalent.

Thus, the proof is complete for the case that the delay δ is short enough so that the states s and $s + \delta$ belong to adjacent regions. As exercise 7.13 establishes, in the general case, a timed action of duration δ can be split into a sequence of timed actions such that the duration of each part is short enough: there exist states $s = s_0, s_1, \ldots s_n = s'$ and delays $\delta_1, \ldots \delta_n$ with $\delta_1 + \cdots + \delta_n = \delta$ such that for each i, $s_i = s_{i-1} + \delta_i$ and for every $0 \le \epsilon \le \delta_i$, the state $s_{i-1} + \epsilon$ is region-equivalent to either s_{i-1} or s_i. Then starting from the state $t_0 = t$ that is region-equivalent to the state $s_0 = s$, by applying the argument above n times, we can find delays $\delta'_1, \ldots \delta'_n$ and states $t_i = t_{i-1} + \delta'_i$ such that each state t_i is region-equivalent to s_i. Thus, the desired duration δ' is the sum $\delta'_1 + \cdots + \delta'_n$, and this ensures that the state $t + \delta'$ is region-equivalent to the state $s' = s_n = s + \delta$. ∎

Let us now establish a bound on the number of clock regions. For a clock variable x, a clock region specifies whether the value of the clock x equals an integer d, where the possible choices of d are $0, 1, \ldots k_x$, whether the value of the clock x exceeds k_x, or whether the value of the clock x is strictly between the integers $d - 1$ and d, where the possible choices of d are $1, 2, \ldots k_x$. This gives a total of $2k_x + 2$ choices in terms of the constraint for the clock x. This means that the total number of partitions due to the axis-parallel constraints is given by the product $\Pi_x(2k_x + 2)$. In terms of the ordering of the fractional parts, if the timed automaton has m clock variables, then we get $m!$ many possible orderings. Finally, for every pair of adjacent clock variables x and y in such an ordering, the clock region makes a distinction based on whether the fractional part of the clock x is strictly smaller than that of the clock y or whether the two coincide. This gives additional 2^m choices. Altogether the number of possible clock regions is at most $2^m \cdot m! \cdot \Pi_x(2k_x + 2)$. This bound is not precise. In particular, this calculation considers orderings of fractional parts of all the clocks in every region, but when a clock x equals an integer d, its fractional part is guaranteed to be 0, and thus the actual number of clock regions is smaller than this bound. However, this bound is useful to get an understanding of how the number of clock regions varies: this number grows exponentially with the number of clocks and grows proportional to the product of the constants used in the description.

Search Using Region Equivalence

We have studied how region equivalence can be used to partition the infinite space of clock valuations into finitely many clock regions. We can now adopt the on-the-fly depth-first search enumerative algorithm for reachability of figure 3.16 to timed automata. Instead of enumerating states, the algorithm now enumerates regions, where each region specifies a discrete state and a clock region.

Figure 7.25 shows some of the regions that are found to be reachable for the timed automaton of figure 7.23. The initial region is described by $[A, x = y = 0]$, which consists of a single state with the mode A and the clock region corresponding to the single clock-valuation $(0, 0)$. In this region, the state can change in two ways: either due to the input event b, leading to the region described by $[B, x = y = 0]$, or due to a timed action of duration at most 2, leading to one of the regions $[A, 0 < x = y < 1]$, $[A, x = y = 1]$, or $[A, 1 < x < 2, y > 1]$. Edges corresponding to successors due to timed actions are labeled with the symbol

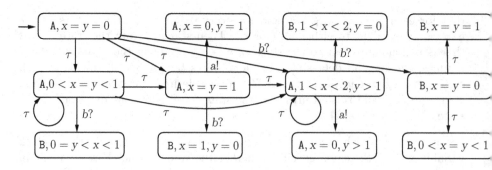

Figure 7.25: Search Using Clock Regions

τ in figure 7.25. This figure shows successors of each of these four regions. For example, consider the region $[A, 1 < x < 2, y > 1]$. If the input event b occurs, then the mode changes to B and the clock y gets reset, leading to the region $[B, 1 < x < 2, y = 0]$. Since the guard $(x \geq 1)$ is enabled, the output event a can be issued, and this resets the clock x, leading to the region $[A, x = 0, y > 1]$. The third possibility is a timed action. Given a state s that satisfies the constraints $(1 < x < 2)$ and $(y > 1)$, we can find a time duration $\delta > 0$ such that the state $s + \delta$ also satisfies the constraints $(1 < x < 2)$ and $(y > 1)$, and this explains the τ-labeled self-loop on the region $[A, 1 < x < 2, y > 1]$. Due to the clock-invariant $(x < 2)$, the region $[A, x = 2, y > 1]$ is not reachable using a timed action.

Recall the two illustrative executions for the timed automaton for figure 7.23 such that at every step, the corresponding states were region-equivalent. Both of these executions correspond to the following execution that records regions instead of concrete states:

$$[A, x = y = 0] \xrightarrow{\tau} [A, 0 < x = y < 1] \xrightarrow{b?} [B, 0 < x < 1, y = 0]$$
$$\xrightarrow{\tau} [B, 0 < y < 1 < x < y+1] \xrightarrow{c!} [A, 0 < y < 1 < x < y+1]$$
$$\xrightarrow{\tau} [A, 0 < y < 1 < x < y+1] \xrightarrow{a!} [A, x = 0 < y < 1]$$
$$\xrightarrow{\tau} [A, 1 < x < 2 < y] \xrightarrow{\tau} [A, 1 < x < 2 < y].$$

In this example, there are two discrete states (the mode can be either A or B), and there are at most 28 clock regions as shown in figure 7.24. This implies that the depth-first search algorithm can explore only 56 possible regions. In general, if the types of variables that are not clocks are finite, then there are only finitely many discrete states, and hence only finitely many regions, and the depth-first search algorithm exploring regions is guaranteed to terminate.

Given a property φ of a timed automaton TP, which in general is a Boolean expression over its state variables, when can we use the search over regions to determine whether the property φ is an invariant of the automaton TP? If the property φ refers only to discrete variables, then since the search over regions keeps track of the discrete state, it is adequate to check if the property φ holds in every reachable region. If the property refers to the clock variables also, then the region-based search is adequate as long as these references are consistent with the partitioning, that is, every atomic constraint involving clock variables in the property φ is of the form $x \leq d$ or $x \geq d$ for some integer constant $d \leq k_x$. In other words, let φ be a property such that whenever two states s and t are region-equivalent, either both states s and t satisfy φ or both do not satisfy φ. Hence, such a property φ is called a *region-invariant* property. For such a region-invariant property φ, to check whether all reachable states of the timed automaton satisfy φ, it suffices to check whether all reachable regions satisfy φ using a search over regions.

The analysis for timed automata can be easily adopted to handle rational constants. In the audio control protocol of section 7.2.2, the model has constraints of the form $(5 - 5\epsilon \leq y \leq 5 + 5\epsilon)$, where ϵ is a rational-valued constant such as $1/15$. To handle such models, we can simply multiply all constants by a factor of 15 to make them integers without changing the executions that are possible in the model.

We conclude this section by noting that defining region-equivalence, and using the resulting equivalence classes for analysis (more specifically, for depth-first search), is an instance of the general concept of *abstraction*. Concrete states such as $\langle A, 0.2, 0.3 \rangle$ are replaced by regions such as $[A, 0 < x < y < 1]$. Such a mapping removes some details and, hence, is called an abstraction, and the regions are called *abstract states*. Since many concrete states get mapped to the same abstract state, searching through abstract states is more efficient. Theorem 7.1 says that, in the case of timed automata, this specific abstraction using regions retains enough information so that a search over the abstract states accurately captures which sequences of input-output events are feasible and whether a region-invariant property is an invariant of the system.

Exercise 7.11: Consider the timed automaton of figure 7.23. List all possible regions that are successors of the region $[A, x = 0, y > 1]$ and all possible regions that are successors of the region $[B, 0 < x = y < 1]$. ∎

Exercise 7.12: Consider a timed automaton with 3 clocks x, y, and z, with $k_x = k_y = k_z = 1$. List all possible clock regions for this automaton. ∎

Exercise 7.13*: Prove that every timed action in a timed automaton can be split into a sequence of timed actions such that during every subaction, the state is region-equivalent to either the start or the end state of this subaction. Formally, consider a state s of a timed automaton and a duration δ. Prove that there exist states $s = s_0, s_1, \ldots s_n = s + \delta$ and delays $\delta_1, \ldots \delta_n$ with $\delta_1 + \cdots + \delta_n = \delta$ such that for each i, $s_i = s_{i-1} + \delta_i$, and for every $0 \leq \epsilon \leq \delta_i$, the state $s_{i-1} + \epsilon$ is region-equivalent to either s_{i-1} or s_i. Find a bound b as a function of the number m of the clock variables of the automaton so that every timed action can be split into at most b subactions of this desired form? ∎

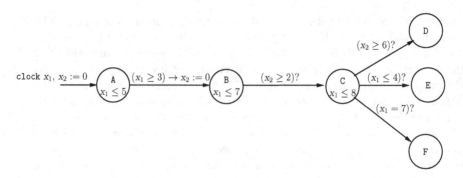

Figure 7.26: Example for Analyzing Timing Feasibility of Executions

Exercise 7.14*: Suppose we modify the definition of a timed automaton so that clock variables can be reset to constant values (that is, for a clock variable x, the allowed assignments are of the form $x := d$, where d is a non-negative integer constant), and tests can also compare differences of clock variables with constants (that is, each atomic expression involving clock variables is of the form $x \leq k$, $x \geq k$, or $x - y \leq k$, where k is an integer constant). How would you modify the definition of the region-equivalence over clock valuations so that there are only finitely many clock regions, and the analog of theorem 7.1 continues to hold? ∎

7.3.3   Matrix-Based Representation for Symbolic Analysis

Region equivalence allows partitioning of the infinite space of clock-valuations into finitely many regions, and we have seen how it can be used for invariant verification of timed automata using a search algorithm that enumerates all reachable regions. While theorem 7.1 implies that it is sufficient to keep track of clock regions to verify region-invariant properties, such a fine partitioning may not be necessary to solve a particular invariant verification problem. For example, consider the search for reachable regions from the initial region $[A, x = y = 0]$ shown in figure 7.25. Instead of enumerating the three regions $[A, 0 < x = y < 1]$, $[A, x = y = 1]$, and $[A, 1 < x < 2, y > 1]$, which are successors of the initial region corresponding to timed actions, we can represent the result of a timed action from the initial region by the single set $[A, 0 < x = y < 2]$. In this section, we consider a symbolic representation called *clock zones* that allows analyzing clock regions in clusters instead of enumerating them in a manner that allows an efficient representation and manipulation.

Example of Timing Analysis Using Clock Zones

To explain the symbolic analysis technique, let us consider the timed automaton shown in figure 7.26. Examination of timing constraints reveals that the mode D is not reachable, that is, the path A, B, C, D cannot be traversed. Similarly,

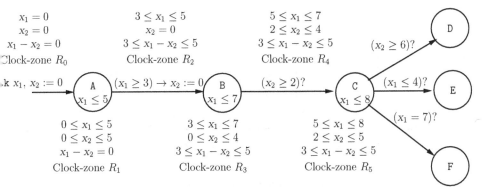

$x_1 = 0$
$x_2 = 0$
$x_1 - x_2 = 0$
Clock-zone R_0

$3 \leq x_1 \leq 5$
$x_2 = 0$
$3 \leq x_1 - x_2 \leq 5$
Clock-zone R_2

$5 \leq x_1 \leq 7$
$2 \leq x_2 \leq 4$
$3 \leq x_1 - x_2 \leq 5$
Clock-zone R_4

$x_1, x_2 := 0$ ——— A $(x_1 \geq 3) \to x_2 := 0$ B $(x_2 \geq 2)?$ C $(x_1 \leq 4)?$ E

$x_1 \leq 5$ $x_1 \leq 7$ $x_1 \leq 8$

$(x_2 \geq 6)?$ D

$(x_1 = 7)?$

$0 \leq x_1 \leq 5$
$0 \leq x_2 \leq 5$
$x_1 - x_2 = 0$
Clock-zone R_1

$3 \leq x_1 \leq 7$
$0 \leq x_2 \leq 4$
$3 \leq x_1 - x_2 \leq 5$
Clock-zone R_3

$5 \leq x_1 \leq 8$
$2 \leq x_2 \leq 5$
$3 \leq x_1 - x_2 \leq 5$
Clock-zone R_5

F

Figure 7.27: Inferring and Propagating Clock Constraints

the mode E cannot be reached, but it is possible to reach the mode F. This is not evident by a local examination of the constraints on the clock values in the clock-invariant of the mode C and the guards associated with the switches out of the mode C but is based on the implied constraints on the values of the clocks x_1 and x_2 when an execution reaches the mode C. This is illustrative of the nature of analysis that is needed to check whether the timing-based mutual exclusion protocol satisfies the mutual exclusion requirement: in the parallel composition of multiple instances of the timed process of figure 7.9, is it possible to reach the state with two processes in the mode Crit while satisfying the timing constraints imposed by the guards and the clock-invariants at every step?

A state in this example consists of the mode that takes values from the enumerated set $\{A, B, C, D, E, F\}$ and the values for the clock variables x_1 and x_2, each of which can be a non-negative real number. We can use finite-state analysis by partitioning the space of clock-valuations into clock regions, but for the automaton of figure 7.26, $k_1 = 7$ and $k_2 = 6$, and as a result, there are many clock regions (140, to be precise), and we wish to avoid considering all such clock regions individually. For this purpose, we generalize the notion of a clock region to a *clock zone*, which is a set of clock-valuations that is represented using constraints of a particular form over the variables x_1 and x_2, namely, bounds on the values of individual clock variables and bounds on the differences between the values of clock variables.

Initially, both clocks are 0, and this leads to the constraint

$$(x_1 = 0) \wedge (x_2 = 0) \wedge (x_1 - x_2 = 0).$$

This is shown as the clock zone R_0 in figure 7.27. Given that the set of clock-valuations upon entry to the mode A is described by the constraints R_0, we can calculate the set of clock-valuations that can be reached using timed actions as the process waits in the mode A. The value of the clock x_1 increases but cannot

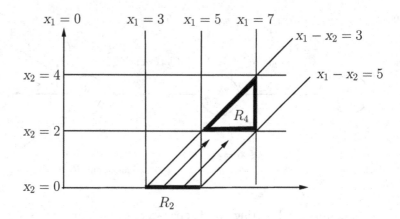

Figure 7.28: Illustrating Clock Zone Manipulations

exceed 5 due to the clock-invariant associated with the mode, and this gives the constraint $0 \leq x_1 \leq 5$. Observe that during timed actions, the difference in the clock values stays unchanged, so the constraint $(x_1 - x_2 = 0)$ from R_0 stays unchanged. These two constraints imply bounds on the value of the clock x_2, and this gives the description of the clock zone R_1:

$$(0 \leq x_1 \leq 5) \wedge (0 \leq x_2 \leq 5) \wedge (x_1 - x_2 = 0).$$

Note that the clock zone R_1 consists of multiple clock regions such as the corner point $x_1 = x_2 = 3$ and the line segment $3 < x_1 = x_2 < 4$. Also observe that the number of clock regions that can be reached from the clock zone R_0 due to a timed action is proportional to the constant 5 appearing in the clock-invariant of the mode A, while there is always a single clock zone that captures all the clock-valuations that are reachable using a timed action starting in the clock zone R_0.

The set R_2 describing the set of clock-valuations upon entry to the mode B is calculated from the clock zone R_1 by intersecting it with the guard condition $(x_1 \geq 3)$ and setting the clock x_2 to 0 to capture the effect of the assignment. The desired clock zone R_2 is described by the constraints:

$$(3 \leq x_1 \leq 5) \wedge (x_2 = 0) \wedge (3 \leq x_1 - x_2 \leq 5).$$

Again notice the implied constraint $(3 \leq x_1 - x_2 \leq 5)$.

This process can be repeatedly applied. The clock zone R_3 describes the set of clock-valuations that are reachable as time elapses in the mode B, and the clock zone R_4 describes the set of clock-valuations upon entry to the mode C. To get some intuition about this calculation, see figure 7.28. The clock zone R_2 is the segment of the line $x_2 = 0$ between $x_1 = 3$ and $x_1 = 5$. As time evolves, this line segment moves diagonally between the lines $(x_1 - x_2 = 3)$ and $(x_1 - x_2 = 5)$. The vertical line $x_1 = 7$ captures the clock-invariant and restricts the reachable clock-valuations in the mode B. The clock zone R_3 is thus the trapezoid between the lines $(x_2 = 0)$, $(x_1 = 7)$, $(x_1 - x_2 = 3)$, and $(x_1 - x_2 = 5)$. The guard condition of the mode-switch from the mode B to the mode C means that this trapezoid should be intersected with the constraint $(x_2 \geq 2)$, leading to the triangular clock zone R_4.

The clock zone R_5 describes the set of clock-valuations that are reachable as time elapses in the mode C (see figure 7.27) and is described by the constraints:

$$(5 \leq x_1 \leq 8) \wedge (2 \leq x_2 \leq 5) \wedge (3 \leq x_1 - x_2 \leq 5).$$

This accurately captures the cumulative effect of timing constraints along the path leading to the mode C. Intersection of this clock zone with the guard condition $(x_2 \geq 6)$ on the switch to the mode D is the empty set, and this establishes that the mode D is unreachable. Similarly, the intersection of the clock zone R_5 with the guard condition $(x_1 \leq 4)$ is also the empty set, and so the mode E cannot be reached. Intersecting the set R_5 with the guard condition $(x_1 = 7)$ gives a non-empty set, namely, $(x_1 = 7) \wedge (2 \leq x_2 \leq 4)$, that describes the set of clock-valuations upon entry to the mode F.

Difference Bounds Matrices

The most natural way of representing the constraints that arise during timing analysis is using a matrix-based representation. Let us assume that the timed automaton has m clock variables: $x_1, x_2, \ldots x_m$. We use a dummy clock x_0 that is assumed to represent the constant 0. Then a clock zone is represented by a square matrix R of dimension $m + 1$: the (i, j)th entry of the matrix gives the upper bound on the difference $(x_i - x_j)$.

We use the symbolic constant ∞ to denote a large value, and this is used to represent absence of a bound. More specifically, let Bounds be the set int of integers together with the symbolic constant ∞. The usual operations of comparison, minimum, and addition over integers are extended to the set Bounds in the following way: for every integer n, $n \leq \infty$ holds and $\min(n, \infty) = n$ and $n + \infty = \infty$.

A clock zone is represented by a square matrix R of dimension $(m + 1)$, with entries in Bounds, which represents the conjunction of constraints

$$\bigwedge_{0 \leq i \leq m, 0 \leq j \leq m} (x_i - x_j) \leq R_{ij}.$$

The column 0 (that is, the entries R_{i0}) gives the upper bounds on the clocks x_i, and the row 0 (that is, the entries R_{0i}) gives the upper bounds on the values of $-x_i$ (and thus, the negations of these entries capture lower bounds on the clocks x_i). Such a matrix representing bounds on the differences of clock values is called a *difference bounds matrix* (DBM).

Going back to our example from figure 7.27, the clock zone R_1 is represented by the following DBM:

$$\begin{bmatrix} 0 & 0 & 0 \\ 5 & 0 & 0 \\ 5 & 0 & 0 \end{bmatrix}$$

and the clock zone R_5 is represented by the following DBM:

$$\begin{bmatrix} 0 & -5 & -2 \\ 8 & 0 & 5 \\ 5 & -3 & 0 \end{bmatrix}.$$

Note that a lower bound of 5 on x_1 shows up as the upper bound -5 on $-x_1$ in the first row, and a lower bound of 3 on $(x_1 - x_2)$ shows up as the upper bound of -3 on the difference $(x_2 - x_1)$.

The representation (and the zone-based analysis of the example) discussed so far assumes that constraints on the clock values do not occur inside negation. A negated constraint such as $\neg(x_1 \geq 2)$ is equivalent to the strict inequality $(x_1 < 2)$. In the presence of such constraints, we need to distinguish between a non-strict upper bound of 2 (generated by the constraint $x_1 \leq 2$) and a strict upper bound of 2 (generated by the constraint $x_1 < 2$). This requires tagging each integral bound with a Boolean flag that indicates whether the associated constraint is strict or non-strict. The representation using DBMs and the techniques for manipulating DBMs can be adopted to handle this distinction (see exercise 7.19).

DBM Manipulation

The key insight regarding algorithmic and efficient inference of constraints (which is necessary to derive the constraint $(x_2 \leq 5)$ from the constraints $(x_1 \leq 5)$ and $(x_1 - x_2 = 0)$ in the description of the clock zone R_1 in our example in figure 7.27) is the following. Since the entry R_{il} is an upper bound on the difference $(x_i - x_l)$ and the entry R_{lj} is an upper bound on the difference $(x_l - x_j)$, the sum $(R_{il} + R_{lj})$ is an *inferred* upper bound on the difference $(x_i - x_j)$. If the entry R_{ij} is larger than the sum $(R_{il} + R_{lj})$, then we can *tighten* the upper bound R_{ij} by replacing it with the inferred bound $(R_{il} + R_{lj})$.

The DBM R is said to be *canonical* if and only if

$$\text{for all } 0 \leq i, j, l \leq m, \quad R_{ij} \leq (R_{il} + R_{lj}).$$

That is, in a canonical matrix, every entry R_{ij} represents the tightest bound that can be inferred on the difference $(x_i - x_j)$. Figure 7.29 shows an algorithm that computes the canonical version of an input DBM.

The tightening of upper bounds and the computation of the algorithm can be readily understood by an alternative view of the DBM as a weighted directed graph. Consider the graph with $m + 1$ vertices $x_0, x_1, \ldots x_m$. For every pair of

```
Input: (m + 1) × (m + 1) DBM R with entries in Bounds.
Output: Empty if R is empty, else canonical version of R.

for l = 0 to m {
    for i = 0 to m {
        for j = 0 to m {
            R[i, j] := min (R[i, j], R[i, l] + R[l, j])
        };
        if R[i, i] < 0 then return Empty
    }
}
return R.
```

Figure 7.29: Algorithm for Canonicalization of DBMs

vertices x_i and x_j, there is an edge from the vertex x_i to the vertex x_j whose cost is equal to the entry R_{ij}. Adding the entries R_{il} and R_{lj} gives the cost of a path from the vertex x_i to the vertex x_j consisting of two edges. If this cost is smaller than the cost of the direct edge from x_i to x_j, then we can replace the cost of this edge by the smaller value. In general, the path with the smallest cost between two vertices gives the tightest upper bound on the difference between the corresponding clocks. Then a DBM can be converted into a canonical one by executing the shortest path algorithm (or, equivalently, the transitive closure construction) on the matrix. The algorithm of figure 7.29 is indeed the classical Floyd-Warshall shortest-path algorithm. In the outer-most loop, the value of the variable l changes from 0 to m, and at every iteration, for all pairs (i, j), the entry $R[i, j]$ captures the shortest path from the vertex x_i to the vertex x_j using only vertices indexed $\leq l$, that is, the tightest upper bound on $(x_i - x_j)$ using constraints that involve variables with indices $\leq l$.

As an example, suppose $m = 3$ and consider the clock zone given by the constraints

$$(1 \leq x_1 \leq 3) \ \wedge \ (x_2 \geq 0) \ \wedge \ (0 \leq x_3 \leq 3) \ \wedge \ (x_2 - x_3 = 1) \ \wedge \ (x_2 - x_1 \geq 2).$$

Translating these constraints to the DBM representation gives the matrix R:

$$\begin{bmatrix} 0 & -1 & 0 & 0 \\ 3 & 0 & -2 & \infty \\ \infty & \infty & 0 & 1 \\ 3 & \infty & -1 & 0 \end{bmatrix}.$$

The corresponding graph representation is shown in figure 7.30 on the left. Note that whenever a matrix entry is ∞, the corresponding edge is not shown as this indicates absence of a constraint. Also, self-loops with cost 0 are not shown.

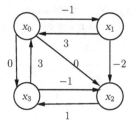 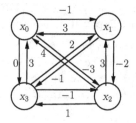

Figure 7.30: Illustrating Canonicalization

The result of canonicalization leads to the matrix

$$\begin{bmatrix} 0 & -1 & -3 & -2 \\ 2 & 0 & -2 & -1 \\ 4 & 3 & 0 & 1 \\ 3 & 2 & -1 & 0 \end{bmatrix}.$$

This matrix corresponds to the graph on the right in figure 7.30. Verify that for the graph on the right, the cost of an edge between a pair of vertices corresponds to the shortest (in terms of total cost) path between those two vertices in the left graph (for example, the shortest path from the vertex x_1 to the vertex x_0 is the path x_1, x_2, x_3, x_0 and has cost $-2 + 1 + 3 = 2$).

The algorithm for canonicalization also needs to address the following question: given a DBM R, is the conjunction of all the constraints represented by R satisfiable? It turns out that the matrix R represents an unsatisfiable set of constraints, and thus the empty set of clock valuations precisely when the corresponding graph has a cycle with a negative cost. For example, consider the constraints $(x_1 \geq 1)$ and $(x_2 \leq 2)$ and $2 \leq (x_1 - x_2) \leq 3$. These constraints are unsatisfiable, corresponding to the empty clock zone. In the DBM representation, due to the first constraint, we set R_{01} to -1; due to the second constraint, we set R_{20} to 2; and the third constraint gives $R_{12} = -2$ and $R_{21} = 3$. Adding up R_{01} and R_{12} implies that R_{02} must be tightened to -3, and adding up R_{02} and R_{20} then implies that R_{00} must be tightened to -1. In the graph view, the edge with cost -1 from x_0 to x_1, the edge with cost -2 from x_1 to x_2, and the edge with cost 2 from x_2 to x_0 form a cycle with a total cost that is negative. In such a case, repeating this cycle lowers the cost further and further, and the DBM cannot be made canonical. In the algorithm of figure 7.29, if some entry R_{ii} is lowered from 0 to some negative value, the algorithm has detected a cycle with a negative cost and returns with the answer that the input DBM corresponds to the empty clock zone. If the input DBM represents a non-empty clock zone, then the algorithm tightens all the entries as much as possible, and the output DBM is the canonical version of the input DBM.

Let us consider some operations that are useful on DBMs.

- **Atomic constraints:** Consider an atomic constraint $(x_i \leq k)$ for a constant k. To obtain the DBM R representing this constraint, we first set all the diagonal entries R_{jj} to 0, for $0 \leq j \leq m$; set the entry R_{i0} to k to reflect the upper bound on the difference $(x_i - x_0)$; set the entry R_{0j} to 0, for $0 \leq j \leq m$, to reflect the implicit assumption that $(x_j \geq 0)$ holds for every clock variable; and set all the remaining entries R_{jl} to ∞ to indicate absence of explicitly stated bounds. We then use the algorithm of figure 7.29 to convert this DBM into a canonical one.

- **Intersection:** Consider two DBMs R and R' both in canonical forms. To compute the intersection of the clock zones represented by these matrices, we simply set the (i, j)th entry of the result to be the minimum of R_{ij} and R'_{ij}. Then, we can test if the resulting matrix is empty, and if not, make it canonical using the algorithm of figure 7.29. The intersection operation is useful for capturing the effect of clock-invariants and of tests in guards appearing on mode-switches.

- **Time elapse:** Given a canonical DBM R representing a clock zone, to compute the set of clock-valuations that can be reached starting in the set R using timed actions (without accounting for the upper bounds imposed by the clock-invariants), we simply set the entry R_{i0}, for $1 \leq i \leq m$, to ∞. As time elapses, clock values increase, so the upper bounds on individual clock values are changed to ∞. Lower bounds on clock values and bounds on differences on clocks do not change because of timed actions.

- **Clock reset:** Given a canonical DBM R and a clock x_i, for $1 \leq i \leq m$, we can define an operation on the DBM so that the result captures the set of clock-valuations that can be obtained by assigning the clock x_i to 0 starting from a clock-valuation in R (see exercise 7.17 to develop details of this operation).

- **Subset test:** If R and R' are two canonical (non-empty) DBMs, then the clock zone represented by the DBM R is a subset of the clock zone represented by the DBM R' precisely when for every $0 \leq i, j \leq m$, $R_{ij} \leq R'_{ij}$. In particular, two canonical (non-empty) DBMs represent the same clock zone precisely when all of their respective entries match.

Reachability Analysis

To verify safety requirements of timed systems, we can now adopt the on-the-fly depth-first search algorithm of section 3.3 using clock zones. A *zone* is now represented as a pair (s, R), where the discrete state s records the values of all the discrete variables and R is a non-empty canonical DBM that captures a set of clock-valuations. The basic search mechanism stays the same. In particular, zones are explored and examined on demand, and the algorithm terminates as soon as a violation of the safety property is encountered. As in the case of the search using clock regions, the clustering of clock-valuations using clock zones is adequate as long as the property being checked is region-invariant.

For a zone (s, R), one possible successor zone is obtained by considering the effect of letting time elapse using a timed action. For this purpose, the algorithm first intersects the DBM R with the clock-invariant corresponding to the discrete state s, updates the matrix R to reflect elapse of time (by setting the entries in the 0-th column to ∞), and then again intersects it with the clock-invariant corresponding to the discrete state s. Note that the clock-invariant corresponding to each discrete state s also needs to be represented as a DBM, and such a DBM can be obtained using the constructions corresponding to atomic constraints and intersection. At every step, the resulting DBM is tested for emptiness and, if non-empty, is made canonical.

For a zone (s, R), the successor corresponding to a discrete transition, that is, execution of either an input, an output, or an internal task, is computed using the following steps. First, we compute the intersection of the DBM R and the DBM that captures the constraints on clock values for the guard condition of the corresponding task. If this intersection is an empty set, then this task is not enabled; otherwise the resulting DBM is made canonical. Then the discrete state s for the discrete variables is updated according to the update description of the task. If the update involves setting a clock variable to 0, then the clock reset operation is applied to the DBM part.

Zones of the form (s, R) are stored in the hash-table *Reach* that contains the zones visited so far. While examining a zone (s, R), the algorithm considers it as visited if a zone of the form (s, R'), where the DBM R is a subset of the DBM R', has been visited before. To implement this check, given a discrete state s, there needs to be an efficient way to access the set of all DBMs R such that the zone (s, R) has been encountered before.

Observe that the search algorithm has a mix of enumerative and symbolic flavors: discrete variables are processed by explicitly enumerating their values and clock variables are manipulated using constraints represented as DBMs. For finite-state timed automata, the number of choices for the discrete states s is bounded a priori, and the zone-based search is guaranteed to terminate.

The search algorithm using clock zones is implemented in tools such as the model checker UPPAAL (see www.uppaal.com) with many optimizations. The same ideas can also be applied to modify the nested depth-first search algorithm of chapter 5 for checking liveness properties of timed systems.

Exercise 7.15: Suppose a timed automaton has two clocks x_1 and x_2. Before entering a mode A, suppose we know that $(3 \le x_1 \le 4)$ and $(1 \le x_1 - x_2 \le 6)$ and $(x_2 \ge 0)$:

1. Show the DBM corresponding to the given constraints.

2. Is the DBM in part (1) canonical? If not, obtain an equivalent canonical form.

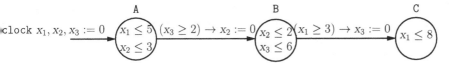

Figure 7.31: DBM Exercise

3. Suppose the clock-invariant of mode A is $(x_2 \leq 5)$. Compute the canonical DBM that captures the set of clock values that can be reached as the process waits in mode A.

4. Consider a mode-switch out of mode A with guard $(x_1 \geq 7)$ and update $x_1 := 0$. Compute the canonical DBM that captures the set of clock values that are possible after taking this transition.

■

Exercise 7.16: Consider the timed process shown in figure 7.31 with three clocks. Compute the sets R_A, R'_A, R_B, R'_B, R_C, and R'_C of clock values represented as canonical DBMs such that each of the DBMs R_A, R_B, R_C captures the possible clock values when the corresponding mode is entered, and each of the DBMs R'_A, R'_B, R'_C captures the possible clock values as the process waits in the corresponding mode. ■

Exercise 7.17: Consider a non-empty canonical DBM R and an index $1 \leq i \leq m$. Describe clearly how to compute the DBM R' that captures the effect of setting the clock x_i to 0. That is, R' should represent the set of all clock-valuations v such that $v = u[x_i \mapsto 0]$ for some $u \in R$. ■

Exercise 7.18: For $m = 3$, consider the constraints given by

$$x_1 \leq 3 \ \wedge \ x_3 \geq 1 \ \wedge \ 4 \leq x_1 - x_2 \leq 10 \ \wedge \ x_1 - x_3 \leq 2 \ \wedge \ x_3 - x_2 \leq 2.$$

Draw the weighted directed graph with four vertices that captures these constraints. Then draw the graph corresponding to the canonical DBM in which the weights reflect the shortest paths in the original graph. ■

Exercise 7.19*: The DBMs we have discussed cannot capture *strict* inequalities such as $(x_1 < 2)$. For this purpose, we can change the set Bounds to contain, in addition to the symbolic constant ∞, pairs of the form (k, b), where k is an integer and b is a Boolean value. The entries of the DBM now range over this new type. Then to capture the constraint $(x_1 < 2)$, we set R_{10} to the value $(2, 0)$, and to capture the constraint $(x_1 \leq 2)$, we set R_{10} to the value $(2, 1)$. The concepts such as tightening of bounds and canonicalization continue to work provided we extend the operations of comparison, minimum, and addition over this new set of bounds. Define these operations precisely. Over two clocks, consider the constraints $(3 < x_1 \leq 6)$ and $(1 \leq x_1 - x_2 < 4)$ and $(x_2 \geq 0)$. Show the DBM corresponding to these constraints. Is this DBM canonical? If not, obtain an equivalent canonical form. ∎

Bibliographic Notes

Since the 1980s, there have been many proposals for incorporating timing constraints in formal models of reactive computation (see, for instance, timed I/O automata as an example of a well-developed model [KLSV10]). The model presented in this textbook is based on timed automata [AD94], which has been studied extensively, resulting in a wealth of theoretical results and practical applications.

The data structure of difference-bounds matrices for analysis of timing constraints was introduced in [Dil89], and the concept of regions for finite partitioning of the state-space of timed models was introduced in [AD94]. Model checkers that implement these analysis techniques include KRONOS [HNSY94], RED [Wan04], and UPPAAL [LPY97], which now supports different forms of efficient analysis tools for real-time systems and has been used in industrial case studies (see www.uppaal.org and [BDL⁺11]).

The mutual exclusion algorithm of figure 7.9 is due to Fischer (see [Lam87] and [Lyn96] for solving distributed coordination problems relying on timing delays). The formal modeling and analysis of the audio control protocol in section 7.2.2 is based on [HW95] (see [BGK⁺96] for an automated analysis of the protocol using UPPAAL). Applying formal methods to the design and verification of a pacemaker is described as a challenge at sqrl.mcmaster.ca/pacemaker.html (see also [LSC⁺12] for a survey of formal modeling and analysis of medical cyber-physical systems). The pacemaker design in section 7.2.3 is based on [JPAM14].

8

Real-Time Scheduling

We have so far studied a model-based approach to the design and analysis of embedded systems. In this chapter, we turn our attention to a key aspect of *implementing* embedded systems so that the implementation exhibits the intended timing behavior. As an example, consider the event-triggered component `MeasureSpeed` of figure 2.30. To implement this component, whenever one of the input events *Second* or *Rotate* is detected, the update code of its task needs to be executed. While defining the execution semantics of synchronous models, we assumed that a task executes instantaneously. Whether such an assumption is justified for a given implementation depends on the answers to a number of questions: How long does it take to execute the code of this task on the underlying processor? Does the task `MeasureSpeed` have its own dedicated processor, or are multiple tasks sharing the same processor? Is the task `MeasureSpeed` independent of the other tasks, or does it have to be executed only after some other task has finished executing? The theory of real-time scheduling focuses on the formalization of demands for processing time by different computational tasks and general policies for allocating processing time so that these demands are met. This subject has a rich history with applications to safety-critical embedded systems as well as signal processing and multimedia systems. In this chapter, we first introduce the most commonly occurring pattern for demand for processing time and then study two classical and widely used algorithms for real-time scheduling.

8.1   Scheduling Concepts

For the purpose of scheduling decisions concerning allocation of processing time, the basic unit of computation is called a *job*. Examples of jobs include execution of the task corresponding to the component `MeasureSpeed` of figure 2.30 and execution of the code corresponding to a mode-switch in the sender process in the timed audio control protocol of figure 7.13. In multimedia applications such as processing of the incoming video stream, a job can correspond to decoding

Scheduler   Job J

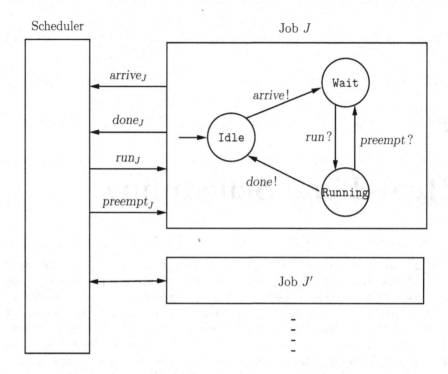

Figure 8.1: Interaction between the Scheduler and the Jobs

of a video frame in an MPEG file, while in a real-time control application such as avionics, a job can correspond to converting an analog signal from a sensor to a discrete value meaningful to the control software.

8.1.1   Scheduler Architecture

Figure 8.1 shows a typical interaction pattern between the scheduler and different jobs. Each job J is an independent process that communicates with the scheduler process responsible for allocation of processing time via events. The illustration also shows an abstract view of how the status of a job changes as a state machine.

Initially, a job J is in the mode `Idle`. When a job needs processing time, it communicates with the scheduler using the event $arrive_J$, and its status changes to `Wait`. When the scheduler decides to allocate the processor to the job J, it notifies the job using the event run_J, and this changes its mode to `Running`. When the current instance of the job J finishes its execution, it communicates with the scheduler using the event $done_J$ and returns to the mode `Idle`. The subsequent instance of the same job can now again request processing time using the event $arrive_J$.

When a job J is running, the scheduler may decide to *preempt* it before its computation is finished and allocate the processor to another job J'. The event *preempt$_J$* switches the mode of the job J from Running to Wait, where it continues to wait for the scheduler to issue another *run$_J$* event.

The scheduler has two requirements. First, the processor can be allocated to only one job at any point in time, and thus at most one job should be in the mode Running at any time. Second, each instance of the job should get "enough" computation time. To formalize this requirement, we need to know the timing pattern of the arrivals of successive instances of each job, how much computation time each job instance needs, and by when each instance needs to finish its execution.

The model of timed processes studied in chapter 7 is rich enough to formalize the arrival pattern and usage requirements for the jobs, the interaction between a job and the scheduler, and the decision logic of the scheduler. However, in a typical implementation, the scheduler is an integral part of the operating system and has tight control over the execution of the jobs. Furthermore, the time needed to execute the decision logic used by the scheduler is much smaller compared with the demands for execution time by the jobs. As a result, we assume that processing time is divided into discrete time slots. All the interaction between the scheduler and all the jobs happens instantaneously at the beginning of each time slot. The demand for processing time by each job is specified in units of time slots (for example, by specifying the bounds on the number of time slots between arrivals of successive instances of a job), and the allocation strategy of the scheduler is completely specified by an assignment of time slots to the jobs. Such an allocation scheme is called *time-triggered* allocation and is an example of a computation model that is synchronous and timed. In this scheme, the choice of the length of a single time slot—whether it is, for example, a second or a millisecond, does not matter for designing resource-allocation policies as long as all the parameters of all the jobs are specified using this length as the basic time unit.

8.1.2 Periodic Job Model

A job model specifies the arrival pattern and usage requirements for the jobs. The most common such job model is the *periodic job model*, in which the demand for processing time is specified using three parameters: a period, a deadline, and a worst-case execution time.

Period

In the periodic model, each job J has an associated period $\pi(J)$, which is a positive number. This specifies that the job J is to be executed periodically every $\pi(J)$ time units, that is, the event *arrive$_J$* is issued every $\pi(J)$ time units starting at time 0. Since a periodic job J is to be executed repeatedly, we use the notation J^a, for every positive integer a, to denote the ath instance of the job. The ath instance of the job J is ready to be executed at time $(a-1)*\pi(J)$, and this time is called the *arrival time* of the ath instance, denoted $\alpha(J, a)$.

Deadline

Each job J has an associated deadline $\delta(J)$, which is a positive number, that specifies that each instance of the job must finish its execution within $\delta(J)$ time units of its arrival. In other words, the delay between the occurrence of an event *arrive$_J$* and the subsequent event *done$_J$* should be bounded by $\delta(J)$ time units. It is required that this relative deadline should not exceed the period: the condition $\delta(J) \leq \pi(J)$ should hold. Since the arrival time of the ath instance of the job J is $(a-1)*\pi(J)$, this instance should finish execution by the deadline $(a-1)*\pi(J)+\delta(J)$, and we denote this absolute deadline for the ath instance of the job J by $\delta(J, a)$. For example, if a job has period 5 and deadline 4, then the arrival time of its third instance is 10, and the deadline for this instance is 14.

When the deadline equals the period, it means that each instance of the job should finish executing before the arrival of the next instance. Such a deadline is called an *implicit deadline*. While implicit deadlines are common, allowing a deadline to be strictly smaller than the period allows specification of more stringent timing requirements, as meeting such explicit deadlines implies improved response time.

Worst-Case Execution Time

While the period specifies how often a job needs to be executed, and the deadline specifies by when each job instance needs to finish executing, the (worst-case) execution time specifies how long it takes to execute an instance of a job. Each job J has an associated worst-case execution time (WCET) $\eta(J)$, which is a positive number, such that the execution of an instance of a job takes at most $\eta(J)$ time units. In other words, the job J is guaranteed to issue the event *done$_J$* if it spends a total cumulative time of $\eta(J)$ time units in the mode **Running** since the last occurrence of its arrival. Note that $\eta(J)$ is an *upper bound* on how long the computation corresponding to an instance of the job J can take to execute on the given platform. The actual execution time may vary, but if the scheduler allocates $\eta(J)$ time units to an instance before its deadline, then this allocation policy is safe. Since $\delta(J)$ specifies the deadline by which a job instance should finish executing, if the WCET $\eta(J)$ exceeds $\delta(J)$, then it is clear that the deadline cannot be met. Henceforth, we will assume that the condition $\eta(J) \leq \delta(J) \leq \pi(J)$ holds.

The three parameters $\eta(J)$, $\delta(J)$, and $\pi(J)$ together specify the requirement that, for every positive integer a, the job J should be allowed to execute for a total of $\eta(J)$ time units between the arrival time $\alpha(J, a) = (a-1)*\pi(J)$ of the instance J^a, and the deadline $\delta(J, a) = (a-1)*\pi(J)+\delta(J)$ for this instance. For example, if a job has period 5, deadline 4, and WCET 3, then it should be allocated three time units between times 0 and 4, three time units between times 5 and 9, three time units between times 10 and 14, and so on.

Estimating WCET

The period and the deadline for a job follow from the design-time requirements of a system. The WCET, however, is an artifact of the software implementation and the execution platform, and we would like an analysis tool to derive this bound automatically by analyzing the code. Deriving WCET bounds is an active area of research with a variety of approaches. Let us review the basic idea underlying the approach based on *statically* analyzing the code, that is, by examining the syntactic structure of the code without actually having to execute it.

Let us assume that the code corresponding to a job is loop-free and consists of atomic assignment statements, conditional statements, and sequences of statements. This is indeed the assumption we have been using for the update code of tasks used in the earlier chapters. Suppose we know how to associate an execution time $\eta(stmt)$ with an atomic statement $stmt$ of the form $x := e$ and an execution time $\eta(e)$ for evaluating a Boolean expression e. Then the following two rules can be used to associate the WCET with a block of code:

1. **Sequencing:** If a statement $stmt$ is the sequence $(stmt_1; stmt_2; \ldots stmt_l)$ of l statements, then the execution time of $stmt$ is the sum of execution times of the component statements:

$$\eta(stmt) \;=\; \eta(stmt_1) + \eta(stmt_2) + \cdots + \eta(stmt_l).$$

2. **Conditional statement:** If a statement $stmt$ is the conditional statement (if e then $stmt_1$ else $stmt_2$), then the execution time of $stmt$ is the sum of the execution time of evaluating the test e and the maximum of the execution times of the statements $stmt_1$ and $stmt_2$:

$$\eta(stmt) \;=\; \eta(e) \;+\; \texttt{max}\,\{\,\eta(stmt_1), \eta(stmt_2)\,\}.$$

To understand the rule for conditional statements, observe that to execute the conditional statement, first the test e is evaluated, and then either of the two statements $stmt_1$ or $stmt_2$ is executed depending on the result of the test. Since we want to estimate an upper bound on the execution time statically, we simply take the maximum of the estimated execution times for $stmt_1$ and $stmt_2$. As an example, consider the code

```
x := y + 1;
if (x > z) then y := z else {y := 0; z := x + 1}.
```

Suppose the execution time of each of the assignment statements is c_1, and the execution time of evaluating the condition $(x > z)$ is c_2, then the execution time of the above code is estimated to be $3c_1 + c_2$.

Thus, a WCET bound for straight-line code can be obtained from execution times for assignment statements and for evaluating Boolean expressions. Let us consider the assignment statement $x := y + 1$. The time it takes to execute such an instruction depends on the specifics of the underlying architecture and, particularly, how memory is organized. If the variables x and y are stored in registers, then such an assignment maps to a single machine instruction and executes in one clock cycle. In contrast, if the variables x and y reside in main memory, then executing such an assignment requires fetching the value of y from the main memory, incrementing a register, and then storing it back in the memory. When an assignment involves such memory operations, its execution time varies over a wide range depending on whether each relevant memory address resides in the local cache or whether it resides in the main memory. If we assume every read or write results in a cache-miss and leads to an access of the main memory, then the resulting upper bound is likely to be too pessimistic for useful analysis. In particular, for the code above, the variables x and y are accessed multiple times, and only the first such access can lead to a cache-miss. As this example illustrates, estimating an upper bound on the execution time that (1) is not too pessimistic, (2) reflects the complexity of the underlying architecture, (3) is guaranteed to be an upper bound on the actual execution time in all cases, and (4) can be computed by statically analyzing the code with reasonable computational effort is a challenging problem.

Periodic Job Model

A periodic job model then consists of a set of periodic jobs, where each job is specified using a period, a deadline, and a bound on execution time. This definition is summarized below.

PERIODIC JOB MODEL

A periodic job model consists of a finite set $\mathcal{J}$ of jobs, where each job J has an associated period $\pi(J)$, deadline $\delta(J)$, and worst-case execution time $\eta(J)$, each of which is a positive integer, such that the condition $\eta(J) \leq \delta(J) \leq \pi(J)$ holds.

As an example, consider the job model consisting of two periodic jobs J_1 and J_2: the job J_1 has period 5, deadline 4, and WCET 3; and the job J_2 has period 3, deadline 3, and WCET 1. The scheduling problem then is to allocate computation time to these two jobs so that for every $a \geq 0$, the job J_1 gets three time units between times $5a$ and $5a + 4$, and the job J_2 gets one time unit between times $3a$ and $3a + 3$.

Exercise 8.1: Consider the code

```
x := y + 1;
if (x > z) then { if y > 1 then y := z} else {y := 0; z := x + 1}.
```

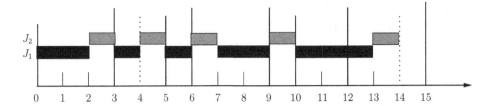

Figure 8.2: Illustrative Schedule for a Job Model with Two Jobs

Assuming that each atomic assignment statement takes c_1 time units and evaluation of each Boolean expression used in the conditional tests takes c_2 time units, estimate the worst-case execution time of the above code. ∎

8.1.3 Schedulability

A schedule specifies allocation of processing time to jobs, and the scheduling problem is to find a schedule that meets the deadlines of all the jobs.

Schedules and Feasibility

As discussed earlier, the scheduler allocates processing time in discrete time slots. A schedule σ for a periodic job model $\mathcal{J}$ specifies for every time t, for $t = 0, 1, 2, \ldots$, the job $J \in \mathcal{J}$ that is allocated the time slot starting at time t. It is possible that a particular time slot is allocated to none of these jobs, and we use $\bot$ to indicate this possibility. Formally, a schedule σ is a function from the set **nat** of natural numbers to the set $\mathcal{J} \cup \{\bot\}$: $\sigma(t) = J$ means that the time slot starting at time t is allocated to the job J, and $\sigma(t) = \bot$ means that the time slot starting at time t is not allocated to any of the jobs in the set $\mathcal{J}$. When $\sigma(t) = \bot$, the processor can stay idle during this time slot or can be used for computation not related to the system corresponding to this job model.

In this scheduling framework, the three parameters $\eta(J)$, $\delta(J)$, and $\pi(J)$ corresponding to a periodic job J specify the requirement that, for every instance a, the job J should be allocated a total of $\eta(J)$ time slots in the interval from time $\alpha(J, a)$ to time $\delta(J, a)$. A schedule σ is *deadline-compliant* for a job J if it indeed allocates the necessary number of slots to each instance of this job. A schedule is deadline-compliant for a periodic job model if it is deadline-compliant for each of the jobs in the model.

Let us reconsider the job model consisting of two periodic jobs J_1 and J_2: the job J_1 has period 5, deadline 4, and WCET 3; and the job J_2 has period 3, deadline 3, and WCET 1. Consider the schedule whose allocation pattern for the first 15 slots is shown in figure 8.2. For each job, vertical lines indicate times when successive instances of the job are ready for execution, and dashed vertical lines indicate the corresponding deadlines (for the job J_2, the deadline coincides with the period, so these dashed lines are missing). When a slot is assigned to a job, the row corresponding to that job is shown as a filled rectangle. Thus, in the schedule of figure 8.2, the first two slots are assigned to the first job, the third slot is assigned to the second job, and the 15th slot is not assigned to any of the two jobs. The same pattern repeats for every 15 slots. Formally, the schedule σ is given by:

$$\sigma(0) = \sigma(1) = \sigma(3) = \sigma(5) = \sigma(7) = \sigma(8) = \sigma(10) = \sigma(11) = \sigma(12) = J_1;$$
$$\sigma(2) = \sigma(4) = \sigma(6) = \sigma(9) = \sigma(13) = J_2;$$
$$\sigma(14) = \bot;$$
$$\text{for every } t \geq 0, \ \sigma(t + 15) = \sigma(t).$$

The abstraction of a schedule as an assignment of time slots to jobs can naturally be implemented in the scheduler architecture of figure 8.1. For instance, to implement the schedule shown in figure 8.2, the scheduler transmits the following events to the job J_1: run_1 at time 0, $preempt_1$ at time 2, run_1 at time 3, run_1 at time 5, $preempt_1$ at time 6, run_1 at time 7, and run_1 at time 10.

The illustration in figure 8.2 should convince you that this schedule is deadline-compliant for both jobs: each instance of the job J_1 is assigned three slots within four time units of its arrival time, and each instance of the job J_2 is assigned one slot before its next instance arrives.

For a periodic job model $\mathcal{J}$, if there exists a deadline-compliant schedule, then the job model is called *schedulable*. Thus, the model consisting of the job J_1 with period 5, deadline 4, and WCET 3 and the job J_2 with period 3, deadline 3, and WCET 1 is schedulable. Now suppose we change the WCET of the job J_1 to 4. Then convince yourself that the job model is not schedulable: during the first 10 slots, the deadlines of at least the first two instances of the job J_1 must be met, and thus it must be given at least eight slots; at the same time, the deadlines of at least the first three instances of the job J_2 must be met, implying that it must be given at least three slots, which is not possible.

These definitions are summarized below. For notational convenience, for a schedule σ, a job J, and time instances t_1 and t_2, such that $t_1 < t_2$, let us denote the number of time slots that the schedule σ allocates to the job J between times t_1 and t_2 by $\sigma(t_1, t_2, J)$, that is,

$$\sigma(t_1, t_2, J) \ = \ |\{t \mid t_1 \leq t < t_2 \text{ and } \sigma(t) = J\}|.$$

SCHEDULABILITY OF PERIODIC JOB MODEL

A schedule σ for a periodic job model $\mathcal{J}$ is a function that maps every natural number $t \geq 0$ to the set $\mathcal{J} \cup \{\bot\}$. Such a schedule is deadline-compliant for a job $J \in \mathcal{J}$ if for every instance $a \geq 1$, $\sigma(\alpha(J, a), \delta(J, a), J) = \eta(J)$. The schedule σ is deadline-compliant for the job model $\mathcal{J}$ if it is deadline-compliant for every job in $\mathcal{J}$. The job model $\mathcal{J}$ is *schedulable* if there exists a deadline-compliant schedule σ for $\mathcal{J}$.

Periodic Schedules

A periodic schedule assigns slots to jobs in a repeating manner. Formally, a schedule σ is a *periodic schedule* with period p, where p is a positive number, if for all time instances $t \geq 0$, $\sigma(t + p)$ equals $\sigma(t)$. The schedule shown in figure 8.2 is a periodic schedule with period 15. A periodic schedule σ can be fully specified by listing its period p and the assignments $\sigma(0), \sigma(1), \ldots \sigma(p-1)$ of slots to jobs for the first p slots.

If we want to check whether a periodic job model is schedulable, then we can limit the search for plausible schedules to only periodic schedules. This is established in the following theorem. Its proof shows that given a periodic job model, it suffices to consider periodic schedules whose period equals the least-common multiple of the periods of all the jobs in the model.

Theorem 8.1 [Periodic Schedules] *A periodic job model $\mathcal{J}$ is schedulable if and only if there exists a periodic schedule that is deadline-compliant for $\mathcal{J}$.*

Proof. Consider a periodic job model $\mathcal{J}$. If there exists a periodic schedule that is deadline-compliant for all the jobs in $\mathcal{J}$, then by definition the job model $\mathcal{J}$ is schedulable. Conversely, suppose the job model $\mathcal{J}$ is schedulable. Then by definition there exists a deadline-compliant schedule σ. The schedule σ need not be periodic, and our goal is to construct a *periodic* schedule σ' that also meets the deadlines of all the jobs.

Let p be the least-common multiple of periods of all the jobs, that is, of all the numbers in the set $\{\pi(J) \mid J \in \mathcal{J}\}$. Define the desired schedule σ' as follows: for $0 \leq t < p$, $\sigma'(t) = \sigma(t)$, and for every $t \geq 0$, $\sigma'(t + p) = \sigma'(t)$. Clearly, the schedule σ' is periodic with period p. Consider a job $J \in \mathcal{J}$. Let $n = p/\pi(J)$ (note that, by the choice of p, it is divisible by $\pi(J)$). Since the schedule σ' is the same as the schedule σ for the first p slots and the schedule σ is deadline-compliant for the job J, the schedule σ' also meets the deadlines of the first n instances of the job J. Since the schedule σ' is periodic, for every $a \geq 1$, the quantity $\sigma(\alpha(J,a), \delta(J,a), J)$ equals the quantity $\sigma(\alpha(J, a + n), \delta(J, a + n), J)$, and thus if it meets the deadline of the instance J^a, then it also meets the deadline of the instance J^{a+n}. This shows that the schedule σ' is deadline-compliant for the job model. ∎

Utilization

Consider the periodic job model consisting of two jobs, the job J_1 with period 5, deadline 4, and WCET 3; and the job J_2 with period 3, deadline 3, and WCET 1. Since the job J_1 requires three time slots in every five slots, it needs $3/5$ of the available processing time. Similarly, since the job J_2 requires one time slot in every three slots, it needs $1/3$ of the available processing time. The sum $3/5 + 1/3$, equal to $14/15$, is called the *utilization* of the job model. Formally,

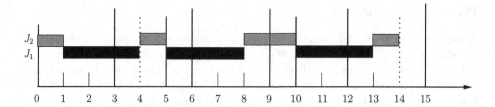

Figure 8.3: A Non-preemptive Schedule for a Job Model with Two Jobs

the utilization of a periodic job model $\mathcal{J}$ is defined as

$$U(\mathcal{J}) \;=\; \sum_{J \in \mathcal{J}} \eta(J)/\pi(J).$$

The utilization indicates what fraction of the available processing time is necessary to meet the demands of all the jobs. Note that the periodic schedule of figure 8.2 for the job model in our example assigns 14 out of 15 slots for scheduling of the two jobs, leaving one slot unassigned.

If the utilization of a job model exceeds 1, then it means that all the jobs together need more processing time than is available. In such a case, the job model is not schedulable. In our example job model, if we change the WCET of the job J_1 to 4, then utilization changes to $4/5 + 1/3$, which exceeds 1, and, as already noted, this leads to non-schedulability. Since the utilization of a job model can be computed easily, checking whether it exceeds 1 is a quick test to detect non-schedulability.

Preemptive vs. Non-preemptive Schedules

Consider the schedule of figure 8.2. The first instance of the job J_1 needs three out of the first four slots, and it is chosen at times 0, 1, and 3. While this meets the deadline of this instance, at time 2, its execution must be interrupted (using the event $preempt_1$) to let the job J_2 execute during the next slot and needs to be resumed in the subsequent slot. The schedule of figure 8.2 is called a preemptive schedule due to its use of such preemptions. A schedule that does not include such preemptions is called non-preemptive. In other words, a non-preemptive schedule assigns each instance of a job J, a chunk of $\eta(J)$ many *consecutive* time slots. Formally, a schedule σ over a set $\mathcal{J}$ of jobs is preemptive if there exists a job J and times $t_1 < t_2 < t_3$, such that (1) $\sigma(t_1) = \sigma(t_3) = J$ and $\sigma(t_2) \neq J$, and (2) there exists a such that $\alpha(J, a) \leq t_1$ and $t_3 < \alpha(J, a + 1)$. This says that there exist three slots, all belonging to an interval during which a single instance of the job J is active, such that the job J is assigned the two extreme slots but not the middle one.

Since preemption of jobs requires switching the context of processing from one job to another, it is expensive to implement and should be avoided whenever possible. For our example job model consisting of the job J_1 with period 5, deadline 4, and WCET 3; and the job J_2 with period 3, deadline 3, and WCET 1, figure 8.3 shows an alternative periodic schedule that is non-preemptive and yet is deadline-compliant.

It is possible that the only way to meet all the deadlines is by relying on preemption: a periodic job model can be schedulable, and yet there may be no non-preemptive schedule that is deadline-compliant. To illustrate this, suppose the model consists of two jobs: the job J_1 with period 2, deadline 1, and WCET 1; and the job J_2 with period 4, deadline 4, and WCET 2. The periodic preemptive schedule, with period 4, that chooses the job J_1 at times 0 and 2, and chooses the job J_2 at times 1 and 3, is deadline-compliant. But it is easy to convince yourself that it is not possible to find a non-preemptive schedule that meets all the deadlines.

Scheduling Policies

Given a periodic job model $\mathcal{J}$, the goal of a scheduling policy is to either produce a deadline-compliant periodic schedule or report failure to find such a schedule. While we will discuss two such policies in detail in sections 8.2 and 8.3, let us consider some of the questions we should ask to understand and evaluate a scheduling policy.

When does the policy succeed in producing a schedule? Ideally, a policy should produce a deadline-compliant periodic schedule whenever the job model is schedulable. If this is not the case, then we should aim to find conditions under which the policy is guaranteed to succeed.

How much computational effort does the policy need to compute the schedule? We already know that if the job model is schedulable, then there exists a periodic deadline-compliant schedule with period equal to the least-common multiple of the periods of all the jobs. If the model has n jobs, then each slot can be assigned only $(n + 1)$ different values (since a schedule maps a slot to either one of the jobs or to $\bot$). Thus, there are p^{n+1} many different periodic schedules possible with period p. A naive scheduling policy analyzes all such possible schedules and chooses a schedule that meets all the deadlines. However, such a policy is too inefficient as its computational cost grows rapidly as the number of jobs and the values of the periods grow. A scheduling policy is required to be *efficient* with time-complexity polynomial in the number of jobs in the model.

How much of the decision logic of the policy is implemented off-line vs. on-line? One possible way to implement a scheduling policy for periodic job models is to compute the desired schedule off-line, that is, before the system starts executing. Then during the execution of the system, that is, on-line, the scheduler simply needs to look up this schedule at the beginning of each time slot (or whenever a scheduling decision regarding allocation of jobs to the processor needs to be made). Alternatively, some scheduling policies only assign priorities to jobs off-line, and when a scheduling decision needs to be made on-line, make the decision based on, say, comparing job priorities. Policies of this latter kind are preferable as there is no need to compute and store a long schedule, and such policies can typically be extended to more complex job models, for instance, to job models in which jobs are added and removed dynamically as the system executes. However, when the decisions are made on-line, it is critical that the overhead associated with such a decision is minimal–no more than a couple of instructions.

Does the policy ensure alternative optimality criteria? Our definition of schedulability requires that the schedule should meet all the deadlines. When there are multiple deadline-compliant schedules, alternative criteria can be used to prefer one schedule over the other. For example, we may want the scheduling policy to compute a schedule with the least number of preemptions (and produce a non-preemptive schedule whenever possible). Another such criterion is *response time*: the response time of an instance of a job is the difference between the arrival time of this instance and the time this instance finishes its execution according to the schedule. We may want the scheduling policy to compute a schedule that minimizes the average response time over all the job instances.

Exercise 8.2: Consider a periodic job model with two jobs: the job J_1 has period 5, deadline 5, and WCET 2; and the job J_2 has period 7, deadline 7, and WCET 4. What is the utilization for this job model? Show a periodic deadline-compliant schedule. ∎

Exercise 8.3: Consider a periodic job model with two jobs: the job J_1 has period 3, deadline 2, and WCET 1; and the job J_2 has period 5, deadline 5, and WCET 3. Argue that this job set is schedulable only if we allow preemptions. Find a deadline-compliant schedule with minimum number of preemptions. ∎

8.1.4   Alternative Job Models

The periodic job model is the simplest model for formalizing the demand for processing time by jobs in a real-time application. Many extensions and variations have been studied in the literature. We close this section with a brief introduction to some of the most significant variants.

Precedence Constraints among Jobs

In a periodic job model with precedence constraints, in addition to the period, deadline, and WCET for each job, precedence constraints among jobs are also specified. The precedence constraint $J_1 \prec J_2$ between two jobs J_1 and J_2 means that for every a, the ath instance of the job J_2 should start executing only after the ath instance of the job J_1 has finished its execution. It is required that whenever there is such a precedence constraint between two jobs, they should have the same period, and the precedence relation should be acyclic. The schedulability problem then is to find a schedule that not only meets the deadlines of all the jobs but also obeys the ordering constraints expressed by the precedence relation. Precedence constraints among tasks in the task-graph-based description of synchronous components in chapter 2 naturally lead to a job model with precedence constraints.

Dynamically Changing Job Set

In the basic periodic job model, it is assumed that the set of jobs is known *a priori* and is fixed. In practice, it is desirable to allow a *dynamically changing* set of jobs, where jobs can be added and removed while the system is executing. In such a scenario, whenever a new job is to be added, the scheduling policy needs to perform a schedulability test on the revised set of jobs and admit the new job only when the schedulability test is successful. The policy also needs to make scheduling decisions dynamically at run-time since the schedule cannot be computed off-line in advance.

Aperiodic Jobs

An *aperiodic* job is a job whose arrival pattern is irregular and not known in advance. For example, in the cruise controller design discussed in chapter 2, while the tasks corresponding to measuring the current speed and controlling the speed (the components `MeasureSpeed` and `ControlSpeed` of figure 2.29) are to be executed in a periodic manner, the task responsible for updating the cruising speed (the component `SetSpeed`) needs to be executed whenever the driver switches on the cruise control or decides to change the cruising speed. An aperiodic job, thus, has an associated deadline and worst-case execution time, but instead of a period, it has an associated triggering event. When the job set has both periodic and aperiodic jobs, the scheduling policy is inspired by the principles and analysis used for design of policies for purely periodic job models and sets aside a fraction of the time slots for allocation to the anticipated but unpredictable demands by the arrival of aperiodic jobs. Such a policy, however, cannot offer guaranteed deadline-compliance, and we can only hope for a "best-effort" policy, where the guarantees of the policy are measured only by comparing them with respect to the guarantees of alternative policies.

Multiprocessor Scheduling

Our definition of a schedule allocates each time slot to at most one job, and this corresponds to the assumption that all the jobs are executing on a single processor. Modern computing platforms consist of multiple computing cores, and in embedded applications, it is even more common to have specialized processors dedicated to executing specific jobs. To formalize the problem of scheduling jobs on multiprocessors, in addition to the period, deadline, and worst-case execution time for each job, we also specify the set of processors, and for each processor, the subset of jobs that can execute on this processor. The multiprocessor schedule then maps each time slot and each processor to a job (or $\perp$ to indicate an idle slot), and the scheduling problem is to compute a deadline-compliant multiprocessor schedule. A job set that is not schedulable on a single processor can become schedulable on multiple processors due to the availability of additional processing time. However, designing an efficient scheduling policy to compute a deadline-compliant schedule in the multiprocessor job model is more challenging since the multiprocessor schedulability problem is computationally intractable (typically NP-complete).

Soft vs. Hard Real-Time Requirements

In the scheduling problem we have defined, a schedule is required to meet the deadlines of all the instances of each job. While such strict deadline-compliance is a necessary requirement in safety-critical and real-time control systems, in applications such as multimedia, it may be appropriate to demand a weaker guarantee. For example, if the job corresponds to refreshing the screen by displaying the next video frame, executing only 95% of all instances of the job may be acceptable. In the literature, systems where it is required that all deadlines must be met, as missing deadlines implies an unacceptable safety violation, are called *hard real-time* systems, and systems where deadlines should be met as frequently as possible, but missing deadlines only causes a degradation of the desired quality, are called *soft real-time* systems. The objective of a scheduling policy for soft real-time systems is then to compute a schedule with a minimum fraction of missed deadlines.

Exercise 8.4*: Consider a set $\mathcal{J}$ of n jobs with a precedence relation $\prec$ such that $\prec$ is acyclic. Suppose every job in $\mathcal{J}$ has period p, deadline p, and WCET c. Let us assume that we have n processors available for scheduling, and each job can be executed on any of the processors. This job set is schedulable if we can find a multiprocessor schedule that meets the precedence constraints (that is, if $J_1 \prec J_2$, then in each period, the job J_1 should finish executing before the job J_2 can start executing, but it's okay if they execute on different processors). Under what conditions is the job set schedulable? Hint: the condition should relate period p, WCET c, and some quantity derived from the precedence relation $\prec$. ∎

8.2   EDF Scheduling

The Earliest Deadline First (EDF) policy is a classical scheduling policy that always selects a job whose deadline is going to expire first. This scheduling policy is applicable to a wide range of job models and is commonly used in practice. We will first describe the EDF policy for the periodic job model and then analyze its performance.

8.2.1   EDF for Periodic Job Model

Given a set of periodic jobs, at every time t, the EDF scheduling policy assigns the next slot to the job that has the earliest (or least) deadline. In a periodic model, for each job, different instances of the job are active at different times, and thus the specific value of the deadline for a job depends on the time t. Also, the policy assigns a slot to a job only if the demand of its active instance has not already been met. The construction of the schedule according to these rules is formalized below.

Scheduling Policy

Consider a periodic job model $\mathcal{J}$, where each job J has an associated period $\pi(J)$, deadline $\delta(J)$, and worst-case execution time $\eta(J)$. For each time $t = 0, 1, 2, \ldots$, the EDF scheduling policy decides allocation of the next slot and builds the EDF schedule σ step by step in the following manner. Consider a job J. Let a be the unique number such that $\alpha(J, a) \leq t < \alpha(J, a + 1)$. The deadline of the job J at time t is the deadline of this instance of J, namely, $\delta(J, a)$. If the schedule σ up to the first t slots has already allocated $\eta(J)$ number of slots to this particular instance of the job J, then it does not need any more processing time. To formalize this, we say that the job J is *ready* at time t according to the schedule σ if $\sigma(\alpha(J, a), t, J) < \eta(J)$. If there is no job that is ready at time t, then the next slot is left unassigned, that is, $\sigma(t) = \bot$. Otherwise it selects a job J such that the job J is ready at time t and has the least deadline at time t among the jobs that are ready at time t, that is, $\sigma(t) = J$, such that the job J is ready at time t and if there is another job K that is also ready at time t, then the deadline of the job J at time t is less than or equal to the deadline of the job K at time t. Note that if multiple ready jobs share the same deadline, then any one of them can be chosen according to the EDF policy, and the choice is made using some alternative criteria in a specific implementation.

The definition of an EDF schedule is summarized below.

EDF SCHEDULE

A schedule σ for a periodic job model $\mathcal{J}$ is called an *EDF-schedule* if for every time $t \geq 0$, if no job is ready at time t in the schedule σ, then $\sigma(t) = \bot$, or else $\sigma(t) = J$, such that the job J is ready at time t in the schedule σ, and for every job $K \in \mathcal{J}$, either the job K is not ready at time t or the deadline of the job J at time t is less than or equal to the deadline of the job K at time t.

Example

Let us revisit the periodic job model consisting of the job J_1 with period 5, deadline 4, and WCET 3; and the job J_2 with period 3, deadline 3, and WCET 1 (see deadline-compliant periodic schedules of figures 8.2 and 8.3 for this model). Suppose we want to construct a schedule according to the EDF policy for this model.

Initially, at time $t = 0$, both jobs J_1 and J_2 are ready (since the demands of their respective first instances have not yet been met), the deadline for the job J_1 is 4, and the deadline for the job J_2 is 3. The EDF policy hence assigns the first slot to the job J_2. As a result, at time 1 as well as at time 2, the job J_2 is no longer ready (the demand of the corresponding active instance has been met), and the policy picks the job J_1 at these times. At time 3, the second instance of the job J_2 arrives, and thus both jobs are ready at time 3. At this point, the deadline for the job J_1 is still 4, but the deadline for the job J_2 is now 6. As a result, the EDF policy assigns the next time slot to the job J_1. At time 4, the job J_1 is no longer ready, and hence the subsequent slot is assigned to the job J_2. At time 5, the job J_2 is no longer ready as the demand for its active instance has been met, but the job J_1 is ready again as its second instance arrives. Thus, the policy chooses the job J_1 at time 5. At time 6, the third instance of the job J_2 arrives, and both jobs are ready. At this time, the deadlines for both jobs equal 9. As a result, the EDF policy is free to choose either of the two jobs. Suppose it chooses the job J_1 since its identifier is smaller than that of the job J_2. By the same reasoning, the next slot is also allocated to the job J_1. At times 8 and 9, only the job J_2 is ready and gets chosen. At times 10 and 11, only the job J_1 is ready, and gets chosen. At time 12, both jobs are ready, the deadline for the job J_1 is 14, and for the job J_2 is 15. Thus, the policy allocates the next slot to the job J_1. At time 13, only the job J_2 is ready and gets the next slot. At time 14, none of the jobs is ready, and hence the following slot is unassigned. The same cycle now repeats with a period of 15. The EDF schedule constructed in this manner is, in fact, identical to the schedule shown in figure 8.3.

Properties of the EDF Policy

In section 8.2.2, we will study under what conditions the EDF policy is guaranteed to produce a deadline-compliant schedule. For now, let us note some of its basic properties.

Whenever a scheduling decision is to be made, the EDF policy picks the job with the earliest current deadline among the ready jobs without explicitly analyzing the global consequences of such a decision. Such a policy is an example of a class of algorithms known as *greedy* algorithms.

Let us assume that whenever the EDF policy needs to choose a job among the ready jobs with identical deadlines, it uses some fixed decision rule (for example, choose the job with the lowest identifier). With this assumption, observe that the schedule produced by the EDF policy is a *periodic* schedule with period equal to the least-common multiple of periods of all the jobs.

In the EDF schedule of figure 8.3, at time 0, the job J_2 is given a priority over the job J_1, while at time 3, the job J_1 is given a priority over the job J_2. Such a scheduling policy whose relative preference among the ready jobs is different at different times is called a *dynamic priority* policy.

In general, the schedule generated by the EDF policy can be *preemptive*. For example, consider the periodic job model with the job J_1 with period 2, deadline 1, and WCET 1; and the job J_2 with period 4, deadline 4, and WCET 2. The

EDF policy chooses the job J_1 at times 0 and 2 and chooses the job J_2 at times 1 and 3. The resulting schedule is deadline-compliant and is preemptive.

In our description of the EDF policy, the policy makes a decision regarding which job is to be scheduled in every slot. However, the decision logic need not be executed in every slot based on the following observation. If a job J is chosen at time t, then the EDF policy is guaranteed to choose the same job at time $t+1$ also if both the following conditions hold: (1) the execution of the currently active instance of the job J is not yet finished, and (2) no new instance of any other job K arrives at time $t+1$. For example, in the schedule of figure 8.3, the job J_1 is chosen at time 1, and at the next time instance, neither the current job finishes its execution nor a new instance of the job J_2 arrives, and this ensures that the job J_1 is chosen at time 2 also. In other words, the scheduler needs to make a decision about allocation of processing time to jobs only when the set of ready jobs changes, which can occur only when either the current instance of the job finishes its execution or a new instance of another job arrives.

This suggests a natural strategy for implementing the EDF policy in the scheduler architecture of figure 8.1. The scheduler maintains a list *WaitingJobs*, which contains all the jobs that are waiting for processing time in the decreasing order of their relative priorities. When the job that is currently running finishes its execution, if the list *WaitingJobs* is non-empty, then the first job from the list *WaitingJobs* is removed and is allocated the processor. When a new instance of a job J arrives, the scheduler performs the following steps. If no job is currently assigned the processor, then the job J is allocated processing time. Otherwise, suppose the job J' is currently assigned the processor. If the deadline of the newly arrived instance of J is less than the deadline of the job J', then the job J' is preempted and added to the front of the queue *WaitingJobs*, and the job J is allocated the processor. If not, the newly arrived instance of the job J is inserted in the queue *WaitingJobs* of ready jobs in the suitable position by comparing its deadline with those of the jobs already in this sorted list.

Exercise 8.5: For the periodic job model consisting of the job J_1 with period 5, deadline 4, and WCET 3; and the job J_2 with period 3, deadline 3, and WCET 1, figure 8.3 shows the schedule constructed by the EDF policy assuming that whenever both jobs are ready with identical deadlines, the job J_1 is chosen. Show the schedule constructed by the EDF policy assuming that whenever both jobs are ready with identical deadlines, the job J_2 is chosen. ∎

Exercise 8.6: Consider a periodic job model consisting of the job J_1 with period 6, deadline 5, and WCET 2; the job J_2 with period 8, deadline 4, and WCET 2; and the job J_3 with period 12, deadline 8, and WCET 4. Construct the schedule according to the EDF policy (if multiple ready jobs have identical deadlines, prefer the job J_1 over the job J_2 over the job J_3). ∎

8.2.2 Optimality of EDF

Given a periodic job model, when is the EDF scheduling policy guaranteed to produce a deadline-compliant schedule? As the next theorem shows, the EDF policy is guaranteed to succeed as long as the model is schedulable. That's why the EDF policy is considered to be an *optimal* algorithm: it is guaranteed to produce a deadline-compliant schedule as long as one such schedule exists.

Theorem 8.2 [Optimality of EDF] *If $\mathcal{J}$ is a schedulable periodic job model and σ is an EDF schedule for $\mathcal{J}$, then σ is deadline-compliant.*

Proof. Let $\mathcal{J}$ be a schedulable periodic job model, and let σ be an EDF schedule for $\mathcal{J}$. We want to prove that the schedule σ meets the deadlines of all instances of all jobs. The proof is by contradiction. That is, we assume that the schedule σ misses some deadline, and we will arrive at a contradiction. Suppose there is an instance of a job such that the schedule σ does not allocate this instance enough slots by its deadline, and let t_0 be the time of this missed deadline.

Since the job model $\mathcal{J}$ is schedulable, there exists a schedule, say σ', that meets all the deadlines. Consider the two schedules σ and σ'. Since the schedule σ' meets all the deadlines and the schedule σ misses at least one deadline, the two schedules cannot be identical. Let $\mathit{diff}(\sigma, \sigma')$ denote the first time instance where the two schedules make different choices, that is, $\mathit{diff}(\sigma, \sigma') = t$, such that $\sigma(t) \neq \sigma'(t)$ and for all $t' < t$, $\sigma(t') = \sigma'(t')$. Such a time instance t must be less than the deadline t_0 that the schedule σ misses but the schedule σ' does not miss.

There can be multiple deadline-compliant schedules for the job model $\mathcal{J}$. Let σ_1 be a schedule among all such deadline-compliant schedules σ', such that $\mathit{diff}(\sigma, \sigma')$ is the largest, that is, the schedule σ_1 is a deadline-compliant schedule for $\mathcal{J}$, such that if the schedule σ' is also a deadline-compliant schedule for $\mathcal{J}$, then $\mathit{diff}(\sigma, \sigma') \leq \mathit{diff}(\sigma, \sigma_1)$. In other words, the schedule σ_1 makes the same choices as the EDF schedule σ for as long as possible without giving up the goal of staying deadline-compliant.

Let $t_1 = \mathit{diff}(\sigma, \sigma_1)$. We know that $\sigma(t_1) \neq \sigma_1(t_1)$, and for all $t < t_1$, $\sigma(t) = \sigma_1(t)$. We will consider different cases based on the values of $\sigma(t_1)$ and $\sigma_1(t_1)$. In each case, we construct another schedule σ_2 such that (1) the schedule σ_2 is deadline-compliant, and (2) $\mathit{diff}(\sigma, \sigma_2) > t_1$. This is a contradiction to the way the schedule σ_1 is chosen, implying that our initial assumption that the schedule σ misses some deadline cannot be true, thus completing the proof.

Consider the case when $\sigma(t_1) = J$ and $\sigma_1(t_1) = K$, with $J \neq K$, such that the job K is ready at time t_1 (see figure 8.4 for illustration). Let the deadlines of the jobs J and K at time t_1 be t_J and t_K, respectively. Since the EDF schedule chooses the job J at time t_1, the job J must be the one with the earliest deadline among all the jobs that are ready at time t_1, and this implies $t_J \leq t_K$. Since

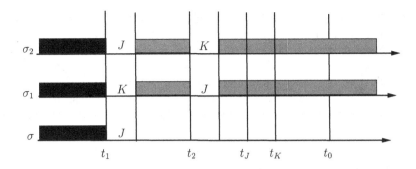

Figure 8.4: Proof of Optimality of EDF

the job J is ready at time t_1, the demand of its active instance at time t_1 has not yet been met, and it needs at least one more slot before the deadline t_J. Since the schedule σ_1 is deadline-compliant, there must be a time instance t_2 such that $t_1 < t_2 < t_J$ and $\sigma_1(t_2) = J$.

Now define the schedule σ_2 such that for all $t \neq t_1$ and $t \neq t_2$, $\sigma_2(t) = \sigma_1(t)$ and $\sigma_2(t_1) = J$ and $\sigma_2(t_2) = K$ (see figure 8.4 for illustration). That is, the schedule σ_2 is obtained from the deadline-compliant schedule σ_1 by swapping the choice of the jobs J and K at times t_1 and t_2. By the choice of these times, the instances t_1 and t_2 belong to the same active instance of the job J and the same instance of the job K, both before their relevant deadlines. As a result, the number of slots allocated to each instance of each job is exactly the same in both the schedules σ_1 and σ_2. It follows that the schedule σ_2 is also deadline-compliant. Now, the EDF schedules σ and σ_2 are identical for times less than t_1 and also at time t_1, and thus the condition $diff(\sigma, \sigma_2) > t_1$ must hold.

For the remaining cases, either (1) $\sigma(t_1) = \perp$ and $\sigma_1(t_1) \neq \perp$, (2) $\sigma(t_1) = J$ and $\sigma_1(t_1) = \perp$, or (3) $\sigma(t_1) = J$ and $\sigma_1(t_1) = K$, such that the job K is not ready at time t_1. In all these cases, define the schedule σ_2 such that for all $t \neq t_1$, $\sigma_2(t) = \sigma_1(t)$ and $\sigma_2(t_1) = \sigma(t_1)$. We leave it as an exercise to verify that the resulting schedule σ_2 is deadline-compliant and the condition $diff(\sigma, \sigma_2) > t_1$ must hold. ■

The proof shows that, at any step, if a deadline-compliant schedule chooses a job K over another job J, whose deadline is earlier than that of K, then choosing the job J, instead of the job K, cannot be the cause of a missed deadline. This core idea of the proof applies to more general job models and shows, for instance, even in the presence of both periodic and aperiodic jobs, that the EDF policy produces a deadline-compliant schedule as long as one exists.

Exercise 8.7: Complete the proof of theorem 8.2: show that the schedule σ_2 defined in the last paragraph of the proof is such that it meets all the deadlines and the condition $diff(\sigma, \sigma_2) > t_1$ holds. ■

Exercise 8.8: We have established that if there exists a deadline-compliant schedule, then the EDF policy produces one such schedule. Show that, however, the following statement is false: if there exists a deadline-compliant schedule with no preemptions, then the EDF policy is guaranteed to produce one such schedule. That is, construct a job model $\mathcal{J}$ (with two jobs) such that (1) there exists a deadline-compliant schedule σ that has no preemptions, and (2) the deadline-compliant schedule produced by the EDF policy involves preemptions.
∎

8.2.3   Utilization-Based Schedulability Test

We know that as long as the periodic job model is schedulable, the EDF policy produces a deadline-compliant schedule. But can we test whether the periodic job model is schedulable or, equivalently, whether the EDF policy is going to succeed without explicitly generating an EDF schedule and checking whether it meets all the deadlines. It turns out that when all the deadlines are implicit, that is, for every job J, the deadline $\delta(J)$ equals its period $\pi(J)$, the test for schedulability is simple: a periodic job model is schedulable exactly when its utilization is 1 or less. We already know that if the utilization exceeds 1, the demand for processing time exceeds the available processing time, and thus the job model is not schedulable. The following theorem proves that if the utilization is 1 or less, then an EDF schedule is deadline-compliant.

Theorem 8.3 [Schedulability test when deadlines equal periods] *Let $\mathcal{J}$ be a periodic job model such that for every job J, $\delta(J) = \pi(J)$. Then the job model $\mathcal{J}$ is schedulable if and only if $U(\mathcal{J}) \leq 1$.*

Proof. Let $\mathcal{J}$ be a periodic job model such that for every job K, $\delta(K) = \pi(K)$. Let

$$U = \sum_{K \in \mathcal{J}} \eta(K)/\pi(K) \qquad (1)$$

be the utilization of the job model.

If $U > 1$, then the total demand for the processing time exceeds the available supply, and there does not exist a deadline-compliant schedule, and the job model is not schedulable.

For the converse, assume that the job model $\mathcal{J}$ is not schedulable. Consider an EDF schedule σ for $\mathcal{J}$. The schedule σ cannot be deadline-compliant. We proceed to prove that the utilization U must exceed 1.

Since σ is not deadline-compliant, there must exist a job J and an instance i of the job J, such that the schedule σ does not allocate enough time slots to the instance J^i. Let t_1 be the arrival time of this instance, that is, $t_1 = \alpha(J, i)$, and let t_2 be the deadline of this instance, that is, $t_2 = \delta(J, i)$, which is the same as $\alpha(J, i + 1)$ (see figure 8.5). We know that the instance J^i misses its deadline, and thus the schedule σ chooses the job J for fewer than $\eta(J)$ time instances in the interval $[t_1, t_2)$: the condition $\sigma(t_1, t_2, J) < \eta(J)$ must hold.

Consider a time instance t with $t_1 \le t < t_2$. We know that the job J is ready at time t and its deadline is t_2. The EDF schedule σ at time t must choose some job keeping the processor busy: $\sigma(t) \ne \perp$. Furthermore, the EDF policy selects a job with the earliest deadline, and hence if $\sigma(t) = K$, then the deadline of the job K at time t must be t_2 or less.

Let t_0 be the least time such that for all time instances t with $t_0 \le t < t_2$, the schedule σ chooses some job K at time t such that the deadline of the job K at time t is $\le t_2$. In other words, the time instance t_0 is chosen so that the interval $[t_0, t_2)$ is the longest interval, such that the processor is busy at all times during the interval and is assigned to a job instance with a deadline t_2 or earlier. From the argument in the preceding paragraph, it follows that the interval $[t_1, t_2)$ does satisfy the desired condition, but it may not be the longest such interval, and thus t_0 is chosen by extending this interval to the left as long as the schedule σ assigns time slots to job instances with deadlines t_2 or less.

By our choice of t_0, the processor is busy throughout the interval $[t_0, t_2)$, and thus it follows that the length of this interval equals the sum of the number of slots the schedule σ allocates to each job during this interval:

$$t_2 - t_0 = \sum_{K \in \mathcal{J}} \sigma(t_0, t_2, K) \qquad (2).$$

The next step in the proof aims to derive a bound on the value of each quantity $\sigma(t_0, t_2, K)$. For this purpose, we show that:

> **Claim:** If the schedule σ allocates a time slot in the interval $[t_0, t_2)$ to a job K, then the corresponding instance of K *lies entirely within the interval* $[t_0, t_2)$.

Consider an arbitrary job K and an instance j of this job. The interval corresponding to the instance K^j is $[\alpha(K, j), \alpha(K, j+1))$. If this interval is not contained within the interval $[t_0, t_2)$, then one of the following three cases can happen.

First, the interval $[\alpha(K, j), \alpha(K, j+1))$ has no overlap with the interval $[t_0, t_2)$. This can happen if $\alpha(K, j) \ge t_2$ or if $\alpha(K, j+1) \le t_0$. In this case, the instance K^j is not relevant to scheduling during the interval $[t_0, t_2)$. In figure 8.5, these are the instances before the ath instance or after the cth instance of the job K.

A second possibility is that there is an overlap, but $\alpha(K, j+1) > t_2$ (see the cth instance of the job K in figure 8.5). We know that throughout the interval $[t_0, t_2)$, the schedule σ picks a job only if its current deadline does not exceed t_2. Throughout the interval $[\alpha(K, j), \alpha(K, j+1))$ corresponding to the instance K^j, the deadline is $\alpha(K, j+1)$, which exceeds t_2, and thus we can conclude that

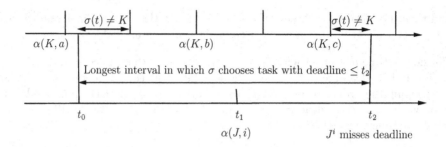

Figure 8.5: Proof of Test for EDF Schedulability

during the interval $[t_0, t_2)$, the schedule does not allocate any time slots to such an instance K^j.

The third and final possibility is that the interval corresponding to the instance K^j does overlap with $[t_0, t_2)$, the deadline $\alpha(K, j+1)$ does not exceed t_2, but $\alpha(K, j) < t_0$ (see the ath instance of the job K in figure 8.5). We claim that, in this case,

the job K is not ready at time t_0 in the schedule σ,

that is, all the demand for the instance K^j has been met during the interval $[\alpha(K, j), t_0)$, and the job K does not need anymore slots during the interval $[t_0, \alpha(K, j+1))$. Since the schedule picks only ready jobs, the claim implies that during the interval $[t_0, t_2)$, the schedule does not allocate slots to the instance K^j.

To prove the claim, assume to the contrary that the job K is ready at time t_0, that is, the instance K^j has not been allocated enough processing time before time t_0. Consider a time instance t such that $\alpha(K, j) \leq t < t_0$. The job K is ready at time t, and its deadline at time t is $\alpha(K, j+1)$. According to the EDF policy, then, $\sigma(t)$ cannot be $\bot$ and must be a job whose deadline at time t is $\leq \alpha(K, j+1)$. Since $\alpha(K, j+1) \leq t_2$, we can conclude that the interval $[\alpha(K, j), t_2)$ is an interval in which at every time the schedule picks a job with deadline t_2 or less. But this contradicts the assumption that $[t_0, t_2)$ is the longest such interval. In other words, if the job K is ready at time t_0, then we should be extending the interval of interest from $[t_0, t_2)$ to $[\alpha(K, j), t_2)$.

Thus, we have shown that whenever the EDF schedule σ picks a job K at a time instance t in the interval $[t_0, t_2)$, the corresponding instance of K must lie entirely within the interval $[t_0, t_2)$ (such as the bth instance of the job K in figure 8.5). The length of the interval $[t_0, t_2)$ is $t_2 - t_0$. The length of the interval corresponding to any instance of K is $\pi(K)$. It follows that the number of instances of K that lie entirely within $[t_0, t_2)$ is at most $(t_2 - t_0)/\pi(K)$. Note that the number of instances of K with a possible overlap with the interval $[t_0, t_2)$ can be the two extreme instances with partial overlap plus the number of instances that lie entirely within it. But we have established that the two extreme instances are not allocated any slots.

To each instance of K that lies entirely inside the interval $[t_0, t_2)$, the schedule allocates at most $\eta(K)$ number of slots. It follows that, for every job K,

$$\sigma(t_0, t_2, K) \leq \eta(K) \cdot (t_2 - t_0)/\pi(K) \qquad (3).$$

The inequality (3) holds for the job J also. However, we know that the instance J^i is allocated strictly less than $\eta(J)$ number of slots by the schedule σ (since this instance misses its deadline t_2). This implies the following strict inequality:

$$\sigma(t_0, t_2, J) < \eta(J) \cdot (t_2 - t_0)/\pi(J) \qquad (4).$$

Summing the inequalities (3) over all the jobs and noting that the inequality is strict at least for the job J leads to:

$$\sum_{K \in \mathcal{J}} \sigma(t_0, t_2, K) < \sum_{K \in \mathcal{J}} \eta(K) \cdot (t_2 - t_0)/\pi(K) \qquad (5).$$

We can substitute U from equation (1) in the above inequality to get

$$\sum_{K \in \mathcal{J}} \sigma(t_0, t_2, K) < (t_2 - t_0) \cdot U \qquad (6).$$

From the equation (2) and inequality (6), we get

$$(t_2 - t_0) < (t_2 - t_0) \cdot U.$$

Since $t_2 - t_0$ is positive, we can conclude that $1 < U$, which is what we wanted to prove. ∎

8.3 Fixed-Priority Scheduling

The EDF policy is an example of a *dynamic priority* policy, where different instances of the same job get assigned different priorities; as a result, at some time instance, a job J is preferred over another job K, while at some other time instance, the job K is preferred over the job J. Now we turn our attention to *fixed-priority* policies, where jobs are statically assigned fixed priorities, and whenever the scheduler has to make a choice, the job with the highest priority is chosen. In particular, we will analyze the properties of the most commonly used such policies, namely, the *deadline-monotonic* policy and the *rate-monotonic* policy.

8.3.1 Deadline-Monotonic and Rate-Monotonic Policies

Fixed-Priority Policies

A priority assignment for a set of jobs assigns each job a number such that no two jobs have the same number. Given two jobs J and K, if the number assigned to the job J is larger than the number assigned to the job K, then the job J has a higher priority than the job K. Given such a priority assignment, a fixed-priority scheduling policy always prefers a job of higher priority over jobs with lower priorities.

More precisely, the fixed-priority schedule is constructed for each time $t = 0, 1, 2 \ldots$ in the following manner. Recall that, given a schedule up to time t, a job J is ready at time t if the instance of J that is active at time t has not already been allocated the necessary $\eta(J)$ number of slots. If there is no job that is ready at time t, then the next slot is left unassigned. Otherwise the fixed-priority schedule selects the job that is ready at time t and has the highest priority among all the jobs ready at that time.

The definition of such policies is summarized below.

FIXED-PRIORITY SCHEDULING POLICY

A *priority assignment* for a periodic job model $\mathcal{J}$ is a function ρ that maps each job $J \in \mathcal{J}$ to a natural number such that for every two distinct jobs J and K, $\rho(J) \neq \rho(K)$. A schedule σ for the periodic job model $\mathcal{J}$ is called a *fixed-priority schedule* with respect to the priority assignment ρ, if for every time $t \geq 0$, if no job is ready at time t in the schedule σ, then $\sigma(t) = \bot$, else $\sigma(t) = J$, such that the job J is ready at time t in the schedule σ, and for every job $K \in \mathcal{J}$, either the job K is not ready at time t in the schedule σ or $\rho(K) < \rho(J)$.

Given a priority assignment ρ, the corresponding fixed-priority schedule depends only on the ordering of the jobs induced by the priorities and not on the numerical values of priorities assigned to jobs. More precisely, consider two priority assignments ρ and ρ' for a periodic job model $\mathcal{J}$ such that for every pair of jobs J and K, $\rho(J) > \rho(K)$ exactly when $\rho'(J) > \rho'(K)$. Then at every time t, the fixed-priority scheduler makes exactly the same decision whether it is based on the priority assignment ρ or is based on the assignment ρ'. Thus, in this case, the fixed-priority schedule with respect to the assignment ρ coincides with the fixed-priority schedule with respect to the assignment ρ'.

Deadline-Monotonic and Rate-Monotonic Priorities

Once we commit to employing a fixed-priority scheduling policy, the only choice for a scheduling policy concerns the priority assignment to jobs. If the period of a job J is smaller than the period of another job K, then the instances of the job J arrive at a faster rate than the instances of the job K, and this suggests that the job J should be assigned a higher priority. This priority assignment rule is called *rate monotonic*: the priority assignment ρ for jobs is a rate-monotonic priority assignment if for every pair of jobs J and K, if $\pi(J) < \pi(K)$, then $\rho(J) > \rho(K)$. Note that if two jobs have the same period, then a rate-monotonic priority assignment still needs to assign different priorities to both, and there is a choice regarding which of the two should be assigned a higher priority.

When all deadlines are implicit, that is, for every job, its deadline equals its period, it turns out that the rate-monotonic policy is an optimal choice for assigning priorities. However, when the jobs have explicitly specified deadlines, then it seems intuitive to prefer a job with an earlier deadline over a job with a later deadline. The resulting priority assignment rule is called *deadline monotonic*: the priority assignment ρ for jobs is a deadline-monotonic priority assignment if for every pair of jobs J and K, if $\delta(J) < \delta(K)$, then $\rho(J) > \rho(K)$. Again, if two jobs have the same deadline, then a deadline-monotonic priority assignment can arbitrarily choose one of them to have a higher priority. In a case where the deadline of each job equals its period, the notions of deadline-monotonic policy and rate-monotonic policy coincide. Note that while the EDF policy also prefers job instances with earlier deadlines, EDF is a dynamic priority policy, whereas deadline monotonic is a fixed-priority policy.

The notions of rate-monotonic and deadline-monotonic schedules are formalized below.

DEADLINE-MONOTONIC AND RATE-MONOTONIC POLICY

A priority assignment ρ for a periodic job model $\mathcal{J}$ is called *deadline monotonic* if for all jobs $J, K \in \mathcal{J}$, if $\delta(J) < \delta(K)$, then $\rho(J) > \rho(K)$; and is called *rate monotonic* if for all jobs $J, K \in \mathcal{J}$, if $\pi(J) < \pi(K)$, then $\rho(J) > \rho(K)$. A schedule σ for the periodic job model $\mathcal{J}$ is called a *deadline-monotonic schedule* if there exists a deadline-monotonic priority assignment ρ, such that the schedule σ is a fixed-priority schedule with respect to ρ; and is called a *rate-monotonic schedule* if there exists a rate-monotonic priority assignment ρ, such that the schedule σ is a fixed-priority schedule with respect to ρ.

Examples

Let us revisit the job model consisting of the job J_1 with period 5, deadline 4, and WCET 3; and the job J_2 with period 3, deadline 3, and WCET 1. We know that this job model is schedulable: see figures 8.2 and 8.3 for deadline-compliant schedules. In particular, the schedule of figure 8.3 is an EDF schedule, and it sometimes prefers the job J_1 over J_2 and other times prefers the job J_2 over J_1. The priority assignment that assigns a higher priority to the job J_2 than to the job J_1 is deadline monotonic (and also rate monotonic). Observe that in the resulting fixed-priority schedule, the job J_2 is selected at time 0 as well as at time 3; as a result, the first instance of the job J_1 gets only two slots before its deadline 4, causing a missed deadline. If the priority assignment were to always prefer the job J_1 over the job J_2, then the first three slots are allocated to the job J_1, causing the first instance of the job J_2 to miss its deadline. Thus, fixed-priority scheduling policies do not produce deadline-compliant schedules for this job model.

Let us change the deadline of the job J_1 to 5: this leads to the job model with implicit deadlines consisting of the job J_1 with period 5 and WCET 3 and the

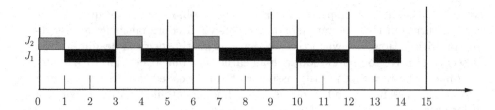

Figure 8.6: Illustrative Rate-Monotonic Schedule

job J_2 with period 3 and WCET 1. The rate-monotonic priority assignment still assigns a higher priority to the job J_2 than to the job J_1. The resulting fixed-priority schedule is shown in figure 8.6. The schedule is periodic with period 15 and meets all the deadlines.

Properties

We have already noted that a deadline-monotonic policy may fail to produce a deadline-compliant schedule even when the job model is schedulable. Conditions under which deadline-monotonic and rate-monotonic policies are guaranteed to succeed are studied in sections 8.3.2 and 8.3.3. For now let us note some basic properties of these policies.

The deadline-monotonic and rate-monotonic policies resolve the choice among the set of ready jobs by a simple local rule and, similar to the EDF policy, are examples of greedy algorithms. We also note that any fixed-priority scheduling policy is guaranteed to produce a periodic schedule with period equal to the least-common multiple of the periods of all the jobs. Furthermore, the schedules generated by these policies can be preemptive (for example, see the rate-monotonic schedule shown in figure 8.6).

The principal benefit of the deadline-monotonic as well as the rate-monotonic policy is that the overhead needed to implement the scheduler is minimal. The scheduler needs to assign a priority to each job off-line, and the computation necessary for this purpose involves ordering jobs according to their deadlines. As in the case of the EDF policy, the deadline-monotonic/rate-monotonic scheduling policy also needs to make a scheduling decision only when either the currently executing instance of a job finishes its execution or a new instance of another job arrives. The priority of an instance of a job equals the statically assigned priority to that job, and thus no computation is needed to determine priorities at run-time. As a result, the deadline-monotonic as well as the rate-monotonic policy can be easily incorporated in any operating system that supports priority-based scheduling of processes.

Exercise 8.9 : Consider a periodic job model of exercise 8.6 consisting of the job J_1 with period 6, deadline 5, and WCET 2; the job J_2 with period 8, deadline

4, and WCET 2; and the job J_3 with period 12, deadline 8, and WCET 4. Is this model schedulable using the deadline-monotonic scheduling policy? ■

Exercise 8.10: Construct a periodic job model with two jobs such that the deadline-monotonic policy leads to a deadline-compliant schedule, but the rate-monotonic policy results in a schedule with missed deadlines. ■

8.3.2   Optimality of Deadline-Monotonic Policy *

We have already noted that, unlike the EDF policy, the deadline-monotonic policy is not an optimal policy for producing deadline-compliant schedules: a deadline-monotonic schedule for a schedulable periodic job model need not be deadline-compliant. In this section, we establish that the deadline-monotonic policy, however, is optimal among all fixed-priority policies: if there exists a priority assignment such that the corresponding fixed-priority schedule meets all the deadlines, then a deadline-monotonic schedule also meets all the deadlines. This implies that if we prefer fixed-priority scheduling for its simplicity and minimal scheduling overhead, then it is desirable to choose a deadline-monotonic priority assignment.

Criticality of First Instances

Before we prove optimality of the deadline-monotonic policy relative to fixed-priority schedules, let us establish a property that all fixed-priority schedules satisfy: if a fixed-priority schedule misses a deadline, then it misses the deadline of the first instance of some job. In other words, in a fixed-priority schedule, the first instances of all the jobs are the critical ones: if all first instances meet their deadlines, then the remaining instances are guaranteed to meet their deadlines. This implies that if we want to check whether a fixed-priority schedule is deadline-compliant, we can explicitly construct the schedule from time $t = 0$ up to time $t = d$, where d is the maximum of the deadlines of all the jobs, ensuring that no deadlines are missed until this time, instead of constructing and examining the entire periodic schedule up to time $t = p$, where p is the least-common multiple of the periods of all the jobs, which is typically much larger than d.

Theorem 8.4 [Criticality of the first instances in fixed-priority schedules] *If σ is a fixed-priority schedule for a periodic job model, and if the deadline of the first instance of every job is met in the schedule σ, then the schedule σ is deadline-compliant.*

Proof. Let $\mathcal{J}$ be a periodic job model, let ρ be a priority assignment for $\mathcal{J}$, and let σ be the fixed-priority schedule for $\mathcal{J}$ with respect to the assignment ρ. Let us assume that the schedule σ allocates enough slots to the first instance of every job. We want to prove that the schedule is deadline-compliant.

For proof by contradiction, assume that the schedule σ is not deadline-compliant. Then there must exist a job instance that misses its deadline in the schedule σ.

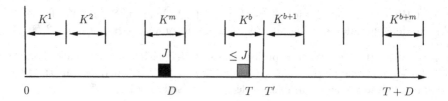

Figure 8.7: Illustration for Proof of Theorem 8.4

Let J be the job with the highest priority that misses a deadline. That is, there exists an instance of the job J that misses its deadline in the schedule σ, and if K is a job such that $\rho(K) > \rho(J)$, then all instances of the job K meet their deadlines in the schedule σ. Let h be the priority of this job J.

By assumption, the first instance of the job J gets $\eta(J)$ number of slots by its deadline. Let D be the time when the first instance finishes its execution, that is, D is the number such that $\sigma(0, D, J) = \eta(J)$ and $\sigma(D-1) = J$ (see figure 8.7 for illustration). By assumption, $D \leq \delta(J)$ holds.

For every $t \geq 0$, let $\theta(t)$ denote the number of slots that the schedule allocates to jobs with priority higher than that of the job J during the time interval of length D starting at time t:

$$\theta(t) \;=\; \sum_{K \in \mathcal{J},\, \rho(K) > h} \sigma(t, t+D, K).$$

For every instance a, the instance J^a of the job J arrives at time $t_a = \alpha(J, a)$. By definition, $\theta(t_a)$ is the number of slots that the schedule allocates to jobs of priority higher than h during the interval of length D starting at time t_a. Any slot not allocated to a job of higher priority is assigned to the job J as long as the job J needs slots, and thus if the condition $\theta(t_a) \leq D - \eta(J)$ holds, then the instance J^a finishes its execution within D units of its arrival and, thus, meets its deadline (since $D \leq \delta(J)$).

We know that the first instance of the job J finishes at time D, and thus $\theta(0) \leq D - \eta(J)$ holds. We also know that there exists an instance of the job J that misses its deadline, and thus for some a, $\theta(t_a) > D - \eta(J)$ must hold, which implies, $\theta(t_a) > \theta(0)$. Let T be the least time for which $\theta(T) > \theta(0)$; that is, if $\theta(t) > \theta(0)$, then $T \leq t$. Note that this particular time instance T cannot exceed the arrival time t_a of the a instance that misses its deadline, but T itself need not be the arrival time of any instance of the job J. By the choice of T, we know that $\theta(T - 1) < \theta(T)$, and thus we can conclude that the schedule σ cannot choose a job of priority higher than h at time $T - 1$.

We proceed to show that the schedule σ does not allocate more slots to a job with priority higher than h during the interval $[T, T+D)$ than during the interval $[0, D)$:

Claim: if $\rho(K) > \rho(J)$ then $\sigma(T, T + D, K) \leq \sigma(0, D, K)$.

Consider a job K such that $\rho(K) > \rho(J)$. By assumption, all instances of such a job meet their deadlines. To prove the claim that $\sigma(T, T + D, K) \leq \sigma(0, D, K)$, our analysis depends on how many instances of the job K overlap with the interval $[0, D)$. Instances of the job K arrive every $\pi(K)$ time units. Let m be the integer such that $(m - 1) \cdot \pi(K) < D \leq m \cdot \pi(D)$, that is, m is obtained by calculating $D/\pi(K)$, and if this quantity is a fraction, then rounding it up to the next integer. Then the first m instances of the job K overlap with the interval $[0, D)$ (see figure 8.7 for illustration).

Observe that the first $(m - 1)$ instances of the job K lie entirely within the interval $[0, D)$, and the mth instance has a partial overlap. Note that since all instances of the job K meet their deadlines, the schedule σ allocates exactly $\eta(K)$ number of slots to each instance of the job K. Each of the first $(m - 1)$ instances of the job K lie entirely within the interval $[0, D)$, and thus each of them contributes $\eta(K)$ to the quantity $\sigma(0, D, K)$. We know that the last slot of the interval $[0, D)$ is allocated to the job J. Since the job K has a higher priority than that of the job J, this implies that at time $(D - 1)$, the job K is not ready, which implies that its corresponding instance has been allocated $\eta(K)$ number of slots by time $(D - 1)$. This means that the contribution of the mth instance of K to the quantity $\sigma(0, D, K)$ is also $\eta(K)$. Thus, $\sigma(0, D, K)$ equals $m \cdot \eta(K)$.

Now let us focus our attention on how different instances of the job K overlap with the interval $[T, T + D)$. Suppose the earliest instance of the job K that arrives at time T or later is the $(b + 1)$-th instance: $\alpha(K, b) < T \leq \alpha(K, b + 1)$. Let T' be the arrival time of the $(b + 1)$-th instance of the job K.

In general, $T' > T$ is possible, and then the bth instance of the job K also has an overlap with the interval $[T, T + D)$ (figure 8.7 illustrates this case). By the choice of T, we know that at time $(T - 1)$, the schedule does not choose a job of priority higher than h and, thus, does not choose the job K or a job of priority higher than that of the job K. This can happen only if the job K is not ready at time $(T - 1)$, which means that all the demand of the bth instance of the job K has been met by time $(T - 1)$. This implies that even though the bth instance can overlap with the interval $[T, T + D)$, it does not contribute to the quantity $\sigma(T, T + D, K)$. This implies that $\sigma(T, T + D, K)$ equals $\sigma(T', T + D, K)$.

The number of instances of K that overlap with the interval $[T', T + D)$ is obtained by dividing its length by $\pi(K)$ and rounding up the answer to the next integer. Since $T' \geq T$, the length of the interval $[T', T + D)$ cannot exceed D, it follows that the number of instances of K that overlap with the interval $[T', T + D)$ is at most m (it is either m or $m - 1$ depending on the difference between T and T'). Since an instance of the job K can be allocated only $\eta(K)$ number of slots, it follows that $\sigma(T', T + D, K) \leq m \cdot \eta(K)$.

We have established the claim that $\sigma(T, T + D, K) \leq \sigma(0, D, K)$ for every job K of priority higher than h. This means that $\theta(T) > \theta(0)$ is not possible, resulting in the desired contradiction. ∎

Proof of Optimality

Now we proceed to prove that a deadline-monotonic policy is guaranteed to produce a deadline-compliant schedule as long as there exists a deadline-compliant fixed-priority schedule.

Theorem 8.5 [Optimality of deadline-monotonic policy] *Given a periodic job model $\mathcal{J}$, if there exists a priority assignment ρ for $\mathcal{J}$, such that the corresponding fixed-priority schedule σ is deadline-compliant, then every deadline-monotonic schedule for $\mathcal{J}$ is deadline-compliant.*

Proof. Let $\mathcal{J}$ be a periodic job model. Let ρ be a priority assignment for $\mathcal{J}$ such that the fixed-priority schedule σ with respect to ρ meets all the deadlines. Suppose the ordering of the jobs in $\mathcal{J}$ according to the priority assignment ρ is $J_1, J_2, \ldots J_n$, where n is the number of jobs in $\mathcal{J}$, with the job J_1 being the job of the highest priority and the job J_n being the job of the lowest priority.

Let ρ' be a deadline-monotonic priority assignment. We want to prove that the fixed-priority schedule with respect to the assignment ρ' is also deadline-compliant. If the ordering of the jobs according the priorities assigned by ρ' equals that according to the assignment ρ, then the schedule σ is also the fixed-priority schedule with respect to ρ' since the choices made by the schedule depend only on the relative priorities of the jobs and not on the numerical values of priorities. Hence, suppose that the ordering of the jobs in decreasing priorities according to the assignment ρ' is not the same as the ordering $J_1, J_2, \ldots J_n$. Then there must exist a pair of adjacent jobs J_a and J_{a+1} in this order such that their relative priorities are different according to the assignment ρ': $\rho'(J_{a+1}) > \rho'(J_a)$. Since the assignment ρ' is deadline-monotonic, it follows that the deadline of the job J_a is not earlier than that of the job J_{a+1}: $\delta(J_a) \geq \delta(J_{a+1})$.

First, we show that if we modify the priority assignment ρ by swapping the priorities of two such adjacent jobs, then the corresponding schedule is still deadline-compliant. That is, let ρ_1 be the priority assignment such that $\rho_1(J_b) = \rho(J_b)$ for $b < a$ and for $b > a+1$, $\rho_1(J_a) = \rho(J_{a+1})$ and $\rho_1(J_{a+1}) = \rho(J_a)$. Thus, the ordering of the jobs with decreasing priorities according to the assignment ρ_1 is $J_1, J_2, \ldots, J_{a-1}, J_{a+1}, J_a, J_{a+2}, \ldots J_n$.

Let σ_1 be the fixed-priority schedule with respect to the priority assignment ρ_1. We want to prove that the schedule σ_1 is deadline-compliant. From theorem 8.4, to prove that the schedule σ_1 meets the deadlines of all instances of all the jobs, it suffices to prove that it meets the deadline of the first instance of each job.

Consider a job J_b for $b = 1, 2, \ldots n$. Let D_b be the time by which the first instance of the job J_b finishes its execution in the deadline-compliant schedule σ according to the original priority assignment ρ. That is, $\sigma(D_b - 1) = J_b$ and $\sigma(0, D_b, J_b) = \eta(J_b)$. Since this is a deadline-compliant schedule, we know that $D_b \leq \delta(J_b)$ holds. We proceed to prove that, in the schedule σ_1 with respect to the revised priority assignment ρ_1, for each $b \neq a$, the first instance of the job J_b finishes its execution by time D_b, and the first instance of the job J_a finishes its execution by time D_{a+1}. Since $D_{a+1} \leq \delta(J_{a+1})$ and $\delta(J_{a+1}) \leq \delta(J_a)$, it follows that the first instances of all the jobs finish their executions by their deadlines, implying deadline-compliance of the schedule σ_1.

To show that the first instance of the job J_a finishes its execution by time D_{a+1}, we prove:

> **Claim:** The schedule σ_1 allocates $\eta(J_a)$ time slots to the job J_a in the time interval $[0, D_{a+1})$.

Let us denote D_{a+1} by D. To prove the claim, let us consider a job J_b of a priority higher than that of the job J_a according to the priority assignment ρ_1 and compute how many time slots are allocated to the job J_b by the schedule σ_1 during the interval $[0, D)$. According to the priority assignment ρ_1, a job J_b has a higher priority than that of the job J_a if either $b < a$ or $b = a + 1$. We consider these two cases one by one.

- **Case $b < a$:** The job J_b has a higher priority than the job J_a in both the schedules σ and σ_1. Suppose the number of instances of the job J_b that overlap with the interval $[0, D)$ is m, that is, m is the integer obtained by rounding up the quantity $D/\pi(J_b)$. Using reasoning similar to the one used in the proof of theorem 8.4, we can conclude that $\sigma(0, D, J_b) = m \cdot \eta(J_b)$: each of the first $(m - 1)$ instances of the job J_b are entirely contained within the interval $[0, D)$, and since this schedule meets all the deadlines, each of them gets $\eta(J_b)$ number of slots; the mth instance can have a partial overlap, but since the priority of the job J_b is higher than that of the job J_{a+1} and the schedule σ chooses the job J_{a+1} at time $(D-1)$, this last instance of J_b must have been allocated all its demand before time D. Since m is the total number of instances of the job J_b that overlap with the interval $[0, D)$, the schedule σ_1 cannot possibly allocate more than $m \cdot \eta(J_b)$ number of slots to the job J_b, and we can conclude that $\sigma_1(0, D, J_b) \leq \sigma(0, D, J_b)$ holds.

- **Case $b = a + 1$:** Now the job J_b has a higher priority than the job J_a in the schedule σ_1 but has a lower priority than the job J_a in the schedule σ. We know that in the schedule σ the first instance of the job J_b finishes its execution by time D, and since this schedule is deadline-compliant, $\sigma(0, D, J_b) = \eta(J_b)$. It also implies that only the first instance of the job J_b has any overlap with the interval $[0, D)$, so it follows that in the schedule σ_1, even though the job J_b has a relatively higher priority, it cannot allocate more then $\eta(J_b)$ number of time slots to the job J_b. We can conclude that $\sigma_1(0, D, J_b) \leq \sigma(0, D, J_b)$ holds.

Thus, we have proved that for a job J_b that has a higher priority than the job J_a in the revised priority assignment ρ_1, the schedule σ_1 does not allocate more slots to the job J_b during the interval $[0, D)$ than the schedule σ does. Since the schedule σ allocates $\eta(J_a)$ number of time slots to the job J_a by time D_a and $D_a \leq D$, it follows that $\sigma(0, D, J_a) = \eta(J_a)$. In the schedule σ_1, during the interval $[0, D)$, no job J_b for $b > a + 1$ will be chosen before the job J_a is allocated $\eta(J_a)$ number of time slots. Hence, the claim follows.

The proof that for each $b \neq a$, the first instance of the job J_b finishes by time D_b in the schedule σ_1 is similar and is left as an exercise.

We have established that the schedule corresponding to the priority assignment ρ_1 obtained by swapping priorities of an adjacent pair of jobs in the ordering given by the original priority assignment ρ is also deadline-compliant. If the ordering of the jobs according to the priority assignment ρ_1 equals the ordering according to the deadline-monotonic assignment ρ', we have already established our goal. If not, we can repeat the argument: in the assignment ρ_1, we can find a pair of jobs that are adjacent according to the ordering given by the assignment ρ_1 but have different relative ordering according to the assignment ρ' and swap their priorities to obtain the assignment ρ_2. By the argument above, deadline-compliance of the fixed-priority schedule σ_1 with respect to the assignment ρ_1 implies the deadline-compliance of the fixed-priority schedule σ_2 with respect to the assignment ρ_2.

To finish the proof, we need to establish that such swapping of jobs will not continue forever. To understand why only a bounded number of swaps suffice, let us consider a concrete example. Suppose the ordering of the jobs according to the original priority assignment ρ is J_1, J_2, J_3, J_4, and the ordering according to the desired (deadline-monotonic) priority assignment is J_3, J_1, J_4, J_2. Then starting from the assignment ρ, we can swap the ordering of the jobs J_2 and J_3 to get the priority assignment ρ_1 with the ordering J_1, J_3, J_2, J_4; then swap the ordering of the jobs J_1 and J_3 to get the priority assignment ρ_2 with the ordering J_3, J_1, J_2, J_4; and finally swap the ordering of the jobs J_2 and J_4 to get the priority assignment ρ'. As this example suggests, this process is the same as an algorithm for sorting a sequence of elements by swapping adjacent out-of-order elements, where each swap makes the current sequence more "similar" to the desired final sequence.

To make this argument precise, let us define the distance between two priority assignments ρ and ρ' to be the number of pairs (a, b), such that the relative ordering of the priorities of the jobs J_a and J_b are different in the two assignments. Such a distance can be at most $n(n-1)$, where n is the number of jobs. The assignment ρ_1 is obtained from the assignment ρ by swapping an adjacent pair of jobs to make the assignment more similar to the target assignment ρ'. More precisely, if the distance between the assignments ρ and ρ' is k, then the distance between the assignment ρ_1 and ρ' cannot be more than $(k-1)$: if the assignment ρ_1 is obtained from the assignment ρ by swapping the pair J_a and J_{a+1}, then the pairs that contribute to the distance between ρ_1 and ρ' are exactly those pairs that contribute to the distance between ρ and ρ', except the swapped pair $(a, a+1)$. Since the distance decreases by at least one at every step, it follows that there can be at most $n(n-1)$ many swaps before the assignment becomes identical to ρ'. ∎

Exercise 8.11: In the proof of theorem 8.5, the schedule σ_1 is the fixed-priority schedule obtained by swapping priorities of two adjacent jobs J_a and J_{a+1}, such that $\delta(J_a) \geq \delta(J_{a+1})$. We proved that in this schedule, the first instance of the job J_a finishes by time D_{a+1}, where for every b, D_b is the time by which the first instance of the job J_b finishes its execution in the original schedule σ. Complete the proof of deadline-compliance of the schedule σ_1 by showing that for every $b \neq a$, the first instance of the job J_b finishes its execution by time D_b in the schedule σ_1. ∎

8.3.3 Schedulability Test for Rate-Monotonic Policy *

Given a periodic job model $\mathcal{J}$, how can we check if the rate-monotonic or the deadline-monotonic scheduling policy is going to succeed in producing a deadline-compliant schedule? One possibility is to explicitly compute the schedule and check if it meets all the deadlines. Theorem 8.4 assures us that it suffices to examine compliance of deadlines only for the first instances of all the jobs, and thus we need to compute the schedule only for the first d time slots, where d is the maximum of the deadlines of all the jobs. We proceed to establish a simpler condition based on the utilization that assures us that for a periodic job model with implicit deadlines, if the utilization is below a certain threshold value, then the rate-monotonic policy is guaranteed to produce a deadline-compliant schedule. Recall that the utilization of a job model specifies the fraction of available processing time that is needed to execute all the jobs and can be computed easily. We also know that the EDF policy is guaranteed to succeed as long as the utilization does not exceed 1. It turns out that the rate-monotonic policy is guaranteed to succeed as long as the utilization does not exceed 0.69. This result is of importance for both practical and theoretical reasons. In practice, if we know that the total demand for processing time is not too high (less than 69%, to be precise), then it suffices to employ the rate-monotonic policy, which has a minimal scheduling overhead. In theory, techniques used to establish this bound illustrate how to analyze algorithms for the worst case.

Analyzing Utilization for Two Jobs

For now, let us suppose that there are only two jobs with implicit deadlines. Let us call the job with the higher priority according to the rate-monotonic priority assignment J_1 and the job with the lower priority J_2. Let σ denote the corresponding fixed-priority rate-monotonic schedule. Let π_1 and π_2 be the periods of the two jobs. By assumption, $\pi_1 \leq \pi_2$. Let η_1 and η_2 be the worst-case execution times of the two jobs. The utilization for this job model is given by

$$U = \eta_1/\pi_1 + \eta_2/\pi_2.$$

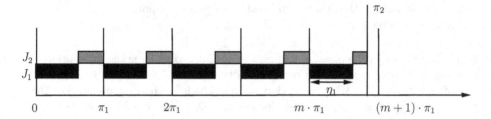

Figure 8.8: Analyzing Rate-monotonic Policy for Two Jobs: Case (a)

Our goal is to come up with a numerical bound B, which should be as high as possible, such that if $U \leq B$ holds, then we are guaranteed that the rate-monotonic schedule σ meets all the deadlines. For this purpose, we eliminate the four parameters in the above expression one by one, and the ensuing analysis reveals relationships among these parameters that cause the *worst-case* scenario for the rate-monotonic policy.

Eliminating WCET η_2

As the first step in the analysis, let us treat the parameters π_1, π_2, and η_1 as given and suppose we want to find out constraints on the fourth parameter η_2, in terms of these three fixed parameters so that the schedule is deadline-compliant.

For checking deadline-compliance, we need to examine only the first instances of the two jobs. Since the job J_1 has the higher priority, its first instance gets the first η_1 slots and is guaranteed to meet its deadline (recall that, by assumption, for every job, its WCET cannot exceed its deadline). To analyze the condition under which the job J_2 gets η_2 slots by its deadline π_2, we need to consider how many instances of the job J_1 are executed in the interval $[0, \pi_2)$. Let m be the number such that the condition $m \cdot \pi_1 \leq \pi_2 < (m + 1) \cdot \pi_1$ holds, that is, m is obtained by rounding down the quantity π_2/π_1 to the nearest integer. Then the first m instances of the job J_1 lie entirely within the interval $[0, \pi_2)$.

To compute the number of slots available for the execution of the job J_2 before its deadline, there are two cases as shown in figures 8.8 and 8.9.

- **Case (a):** If $m \cdot \pi_1 + \eta_1 < \pi_2$, then the $(m + 1)$th instance of the job J_1 finishes its execution before the deadline π_2 of the first instance of the job J_2. This is the case shown in figure 8.8. In this case, the total time allocated to the first job in the interval $[0, \pi_2)$ is $(m + 1) \cdot \eta_1$. Then the first instance of the job J_2 can be allocated up to $\pi_2 - (m + 1) \cdot \eta_1$ time slots. Thus, the schedule meets the deadlines as long as the condition $\eta_2 \leq \pi_2 - (m + 1) \cdot \eta_1$ holds. This implies that the rate-monotonic policy succeeds as long as

$$U \leq \eta_1/\pi_1 + [\pi_2 - (m + 1) \cdot \eta_1]/\pi_2.$$

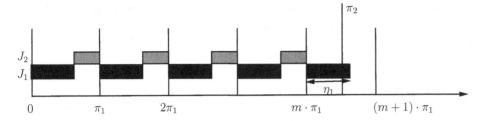

Figure 8.9: Analyzing Rate-monotonic Policy for Two Jobs: Case (b)

- **Case (b):** If $m \cdot \pi_1 + \eta_1 \geq \pi_2$, then the $(m + 1)$th instance of the job J_1 does not finish its execution before the deadline π_2 of the first instance of the job J_2. In this case, the job J_2 meets its deadline only if it finishes execution before the $(m+1)$th instance of the job J_1 arrives (see figure 8.9), and the maximum number of slots that can be allocated to the job J_2 by its deadline is $m \cdot (\pi_1 - \eta_1)$. Thus, the schedule meets the deadlines as long as the condition $\eta_2 \leq m \cdot (\pi_1 - \eta_1)$ holds. This implies that the rate-monotonic policy succeeds as long as

$$U \leq \eta_1/\pi_1 + [m \cdot (\pi_1 - \eta_1)]/\pi_2.$$

Thus, in each case, we have obtained a bound B such that if the utilization is below that bound, the schedule is guaranteed to be deadline-compliant. This bound B can be viewed as a function of the three parameters, π_1, π_2, and η_1, and can be summarized as

```
if (m·π₁+η₁ < π₂) then η₁/π₁+[π₂−(m+1)·η₁]/π₂ else η₁/π₁+[m·(π₁−η₁)]/π₂.
```

Eliminating WCET η_1

The next step is to eliminate the parameter η_1 by *minimizing* this function B over all possible choices of η_1: if the utilization is below this minimized value, then we are guaranteed that the utilization is below the desired bound no matter what value of η_1 is chosen, and hence the schedule is deadline-compliant.

For the case $m \cdot \pi_1 + \eta_1 < \pi_2$, we have

$$
\begin{aligned}
B &= \eta_1/\pi_1 + [\pi_2 - (m + 1) \cdot \eta_1]/\pi_2, \\
&= 1 + \eta_1/\pi_1 - (m + 1) \cdot \eta_1/\pi_2, \\
&= 1 - \eta_1 \cdot (m + 1 - \pi_2/\pi_1)/\pi_2.
\end{aligned}
$$

Since $\pi_2 < (m + 1) \cdot \pi_1$, the quantity $(m + 1 - \pi_2/\pi_1)$ is positive. Hence, for given values of the parameters π_1 and π_2, the value of the bound B *decreases* as η_1 increases. Hence, the minimum occurs when η_1 has the highest possible value, which due the condition $m \cdot \pi_1 + \eta_1 < \pi_2$, equals $\pi_2 - m \cdot \pi_1$.

For the case $m \cdot \pi_1 + \eta_1 \geq \pi_2$, we have

$$
\begin{aligned}
B &= \eta_1/\pi_1 + m \cdot (\pi_1 - \eta_1)/\pi_2, \\
&= m \cdot \pi_1/\pi_2 + \eta_1/\pi_1 - m \cdot \eta_1/\pi_2, \\
&= m \cdot \pi_1/\pi_2 + \eta_1 \cdot (\pi_2/\pi_1 - m)/\pi_2.
\end{aligned}
$$

Since $m \cdot \pi_1 \leq \pi_2$, the quantity $(\pi_2/\pi_1 - m)$ is positive. As a result, for given values of the parameters π_1 and π_2, the value of the bound B *increases* as η_1 increases. Hence, the minimum occurs when η_1 has the least possible value, which due the condition $m \cdot \pi_1 + \eta_1 \geq \pi_2$, equals $\pi_2 - m \cdot \pi_1$.

Thus, we have shown that the bound B is minimal when $\eta_1 = \pi_2 - m \cdot \pi_1$. Substituting this value of η_1 in the expression for B in the second case gives us the desired bound as a function of the parameters π_1 and π_2:

$$
B = m \cdot \pi_1/\pi_2 + (\pi_2 - m \cdot \pi_1) \cdot (\pi_2/\pi_1 - m)/\pi_2.
$$

We want to choose the parameters π_1 and π_2 to minimize this function. Note that m is the largest integer less than or equal to the ratio π_2/π_1. Let π_2/π_1 be $m + f$, where $f \in [0, 1)$ is the fractional part of this ratio. Substituting $m + f$ for the ratio π_2/π_1 in the above expression for the bound B leads to:

$$
B = m/(m+f) + (1 - m/(m+f)) \cdot f = (m + f^2)/(m + f).
$$

Computing the Numerical Bound

We have expressed the desired bound as a function of the two parameters m and f. Now we want to minimize this quantity over all choices of m and f, where m is a positive integer and f is a fraction in the interval $[0, 1)$.

$$
B = (m + f + f^2 - f)/(m + f) = 1 + (f^2 - f)/(m + f).
$$

Since $0 \leq f < 1$, the quantity $f^2 - f$ is always *negative*, and thus for a given value of the fraction f, the value of the bound B increases as m increases. Thus, the minium occurs at the smallest possible value of m. Note that, by assumption, $\pi_1 \leq \pi_2$, and thus m cannot be 0. This means that the smallest possible value of m is 1. Substituting this value in the expression for the bound B gives us

$$
B = (1 + f^2)/(1 + f).
$$

This says that, for a given f, which is the fractional part of the ratio of the two periods, the schedule is deadline-compliant as long as the utilization does not exceed $(1 + f^2)/(1 + f)$.

The last step of the analysis is to minimize this function with respect to the parameter f, where $0 \leq f < 1$. For this purpose, let us *differentiate* the expression B with respect to f:

$$
\begin{aligned}
dB/df &= [2f(1+f) - (1+f^2)]/(1+f)^2, \\
&= (f^2 + 2f - 1)/(1+f)^2.
\end{aligned}
$$

Thus, the derivative dB/df is 0 exactly when $f^2 + 2f - 1 = 0$. This quadratic equation has two roots, $f = -1 - \sqrt{2}$ and $f = -1 + \sqrt{2}$, of which only $-1 + \sqrt{2}$ lies in the range $[0, 1)$ for f. Thus, the bound B is minimal when $f = -1 + \sqrt{2}$. Substituting this value in the expression B gives:

$$
B = [1 + (-1 + \sqrt{2})^2]/(1 - 1 + \sqrt{2}) = 2(\sqrt{2} - 1).
$$

This leads to the main claim for the case of two jobs:

> For a periodic job model with two jobs with implicit deadlines, a rate-monotonic schedule is guaranteed to be deadline-compliant if the utilization does not exceed $2(\sqrt{2} - 1)$.

Note that the quantity $2(\sqrt{2} - 1)$ is 0.828. This means that with two jobs, if the utilization does not exceed 0.828, then we know that the rate-monotonic policy will succeed no matter what values of the periods and WCETs are chosen.

Understanding the Worst Case for Two Jobs

To recap the proof of the bound for two jobs, the worst case for the rate-monotonic policy, that is, the case where the utilization is as small as possible, while the resulting schedule is just barely deadline-compliant, occurs when (1) the period π_2 is $\sqrt{2}$ times the period π_1, (2) the WCET η_1 equals the difference $\pi_2 - \pi_1$, and (3) the WCET η_2 equals the difference $\pi_1 - \eta_1$.

Figure 8.10 shows this critical scenario for two jobs. For the job J_1, the period is 100, whereas for the job J_2, the period is 141. The WCET for the job J_1 is 41, whereas the WCET of the job J_2 is 59. Observe that the resulting schedule is deadline-compliant, and the utilization is $41/100 + 59/141$, which is about 0.828.

Suppose we increase the WCET of the first job to 42. The utilization for this updated job model is $42/100 + 59/141$, which is about 0.838, which exceeds the bound of the schedulability test. Observe that in the rate-monotonic schedule for this revised model, the second job gets only 57 slots by its deadline, and thus the schedule is not deadline-compliant.

Observe that the bound we have calculated is only a *sufficient* test for deadline-compliance. It may happen that the utilization for a model with two jobs exceeds the bound 0.828, and yet the rate-monotonic policy produces a deadline-compliant schedule. This can happen if the specific values of the periods and

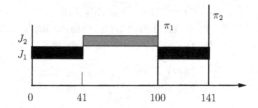

Figure 8.10: Worst Case for Rate-monotonic Policy with Two Jobs

WCETs in this model do not correspond to the worst-case scenario for the rate-monotonic policy. For example, for the job model consisting of the job J_1 with period 5 and WCET 3 and the job J_2 with period 3 and WCET 1, the utilization is 0.93, and yet as figure 8.6 illustrates, the rate-monotonic schedule meets all the deadlines.

Schedulability Test for n Jobs

The analysis for two jobs can be generalized to a job model with n jobs. We only state the worst-case scenario that the analysis reveals. Suppose the jobs are ordered $J_1, J_2, \ldots J_n$ in a decreasing order of priorities (that is, in an increasing order of periods) according to the rate-monotonic policy. For each job J_a, let us denote its period (and deadline) by π_a and its WCET by η_a. Then the worst case for the rate-monotonic policy occurs when the following relationships hold among the different parameters:

$$\pi_1 \;<\; \pi_2 \;<\; \cdots \;<\; \pi_n \;<\; 2 \cdot \pi_1,$$
$$\eta_1 \;=\; \pi_2 - \pi_1,$$
$$\eta_2 \;=\; \pi_3 - \pi_2,$$
$$\cdots$$
$$\eta_n \;=\; \pi_1 - (\eta_1 + \eta_2 + \cdots + \eta_{n-1}) \;=\; 2 \cdot \pi_1 - \pi_n,$$
$$\pi_2/\pi_1 \;=\; \pi_3/\pi_2 \;=\; \cdots \;=\; \pi_n/\pi_{n-1} \;=\; 2^{1/n}.$$

If we calculate the utilization for these values of the parameters, then we get the bound $B_n = n(2^{1/n} - 1)$.

This scenario for the case of $n = 3$ is shown in figure 8.11. The WCETs and periods are chosen so that the first instances of all three jobs finish by the time the second instance of the first job arrives, the second instance of the first job finishes by the time the second instance of the second job arrives, which finishes by the time the second instance of the third job arrives, which finishes by the time the third instance of the first job arrives. The relationship among the three periods π_1, π_2, and π_3 and the three WCETs η_1, η_2, and η_3 is given by:

$$\pi_2 \;=\; 1.26 \cdot \pi_1,$$

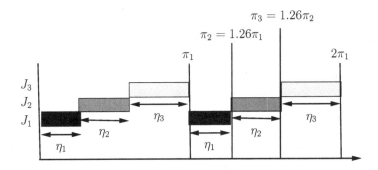

Figure 8.11: Worst Case for Rate-monotonic Policy with Three Jobs

$$
\begin{aligned}
\pi_3 &= 1.26 \cdot \pi_2 = 1.59 \cdot \pi_1, \\
\eta_1 &= \pi_2 - \pi_1 = 0.26 \cdot \pi_1, \\
\eta_2 &= \pi_3 - \pi_2 = 0.33 \cdot \pi_1, \\
\eta_3 &= 2 \cdot \pi_1 - \pi_3 = \pi_1 - (\eta_1 + \eta_2) = 0.41 \cdot \pi_1, \\
B_3 &= \eta_1/\pi_1 + \eta_2/\pi_2 + \eta_3/\pi_3 = 0.78.
\end{aligned}
$$

The following theorem summarizes the bound for n jobs:

Theorem 8.6 [Schedulability test for Rate-Monotonic Policy] *Given a periodic job model with n jobs with implicit deadlines, if the utilization does not exceed the quantity $B_n = n(2^{1/n} - 1)$, then every rate-monotonic schedule is guaranteed to be deadline-compliant.* ∎

Note that the bound B_n decreases as n increases. The table in figure 8.12 shows the values of these bounds for $n = 1, 2, \ldots 10$. This means, for instance, when we have six jobs, if the utilization is 0.735 or less, then we are guaranteed that the rate-monotonic policy produces a deadline-compliant schedule.

Finally, let us consider the *limit* of the expression $n(2^{1/n} - 1)$. This limit turns out to be $ln\ 2$, the natural logarithm of 2, and equals 0.69. That is, for every number n, the value of the expression $n(2^{1/n} - 1)$ is at least 0.69. This means:

If the utilization of a periodic job model with implicit deadlines is 0.69 or less, then a rate-monotonic schedule is deadline-compliant.

Exercise 8.12: Consider a periodic job model with three jobs with implicit deadlines: the job J_1 with period 4 and WCET 1, the job J_2 with period 6 and WCET 2, and the job J_3 with period 8 and WCET 3. Can we conclude that the rate-monotonic policy results in a deadline-compliant schedule using the utilization-based schedulability test? Does the rate-monotonic policy result in a deadline-compliant schedule? ∎

n	1	2	3	4	5	6	7	8	9	10
B_n	1	0.828	0.780	0.757	0.743	0.735	0.729	0.724	0.721	0.718

Figure 8.12: Utilization Bound for Rate-monotonic Policy

Exercise 8.13 *: Let $\mathcal{J}$ be a periodic job model, let ρ be a priority assignment for $\mathcal{J}$, and let σ be the fixed-priority schedule for $\mathcal{J}$ with respect to the assignment ρ. Prove that the schedule σ is deadline-compliant if the following condition is satisfied for every job J:

$$\delta(J) \geq \eta(J) + \sum_{K \in \mathcal{J}:\rho(K)>\rho(J)} \lceil (\delta(J)/\pi(K)) \rceil \cdot \eta(K)$$

In this formula, for a rational number f, $\lceil f \rceil$ denotes the integer obtained by rounding up f, that is, the smallest integer that is greater than or equal to f.
∎

Bibliographic Notes

Scheduling algorithms for real-time systems is a well-studied topic: see [SAÅ+04] for a survey. The key result developed in this chapter, namely, the optimality and the analysis of the rate-monotonic scheduling policy, is due to Liu and Layland [LL73]. The presentation in this chapter is based on [But97] (see also [Liu00]). We also refer the reader to [FMPY06, BDL+11] for formalization of job models and schedulability analysis using the computational model of timed processes and reachability analysis for timed automata.

We have only briefly discussed the problem of estimating the worst-case execution time of tasks, which is also a well-studied problem with multiple theoretical approaches and tools (see [WEE+08] for a survey).

Real-time scheduling is supported by a number of operating systems (see [RS94] and [Kop00]).

9

Hybrid Systems

In chapter 6, we studied continuous-time models of physical plants and controllers. In this chapter, we turn our attention to systems whose dynamics
consists of both continuous evolution as time elapses and discrete instantaneous
updates to state. In fact, the timed processes of chapter 7 already exhibit this
mix of discrete and continuous updates in their dynamics: the values of the
clock variables of a timed process increase as time elapses while the process
waits in a mode, and state variables are updated in a discrete manner during
a mode switch. Hybrid systems admit more general forms of continuous-time
evolution for state variables described using differential and algebraic equations.
Such models provide a unified framework for designing and analyzing systems
that integrate computation, communication, and control of the physical world.

9.1 Hybrid Dynamical Models

The model of computation for hybrid processes is a generalization of the model
for timed processes studied in chapter 7.

9.1.1 Hybrid Processes

We describe a hybrid process using an extended-state machine with modes and
mode-switches. Each process has input, output, and state variables, and some
of these variables are of type cont. A variable of type cont takes values from
the set of real numbers (or an interval of real numbers) and is updated continuously as time progresses while a process waits in a mode. A mode-switch
is executed discretely and takes zero time. As usual, such a switch is guarded
with a condition over state and input variables, can update state and output
variables, and describes either an input, an output, or an internal action. A
mode is annotated with differential and algebraic equations that specify how
state and output variables of type cont evolve. In addition, each mode also
specifies a constraint on how long the process can wait in that mode using a

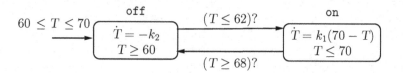

Figure 9.1: A Thermostat Model Switching between Two Modes

Boolean expression over the state variables.

Switching Thermostat

As an example of a dynamical system switching between two modes, consider a simple model of a self-regulating thermostat shown in figure 9.1. The process Thermostat can be in two modes, off and on. The temperature T is a continuous-time variable. When the mode is on, the dynamics of the system is given by the differential equation $\dot{T} = k_1(70 - T)$, where k_1 is a constant. The temperature changes continuously according to this differential equation. Note that this dynamics is linear, and for a given initial value of the temperature, there is a unique response signal that captures how the temperature evolves with time. The constraint $(T \leq 70)$ associated with the mode specifies that the process can stay in this mode only as long as this condition holds: the switch to the mode off must occur before this constraint is violated. The switch to the mode off is guarded by the condition $(T \geq 68)$. This implies that the switch may happen any time after the temperature exceeds 68.

In the mode off, the dynamics of how the temperature changes is described by the differential equation $\dot{T} = -k_2$, where k_2 is a constant. Thus, the temperature falls linearly with time when the thermostat is off. The constraint $(T \geq 60)$ associated with the mode off specifies that the process must switch to the mode on before the temperature falls below 60, and the guard $(T \leq 62)$ associated with the switch from the mode off to the mode on implies that this switch may occur at any time after the temperature drops below 62.

Initially the mode is off and the temperature is T_0. The set of choices for the initial temperature is described by the initialization constraint $60 \leq T \leq 70$. Figure 9.2 shows a possible execution of the thermostat process for the initial temperature $T_0 = 66$ with the constants $k_1 = 0.6$ and $k_2 = 2$. The execution proceeds in phases: during each phase, the mode stays unchanged, and the temperature changes as a continuous function of time according to the differential equation of the current mode. When a mode-switch occurs, state changes discontinuously. In this model, there is an uncertainty in the times at which the mode-switches occur, and thus the model is non-deterministic, and even when we fix the initial temperature, the process has many possible executions.

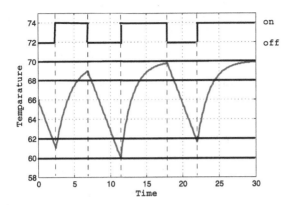

Figure 9.2: A Possible Execution of the Process `Thermostat`

The behavior of the process `Thermostat` while it stays in a given mode can be analyzed using techniques discussed in chapter 6. If the process switches to the mode `off` at time t^* with the temperature equal to T^*, then until the next mode-switch, the value of the temperature at time t is given by the expression $T^* - k_2 (t - t^*)$. Assuming that the temperature T^* upon entry is at least 62, the process spends at least $(T^* - 62)/k_2$ seconds and at most $(T^* - 60)/k_2$ seconds in the mode `off`.

If the process switches to the mode `on` at time t^* with the temperature equal to T^*, then until the next mode-switch, the value of the temperature at time t is given by the expression $70 - (70 - T^*) e^{-k_1(t-t^*)}$. Assuming that the entry temperature T^* does not exceed 68, the process spends at least $-\ln (2/(70 - T^*))/k_1$ seconds in the mode `on`. It can stay there for an indefinite period as the value of the temperature is guaranteed not to exceed 70 according to this equation.

Bouncing Ball

Consider a ball dropped from an initial height $h = h_0$ with initial velocity $\dot{h} = v = v_0$. The ball drops freely with its dynamics given by the differential equation $\dot{v} = -g$, where g is the gravitational acceleration. When it hits the ground, that is, when the value of the variable h becomes 0, there is a discontinuous update in its velocity. This discrete change can be modeled as a mode switch with the update given by $v := -a\, v$. This assumes that the collision is inelastic, and the velocity decreases by a factor a, for some appropriate constant $0 < a < 1$. This behavior is captured by the hybrid process `BouncingBall` of figure 9.3. It has a single mode and two state variables of type `cont`. Whenever the condition $(h = 0)$ holds, a mode-switch is triggered. The output event *bump* is issued by the process, and the assignment is executed reflecting the change in the direction

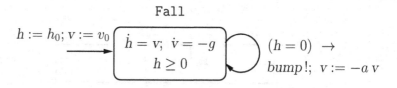

Figure 9.3: A Bouncing Ball as a Single-mode Hybrid System

of the velocity. The invariant constraint ($h \geq 0$) ensures that the switch is executed whenever the height h becomes 0. Figure 9.4 shows an execution of the process BouncingBall with the initial velocity $v_0 = 0$, initial height $h_0 = 5m$, gravitational acceleration $g = 9.8 \, m/s^2$, and damping coefficient $a = 0.8$.

Note that the model capturing the continuous-time evolution of the state is a two-dimensional linear system: for each state variable, its rate of change with time is given as a linear function of the state variables. However, due to the discontinuous update of the velocity during the discrete mode-switch, the value of a state variable at time t cannot be expressed as a nice closed-form function of the initial state.

Formal Model

We define the formal model for hybrid processes along the lines of the formal model for timed processes. A hybrid process consists of an asynchronous process that is defined by listing its input channels, output channels, state variables, initialization, input tasks, output tasks, and internal tasks. Input, output, and internal actions are defined as in the case of asynchronous processes and can discretely update any of the variables. A variable can be of type cont, indicating that the variable evolves continuously during a timed action. A variable of a type other than cont is updated only discretely, and let us call such variables *discrete* variables.

To execute a timed action of duration δ, for each continuously updated input variable u, we need a continuous signal $\bar{u}$ that specifies the value of the input u over the interval $[0, \delta]$. As in the case of continuous-time components, the continuous-time evolution is specified by a real-valued expression h_y for every continuously updated output variable y and a real-valued expression f_x for every continuously updated state variable x. Each of these expressions is an expression over the continuously updated input variables and state variables. The value of the output variable y of type cont at time t is obtained by evaluating the expression h_y using the values of the state and input variables at time t. The signal for a continuously updated state variable x should be a differentiable function such that its rate of change at time t equals the value of the expression f_x evaluated using the values of the state and input variables at time t. Note that the discrete input and discrete output variables are not relevant during a

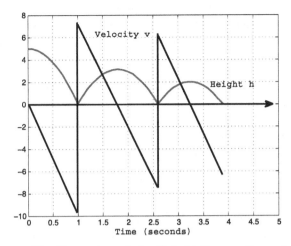

Figure 9.4: The Execution of the Hybrid Process `BouncingBall`

timed action, and the value of a discrete state variable stays unchanged during a timed action. These rules define the state signal and the output signal for the continuously updated output variables for the duration $[0, \delta]$. If the expressions h_y and f_x used to define the dynamics are Lipschitz-continuous, then the state and the output signals corresponding to a given continuous input signal during a timed action are uniquely defined. The *continuous-time invariant* is specified as a Boolean expression over the state variables, and it is required that at every time instance during the timed action, the state signal satisfies this invariant. During a timed action of duration δ, the process and its environment synchronize on the evolution of the continuously updated variables as in the case of dynamical models and also agree on not executing a discrete action for this duration.

Let us revisit the process `Thermostat` shown in figure 9.1. Let us assume that the output of the process is the temperature. Then the hybrid process corresponding to this state machine consists of the following components:

- it has no input variables;

- it has a single output variable T of type `cont`;

- it has a discrete state variable *mode* of enumerated type {`off, on`}, and a state variable T of type `cont`;

- the variable *mode* is initialized to `off`, and the initialization of the variable T is given by the nondeterministic choice $60 \leq T \leq 70$;

- it has no output tasks, which means that the value of the temperature is not transmitted during the discrete actions;

- it has two internal tasks corresponding to the two mode-switches: one task has the guard $(mode = \texttt{off} \ \wedge \ T \leq 62)$ and the update $mode := \texttt{on}$, and the second task has the guard $(mode = \texttt{on} \ \wedge \ T \geq 68)$ and the update $mode := \texttt{off}$;

- the expression defining the value of the output variable T equals the state variable T;

- the expression defining the derivative of the state variable T is given by the *conditional* expression

$$\texttt{if } (mode = \texttt{off}) \texttt{ then } - k_2 \texttt{ else } k_1 \, (70 - T);$$

- the continuous-time invariant CI is given by the expression:

$$(mode = \texttt{off}) \ \rightarrow \ (T \geq 60)] \ \wedge \ [(mode = \texttt{on}) \ \rightarrow \ (T \leq 70)].$$

The formal definition is summarized below:

HYBRID PROCESS

A *hybrid process* HP consists of (1) an asynchronous process P where some of the input, output, and state variables are of type $\texttt{cont}$; (2) a *continuous-time invariant* CI, which is a Boolean expression over the state variables S; (3) for every output variable y of type $\texttt{cont}$, a real-valued expression h_y over the state and input variables of type $\texttt{cont}$; and (4) for every state variable x of type $\texttt{cont}$, a real-valued expression f_x over the state and input variables of type $\texttt{cont}$. Inputs, outputs, states, initial states, internal actions, input actions, and output actions of the hybrid process HP are the same as those of the asynchronous process P. Given a state s, a real-valued time $\delta > 0$, and an input signal $\overline{u}$ for every input variable u of type $\texttt{cont}$ over the interval $[0, \delta]$, the corresponding timed action of the process HP is the differentiable state signal $\overline{S}$ over the state variables, and the signal $\overline{y}$ for every output variable y of type $\texttt{cont}$ over the interval $[0, \delta]$ such that (1) for every state variable x, $\overline{x}(0) = s(x)$; (2) for every discrete state variable x and time $0 \leq t \leq \delta$, $\overline{x}(t) = s(x)$; (3) for every output variable y of type $\texttt{cont}$ and time $0 \leq t \leq \delta$, $\overline{y}(t)$ equals the value of h_y evaluated using the values $\overline{u}(t)$ and $\overline{S}(t)$; (4) for every state variable x of type $\texttt{cont}$ and time $0 \leq t \leq \delta$, the time derivative $(d/dt)\,\overline{x}(t)$ equals the value of f_x evaluated using the values $\overline{u}(t)$ and $\overline{S}(t)$; and (5) for all $0 \leq t \leq \delta$, the continuous-time invariant CI is satisfied by the values $\overline{S}(t)$ of the state variables at time t.

Executions

The execution of a hybrid process starts in an initial state. At each step, either an internal, an input, an output, or a timed action is executed. For example,

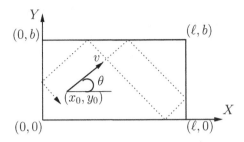

Figure 9.5: Motion of a Billiard Ball

the execution of the process **Thermostat** shown in figure 9.2 corresponds to the following sequence of alternating timed and internal actions:

$$(\text{off}, 66) \xrightarrow{2.5} (\text{off}, 61) \xrightarrow{\varepsilon} (\text{on}, 61) \xrightarrow{3.7} (\text{on}, 69.02) \xrightarrow{\varepsilon}$$
$$(\text{off}, 69.02) \xrightarrow{4.4} (\text{off}, 60.22) \xrightarrow{\varepsilon} (\text{on}, 60.22) \xrightarrow{7.6} (\text{on}, 69.9) \xrightarrow{\varepsilon}$$
$$(\text{off}, 69.9) \xrightarrow{4.1} (\text{off}, 61.7) \xrightarrow{\varepsilon} (\text{on}, 61.7) \xrightarrow{7.7} (\text{on}, 69.92).$$

During each timed action, the process continuously outputs the value of the temperature. For example, during the first timed action of duration 2.5, the temperature signal is given by $\overline{T}(t) = 66 - 2t$, while during the second timed action of duration 3.7, the temperature signal is given by $70 - 9e^{-0.6t}$.

Note that discrete and timed actions need not strictly alternate during an execution: two timed actions can appear consecutively and so can two discrete actions. In particular, the first timed action of duration 2.5 in the execution above can be split into multiple timed actions:

$$(\text{off}, 66) \xrightarrow{1.5} (\text{off}, 63) \xrightarrow{0.8} (\text{off}, 61.4) \xrightarrow{0.2} (\text{off}, 61).$$

A state s of a hybrid process is reachable if there is an execution that ends in the state s. Given a hybrid process HP and a property φ over its state variables, the property φ is said to be an invariant of the process HP if every reachable state of the process HP satisfies the property φ. For example, the property $60 \leq T \leq 70$ is an invariant of the process **Thermostat**.

Exercise 9.1: Specify the model of the bouncing ball shown in figure 9.3 as a formal hybrid process by listing all its components. Assume that the outputs of the process are the discrete bump events and the continuously updated height.
∎

Exercise 9.2: In this problem, we want to construct a hybrid systems model of the motion of a ball on a billiards table with perfect collisions (see figure 9.5). The table has length ℓ units and breadth b units. The ball is initially hit from the position (x_0, y_0) with an initial speed v in the direction θ. Whenever the

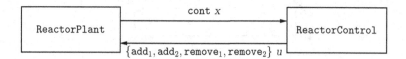

Figure 9.6: The Block Diagram for the Reactor Example

ball hits a side parallel to the X-axis, its velocity in the Y-direction flips sign, and the velocity in the X-direction stays unchanged. Symmetrically, whenever the ball hits a side parallel to the Y-axis, its velocity in the X-direction flips sign, and the velocity in the Y-direction stays unchanged. Thus, we are ignoring friction and collision impact. When the ball reaches one of the corner points, it drops and stops. Describe a precise hybrid state machine corresponding to this description. ∎

Exercise 9.3*: Consider a mobile robot that moves in a two-dimensional world corresponding to the positive quadrant of the X-Y plane. The robot is initially at the origin and is stationary. The input command to the robot consists of a target location to go to. Assume that there are no obstacles. The robot can move in the horizontal direction at speed 6 m/s, in the vertical direction at speed 8 m/s, or along any other arbitrary direction at speed 5 m/s. The robot plans its trajectory to the target to minimize the time taken. Once it reaches the target, it waits there to receive another input command to move to a new target and repeats the same behavior. Construct a hybrid process (using the extended-state machine notation) to model the behavior of the robot. Clearly specify input and state variables along with their types. For the purpose of this question, you can assume that the time needed to change the velocity is negligible (that is, the robot can change its speed from, say, 0 to 5, instantaneously). ∎

9.1.2   Process Composition

Hybrid processes can be put together using block diagrams. Operations such as instantiation, variable renaming, and output hiding are defined in the usual way. To compose two hybrid processes, we compose the corresponding asynchronous processes using the composition operation for asynchronous processes and compose the dynamics during timed actions as in the case of continuous-time components. Thus, two hybrid processes are compatible and can be composed, provided the state variables of the two are disjoint, the output variables of the two are disjoint, and there are no cyclic await dependencies among the continuously updated common input/output variables of the two. The continuous-time invariant for the composed process is simply the conjunction of the continuous-time invariants of the component processes. Thus, the discrete actions of the composite process are obtained by the asynchronous composition of the discrete actions of the component processes, and the timed actions of the composite

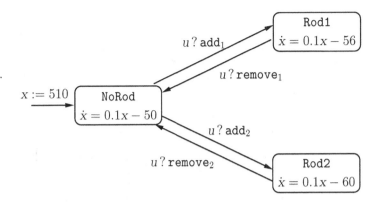

Figure 9.7: The Hybrid Model of the Reactor Plant

process are obtained by the synchronous composition of the timed actions of the component processes. In particular, a timed action of a duration δ is possible in the composite process only when both components are willing to evolve continuously for the duration of δ without an interrupting discrete action.

To illustrate process composition, we consider a toy model of a nuclear reactor with two control rods. The reactor and the controller are modeled by the hybrid processes ReactorPlant and ReactorControl, respectively. The interaction pattern is shown in the block diagram of figure 9.6. The output of the process ReactorPlant is the continuously updated variable x that captures the reactor temperature, and this variable is monitored by the controller. The output of the process ReactorControl is the event variable u that can take values add_1 (a control command to insert the first rod), $remove_1$ (a control command to remove the first rod), add_2 (a control command to insert the second rod), and $remove_2$ (a control command to remove the second rod).

The hybrid process corresponding to the plant model is shown in figure 9.7. The process has three modes NoRod, Rod1, and Rod2, corresponding, respectively, to whether there is no rod in the reactor, the first rod is in the reactor, or the second rod is in the reactor. Initially, the temperature is 510 degrees, and no rods are in the reactor. The dynamics for the change in the temperature is described by the differential equation $\dot{x} = 0.1x - 50$. When the controller issues the event add_1, the plant switches to the mode Rod1. The rod has a dampening effect, which slows down the rate of increase in temperature, and the dynamics is given by the differential equation $\dot{x} = 0.1x - 56$. On receiving the control command $remove_1$, the plant switches back to the mode NoRod. The mode Rod2 is similar except that the second rod causes a stronger dampening with the dynamics given by the differential equation $\dot{x} = 0.1x - 60$.

The controller for the plant is shown in figure 9.8. Once a rod is removed, it cannot be reinserted for a time period of c time units. To capture this restriction,

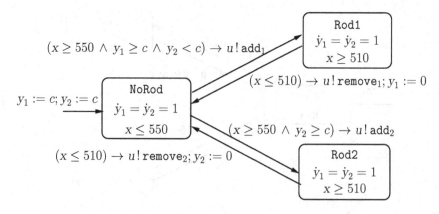

Figure 9.8: The Hybrid Model of the Reactor Controller

we introduce two variables: y_1 and y_2. The rate of change of these continuously updated variables is 1 in all the modes, and thus they are the same as the clock variables of timed models. Initially, the clock y_1 equals c and is reset to 0 every time the first rod is removed. The controller can issue the output $\mathbf{add}_1$ only when the clock y_1 is at least c. This ensures that the delay between an event $\mathbf{remove}_1$ and the subsequent $\mathbf{add}_1$ is at least c time units. The variable y_2 is updated similarly and ensures that the delay between an event $\mathbf{remove}_2$ and the subsequent $\mathbf{add}_2$ is at least c time units.

Initially, the controller is in the mode NoRod. The continuous-time invariant $(x \leq 550)$ ensures that when the plant temperature rises to 550 degrees, a mode-switch is triggered. Note that in this example, the variable x is updated by the process ReactorPlant, which specifies the differential equations regarding how the variable evolves. It is monitored by the process ReactorControl, which constrains the durations of timed actions via continuous-time invariants that refer to x. When the temperature reaches 550, depending on the values of the clock variables y_1 and y_2, the controller process can decide to insert the first rod by issuing the output $\mathbf{add}_1$ or the second rod by issuing the output $\mathbf{add}_2$. If both choices are available, then the controller prefers the second rod with the stronger dampening effect. The controller process stays in the mode Rod1 or Rod2 as long as the temperature is above 510 degrees, and if it falls below that, it switches back to the mode NoRod by issuing a command to remove the corresponding rod.

When the mode equals NoRod, if the temperature rises to 550 and both the clock variables y_1 and y_2 are smaller than c, thereby disabling both the control actions $\mathbf{add}_1$ and $\mathbf{add}_2$, the temperature of the core can rise to an unacceptable level, causing an alarm. Formal analysis can reveal the range of values for the parameter c for which such an alarm condition is not feasible.

Exercise 9.4: Consider the composition of the hybrid processes `ReactorPlant` and `ReactorControl`. Show executions of the system using a simulation tool (such as MATLAB) for the following values of the constant c: 10, 20, 30, 40, 50, and 60. ■

9.1.3 Zeno Behaviors

Execution of the Bouncing Ball

Let us reconsider the hybrid process `BouncingBall` corresponding to the bouncing ball of figure 9.3. Let us assume that the initial velocity v_0 equals 0. Then before the first bump, the change in height is described by the equation $\overline{h}(t) = h_0 - g\,t^2/2$. The guard condition ($h = 0$) for the mode-switch becomes true at time $\delta_1 = \sqrt{2\,h_0/g}$. The first timed action is of this duration δ_1, and during this action, the height decreases from h_0 to 0, and the velocity changes from 0 to $-v_1$, where $v_1 = g\,\delta_1 = \sqrt{2\,g\,h_0}$. At this instance, a discrete output action is executed, the event *bump* is issued, and the velocity changes its direction while its magnitude decreases by a factor of a. Thus, the new velocity is $v_2 = a v_1$. After the first bump until the next bump, the height signal during the timed action is given by $\overline{h}(t) = v_2\,t - g\,t^2/2$ and the velocity signal is given by $\overline{v}(t) = v_2 - g\,t$. This timed action is of duration $\delta_2 = 2\,v_2/g$, and captures one bounce corresponding to the parabolic motion of the ball. At the end of this action, the height becomes 0 again, and the velocity is $-v_2$. As a result of the bump, the velocity is updated to $v_3 = a\,v_2 = a^2\,v_1$, and the entire cycle repeats.

If we concatenate a sequence of successive timed actions into one single timed action, then the model has a single infinite execution that can be described as

$$(h_0, 0) \xrightarrow{\delta_1} (0, -v_1) \xrightarrow{bump!} (0, v_2) \xrightarrow{\delta_2} (0, -v_2) \xrightarrow{bump!} (0, v_3) \xrightarrow{\delta_3} \cdots$$

where for each i, $v_{i+1} = a\,v_i = a^i\,v_1$ and $\delta_{i+1} = 2\,v_{i+1}/g = 2\,a^i\,v_1/g$. After k bumps, the velocity of the ball is $a^k\,v_1$. Since $a < 1$, this sequence converges to 0. Similarly, the sequence of durations $\delta_1, \delta_2, \ldots$ of timed actions corresponding to the successive bounces of the ball is decreasing, converging to 0. However, at no point during this infinite execution of the ball, it is stationary, and there is always one more bounce possible. An inductive reasoning about such a behavior of the ball can lead to the flawed conclusion that at no point in time the ball is at rest.

Zeno's Paradox

The phenomenon exhibited by the bouncing ball was noted by the Greek philosophers many centuries ago and is known as the *Zeno's Paradox*. This was originally posed in the context of a race between Achilles and the tortoise, in which the tortoise has a head start. Suppose the tortoise is ahead of Achilles by d_1

meters at the beginning of a round. By the time Achilles has covered this distance d_1, the tortoise has moved a little bit ahead, say d_2 meters, with $d_2 < d_1$, and for the next round, Achilles has now to cover another d_2 meters, during which the tortoise has moved farther by d_3 meters with $d_3 < d_2$. By inductive reasoning, for every natural number n, after n rounds, the tortoise is ahead of Achilles by a non-zero distance, and thus Achilles will never be able to catch up with the tortoise! The source of paradoxical behavior lies in the fact that executions in which the total elapsed time does not grow in an unbounded manner do not adequately describe the system state with progression of time.

The total time elapsed along an execution is the sum of the durations of all the timed actions in the execution. In our bouncing ball example, this sum is $\Sigma_{i \geq 1} \delta_i$. Since the sequence $\delta_1, \delta_2, \delta_3 \ldots$ converges to 0, the sum is bounded by a constant K. Plugging in the expressions for the durations δ_i and velocities v_i discussed above, and using the fact that the geometric series $\Sigma_{i \geq 1} a^i$ equals $a/(1 - a)$, we can compute the expression for the constant K:

$$\sum_{i \geq 1} \delta_i \;=\; \sqrt{2\,h_0/g}\,(1 + a)/(1 - a).$$

Thus, the execution, even though it contains infinitely many output and timed actions, does not describe what happens at time K (and beyond). In reality, the ball would be stationary on the ground at time K, but the execution never gets to time K.

Zeno Executions and Zeno States

An infinite execution of a hybrid process *HP* is said to be a *Zeno execution* if the sum of the durations of all the timed actions in the execution is bounded by a constant. Thus, a non-Zeno execution is an execution in which time diverges. Zeno executions are an artifact of mathematical modeling, and proving properties using Zeno executions will lead to misleading conclusions.

In the bouncing-ball model, from the initial state, every possible infinite execution is a Zeno execution. Such a state is called a *Zeno state*. It is the analog of the deadlock states discussed in chapter 4. From a deadlock state there is no enabled action, and thus there is no way to continue the execution even for one step. From a Zeno state, there is no way to produce an infinite execution on which time keeps increasing unboundedly.

Note that the existence of a Zeno execution starting from a state does not imply that the state is Zeno. For example, consider the initial state (off, 66) of the process Thermostat of figure 9.1. Consider the execution in which we first execute a timed action of duration $\delta_1 = 0.5$, then a timed action of duration $\delta_2 = 0.25$, and repeat this pattern: the ith action in the execution is a timed action of duration $1/2^i$. This is an infinite execution, where the total sum of durations of all the timed actions is bounded by 1. Thus, the execution is a Zeno execution. However, the initial state is not a Zeno state: we could have chosen durations of timed actions so that the resulting execution is non-Zeno (in particular, see the execution corresponding to the one illustrated in figure 9.2).

A hybrid process is called Zeno if some reachable state of the system is a Zeno state. For a Zeno process, the execution can end up in a state from which every possible way of continuing the execution causes convergence in the sum of the durations of timed actions. The process BouncingBall of figure 9.3 is a Zeno process. However, the process Thermostat of figure 9.1 is a non-Zeno process, and so is the process obtained by composing the processes ReactorPlant and ReactorControl (see section 9.1.2).

These definitions are summarized below.

ZENO EXECUTIONS, STATES, AND PROCESSES

An infinite execution of a hybrid process *HP* is said to be a *Zeno execution* if the sum of the durations of all the timed actions in the execution is bounded by a constant. A state *s* of a hybrid process *HP* is said to be a *Zeno state* if every infinite execution that contains the state *s* is a Zeno execution. A hybrid process *HP* is said to be a *Zeno process* if there exists a state *s* that is reachable and is a Zeno state.

Revising Zeno Models

The Zenoness of the hybrid process BouncingBall is perhaps not a serious problem if we were analyzing it in isolation. However, consider the composite process

$$HP = \texttt{BouncingBall} \mid \texttt{Thermostat}$$

obtained by the parallel composition with the thermostat model. Even though the two processes BouncingBall and Thermostat do not communicate, they synchronize on the passage of time. In particular, the composite process *HP* is a Zeno process, and its initial state is a Zeno state. This can lead to absurd conclusions. For instance, suppose that the parameters of the two processes are chosen so that the expression $\sqrt{2\,g\,h_0}\,(1+a)/(1-a)$ that gives the bound on the sum of the durations of all the timed actions of the bouncing ball is less than the earliest time, given by the expression $(T_0 - 62)/k_2$, by which the thermostat can switch to the mode on. Then on every execution of the composite process *HP*, the mode of the process Thermostat stays off, and its temperature variable keeps decreasing from the initial value T_0 without ever crossing 62. By definition, a state is considered reachable only when it appears on some execution of the system. Hence, the following property is an invariant of the composite process *HP*:

$$(mode = \texttt{off}) \wedge (62 \leq T \leq T_0).$$

This property of course is not an invariant of the process Thermostat and will be violated in the physical world regardless of whether a ball is bouncing next to the thermostat.

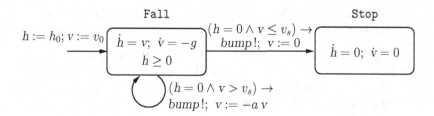

Figure 9.9: A Non-Zeno Model of the Bouncing Ball

Thus, in a system consisting of multiple components, the presence of a single Zeno process can affect the analysis of the entire system in unexpected ways. This suggests that Zeno components should be avoided during formal modeling.

A Zeno process can be converted into a non-Zeno process by modifying the model so that the model does not force mode-switches after shorter and shorter durations. For example, figure 9.9 shows the non-Zeno process NonZenoBall obtained from the process BouncingBall by adding a new mode called Stop in which the ball is stationary and adding a mode-switch from the initial mode Fall to the mode Stop when the velocity during a bump is smaller than some threshold value v_s. Starting from the initial state, if we execute a timed action of maximum possible duration, followed by the output action corresponding to the discrete bump, then it is guaranteed that after finitely many actions, the magnitude of the velocity becomes smaller than this threshold value, and the process switches to the mode Stop. Once in the mode Stop, timed actions of arbitrary durations are possible, and in particular one can generate a non-Zeno execution.

If we compose the process Thermostat and the modified bouncing ball process NonZenoBall, then the behavior of the process Thermostat is not influenced in any meaningful way. In particular, verify that for any property φ that refers to the mode and the temperature of the thermostat, the property φ is an invariant of the composite process NonZenoBall | Thermostat if and only if it is an invariant of the process Thermostat.

Exercise 9.5: Consider the following scenario. Two trains are heading toward each other on a single track at constant speeds: the train E is traveling east at a fixed speed v_e, and the train W is traveling west at a fixed speed v_w. A bee B is initially traveling west at a fixed speed v_b along the line joining the two trains. When the bee reaches the train E, it reverses its direction, heads east at the same speed v_b, and reverses its direction again when it reaches the train W. This cycle repeats. Model this scenario as a hybrid process. The state machine can have two modes, one each corresponding to the direction in which the bee is traveling, and three state variables that capture the positions of the train E, the train W, and the bee B. Show that the process is Zeno. Compute the formula that expresses the distance between the two trains as a function of

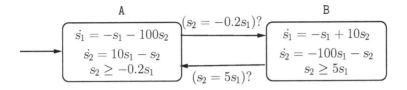

Figure 9.10: Instability Due to Mode Switching

the speeds v_e, v_w, and v_b and the initial values of the three position variables. ∎

Exercise 9.6*: In this exercise, we establish that the property of being non-Zeno is not preserved by parallel composition. Consider the following specification of the hybrid process HP_1. It has a single input event x and a single output event y. Whenever it receives an input, it waits for a duration of $1/2^i$ time units, if this is the ith input event it has received so far, and then issues an output event (it does not accept any inputs while it is waiting to issue its output). Design the process HP_1 as an extended-state machine. It suffices for the process HP_1 to use a single clock variable, and thus the process HP_1 is a timed process. Argue that the process HP_1 is non-Zeno.

Now consider the following specification of another process HP_2. It has a single input event y and a single output event x. It first issues an output event after a delay of 1 second; subsequently, whenever it receives an input, it waits for a duration of $1/2^i$ time units, if this is the ith input event it has received so far, and then issues an output event. Design the timed process HP_2 as an extended-state machine. Argue that the process HP_2 is non-Zeno.

Now consider the parallel composition $HP_1 | HP_2$. Show that the composite process is Zeno. ∎

9.1.4 Stability

As discussed in chapter 6, stability is a desirable feature for dynamical systems. Recall that a state s_e of a dynamical system is an equilibrium state if the system, starting in the state s_e, continues to stay in that state in the absence of external inputs. Such an equilibrium state is stable if we perturb the system state slightly, that is, choose the initial state s such that the distance $\|s - s_e\|$ is small, then at all times the state of the system stays within a bounded distance from the equilibrium, and is asymptotically stable if the state converges to the equilibrium with the passage of time.

We can use these same definitions to understand stability of hybrid processes. However, due to the presence of mode-switches, the mathematical analysis used to characterize stability for linear systems, and the associated techniques for

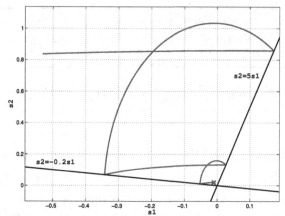

Figure 9.11: Unstable Response due to Mode-switching

designing stabilizing controllers, are not applicable to hybrid systems. We will illustrate the difficulties introduced due to mode-switching using an example.

Consider the hybrid process shown in figure 9.10. The dynamics in the mode A is specified by the linear differential equations $\dot{s}_1 = -s_1 - 100s_2$ and $\dot{s}_2 = 10s_1 - s_2$. Observe that the origin, that is, the state 0, is an equilibrium state. The eigenvalues of the dynamics matrix are $-1 + \sqrt{1000}j$ and $-1 - \sqrt{1000}j$, and from theorem 6.3, we can conclude that the continuous-time system with this dynamics is asymptotically stable.

However, the hybrid process of figure 9.10 stays in the mode A only as long as the invariant $(s_2 \geq -0.2s_1)$ holds, and when the state satisfies the switching condition $(s_2 = -0.2s_1)$, it switches to mode B. The dynamics associated with the mode B is specified by the linear differential equations $\dot{s}_1 = -s_1 + 10s_2$ and $\dot{s}_2 = -100s_1 - s_2$. Observe that this dynamics matrix is the transpose of the dynamics matrix in mode A with identical eigenvalues, and thus a system that evolves according to this dynamics is also asymptotically stable. The process can stay in the mode B only as long as the invariant $(s_2 \geq 5s_1)$ holds, and when the condition $(s_2 = 5s_1)$ is satisfied, it switches back to mode A.

Although the dynamics in the individual modes A and B are asymptotically stable, the switching causes instability. Figure 9.11 shows the execution of the system from the initial state $(-0.01, 0.02)$ in mode A. Indeed the origin is unstable: if the initial state s in mode A is not the origin, then no matter how close it is to the origin, the system state diverges from the origin as time passes.

Analyzing stability of hybrid systems turns out to be a difficult problem. Generalizing analysis techniques from the theory of continuous-time systems to hybrid systems remains an active area of research and is beyond the scope of this textbook.

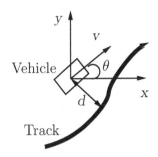

Figure 9.12: The Design Problem for the Automated Guided Vehicle

9.2 Designing Hybrid Systems

We illustrate the modeling and design of controllers for hybrid systems using three examples. The first example illustrates design of a controller that switches between different modes of operation, the second example illustrates multi-agent collaboration for improved planning, and the third example of multi-hop control networks shows how to model a system that integrates control, computation, and communication.

9.2.1 Automated Guided Vehicle

Consider an autonomous vehicle that needs to be programmed to move along a track as closely following the track as possible. The track is not known to the vehicle in advance but is equipped with sensors. In particular, assuming that the vehicle has not strayed too far from the track, the sensors can measure the distance d of the vehicle from the center of the track. Such information can be provided, for instance, by placing photodiodes along the track.

The vehicle dynamics is modeled as a planar rigid-body motion with two degrees of freedom. It can move forward along its body axis with a maximum possible speed v, and it can rotate about its center of gravity with an angular speed w, which can range over the interval from $-\pi$ to $+\pi$ radians/second. The variables (x, y) model the position of the vehicle, and θ gives the relative angle with respect to some fixed planar global frame in which the vehicle is headed.

Figure 9.12 shows the design problem for the automated guided vehicle. Based on the current measurement of the distance d, the controller must adjust the control inputs v and w so as to keep the value of the distance d as close to 0 as possible. The control design problem is additionally constrained by the requirement that the vehicle hardware provides only three discrete settings for the angular speed w: it can be 0, $+\pi$, or $-\pi$. When $w = 0$, the direction θ stays unchanged, and the vehicle is going straight. When $w = -\pi$, the direction θ is decreasing as fast as possible, and thus the vehicle is attempting to turn right.

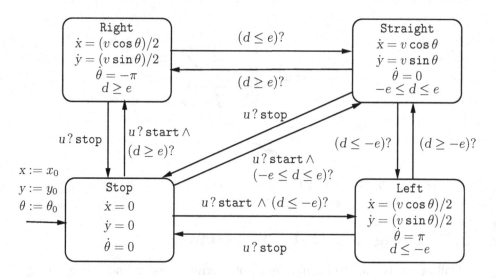

Figure 9.13: The Hybrid Controller for the Automated Guided Vehicle

Conversely, when $\omega = \pi$, the direction θ is increasing as fast as possible, and thus the vehicle is attempting to turn left.

The control designer makes a further design decision that when the vehicle is headed straight, it moves as fast as possible, with speed v. When the vehicle is turning left or turning right, it attempts to do so at half the maximum possible speed. This leads to four modes of operation as shown in the hybrid state machine of figure 9.13. In the mode Stop, the vehicle is stationary; in the mode Straight, the vehicle is moving with speed v and $\omega = 0$; in the mode Left, the vehicle is turning left with speed $v/2$ and $\omega = \pi$; and in the mode Right, the vehicle is turning right with speed $v/2$ and $\omega = -\pi$. In this model, we have assumed that the vehicle can change its speed and direction instantaneously during a mode-switch. This assumption is justifiable if the time needed to switch from one mode to another is insignificant compared with time spent in each mode.

The input variables to the process are the discrete input channel *in* that carries the commands start and stop to start and stop the vehicle and the continuously updated signal d that captures the error of the trajectory from the desired track. By design, a positive value of d means that the vehicle is off to the left of the track, and a negative value of d means that the vehicle is off to the right of the track.

The switching laws for the controller are designed using the parameter e. Whenever the current distance is in the interval $[-e, +e]$, the vehicle is assumed to be close enough to the track, and the controller decides to move straight. Whenever the current distance exceeds the threshold value e, the controller decides

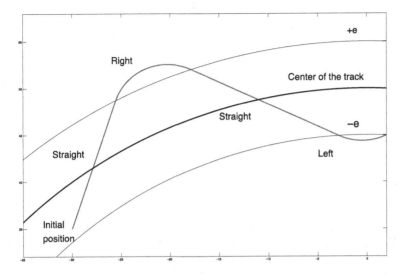

Figure 9.14: A Sample Vehicle Trajectory along a Curved Track

that the vehicle has strayed too far to the left and must be steered to the right by switching to the mode `Right`. Symmetrically, whenever the current distance is smaller than the threshold value $-e$, the controller concludes that the vehicle has strayed too far to the right and must be steered to the left by switching to the mode `Left`. The continuous-time invariants and the guards on the switches of the hybrid process of figure 9.13 capture this logic.

The behavior of the hybrid controller can be understood by considering tracks of different shapes. Figure 9.14 shows a sample trajectory followed by the vehicle along a curve in the track. In the illustrated scenario, the track is a circle centered at the origin with the radius 50. The following values of the parameters are used: $v = 35$, $e = 5$, $x_0 = -30$, $y_0 = 35$, and $\theta_0 = 0.4\pi$. The vehicle is initially going straight. As the distance from the center of the track exceeds the threshold value e, it starts moving right, and when the distance from the center of the track decreases below e, it decides to move in a straight line. In this example, this maneuver causes an overshoot, eventually causing the distance to be less than the threshold $-e$, and this triggers the vehicle to switch to the mode `Left`.

Exercise 9.7: For the automated guided vehicle, consider the following additional constraint: once the vehicle starts moving straight, left, or right, it cannot change its direction before Δ minutes for a given constant Δ. How will you modify the hybrid process of figure 9.13 to capture this additional constraint? How will the trajectory of the vehicle be affected by this change? ∎

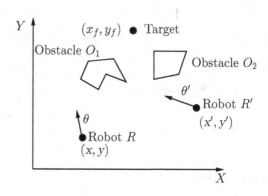

Figure 9.15: Path Planning in Presence of Obstacles for Two Robots

9.2.2 Obstacle Avoidance with Multi-robot Coordination

A challenging application domain for modeling and analysis of hybrid systems is the design of multi-robot coordination for a system of autonomous mobile robots. A typical surveillance task involves identifying a target and exploring a room with unknown geometry, possibly with obstacles, to reach the target. The sensory capabilities of each robot yield only imperfect information about the surroundings, and in particular, each robot has only estimates about the obstacle positions. The robots can send information to one another over wireless links and use this information to improve the accuracy of their estimates for better motion planning. The robots also need to coordinate with one another to achieve collaborative goals. For instance, it may be required that a team of robots should arrive at the same target, or the robots may be required to partition a set of targets among themselves. The solution to the design problem should ideally also be optimal among the space of solutions. For example, the objective can be to minimize either the total distance traveled by the robots or the time by which all the targets are reached. Thus, the design problem involves coordination, planning, and control in an optimal manner while satisfying the safety requirements and is representative of design problems in intelligent vehicle systems and flight management systems.

Illustrative Scenario

To illustrate modeling in a concrete scenario, suppose there are two autonomous mobile robots R and R'. We assume a two-dimensional world in which each robot is modeled as a point in the two-dimensional X-Y plane (see figure 9.15). The initial position of the robot R is (x_0, y_0), and the initial position of the robot R' is (x_0', y_0'). The goal of each robot is to reach the target located at the position (x_f, y_f). Both the robots want to reach the target while minimizing the total distance traveled. Assume that both robots travel at a fixed speed, say v. Then the sole control input for each robot is the direction in which it is moving. If the state variables (x, y) specify the coordinates of the robot R, the variables (x', y') specify the coordinates of the robot R', the variable θ specifies the direction in which the robot R is headed, and the variable θ' specifies the direction in which the robot R' is headed, then the dynamics of the system is captured by the differential equations:

$$\dot{x} = v \cos \theta; \quad \dot{y} = v \sin \theta; \quad \dot{x}' = v \cos \theta'; \quad \dot{y}' = v \sin \theta'.$$

The room has obstacles, and this can prevent each robot from traveling in a straight line from its initial position to the target. More specifically, suppose the room has two obstacles and the areas occupied by the two obstacles are O_1 and O_2, respectively (see figure 9.15). Each robot has a camera that can detect the approximate position of each obstacle. The robots can also communicate their current estimates so that their joint knowledge can be used for more accurate information.

The safety requirement for the problem is that no robot should ever collide with an obstacle. That is, the following property should be an invariant of the system:

$$[(x, y) \notin O_1 \wedge (x, y) \notin O_2 \wedge (x', y') \notin O_1 \wedge (x', y') \notin O_2].$$

The liveness requirement is that each robot should eventually reach its target:

$$\Diamond[(x, y) = (x_f, y_f)] \wedge \Diamond[(x', y') = (x_f, y_f)].$$

Estimating Obstacles

Mapping obstacles accurately using images from a camera is a computationally expensive task. Furthermore, optimal path planning to a target location given complex descriptions of obstacles is also computationally expensive. To address these difficulties, let us estimate each obstacle using a circle. In figure 9.16, the actual obstacle is a concave polygon occupying the area O. The circle of radius r contains this area entirely and is the best possible circular approximation of the obstacle. The image-processing algorithm on board a robot then simply needs to return the parameters of the circle, and this does not require detecting edges of the obstacle accurately. The planning algorithm has to compute a path to the target that avoids the circular shapes, and such a path is guaranteed to avoid collisions with the actual obstacles, thereby satisfying the safety requirement, although it may not be the shortest such path to the target.

In vision applications, the accuracy of obstacle estimation is limited by several factors, and the estimates improve as the distance to the target decreases. This is particularly true when sonars are used for obstacle detection. In figure 9.16, the obstacle is a circle with the center (x_o, y_o) and radius r. The estimate by a robot with the current position (x_1, y_1) is a circle that is concentric with the obstacle circle but with radius e_1 that exceeds r. This estimate depends on the distance d_1 between the obstacle center (x_o, y_o) and the robot position (x_1, y_1).

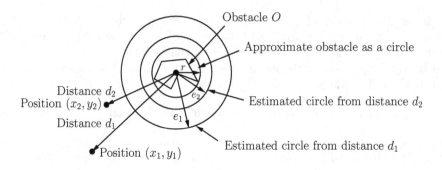

Figure 9.16: Approximate Estimation of an Obstacle

As the robot moves to the position (x_2, y_2), which is distance d_2 away from the obstacle center, the estimate improves to another concentric circle of radius e_2. As $(d_2 < d_1)$, we have $(e_2 < e_1)$. As the robot approaches the obstacle boundary, the distance between the robot position and the center of the obstacle decreases, and the estimate becomes more and more accurate, converging to the correct value of the obstacle radius. We assume the dependence of the estimate on the distance to be linear. If d is the current distance between the robot position and the obstacle center and r is the obstacle radius, then the estimated radius e is given by the equation

$$e = r + a(d - r)$$

where $0 < a < 1$ is a constant.

In our example scenario of figure 9.15, we have two obstacles. The first obstacle is modeled as a circle with the center (x_o^1, y_o^1) and radius r_1, and the second obstacle is modeled as a circle with the center (x_o^2, y_o^2) and radius r_2. Each robot can compute the estimates of each of the two obstacles based on the distance of its current position from the centers of the two obstacles. Furthermore, obstacle estimation is a computationally expensive process. As a result, the estimates are updated only discretely, every t_e seconds.

Path Planning

Consider the robot R with the current position (x, y). Its goal is to reach the target (x_f, y_f) while avoiding the two obstacles. For the obstacle O_1, the area it occupies according to the robot R is a circle centered at (x_o^1, y_o^1) with the current estimated radius e_1. Similarly, for the obstacle O_2, the area it occupies according to the robot R is a circle centered at (x_o^2, y_o^2) with the current estimated radius e_2. The objective of the planner is to compute a shortest path from the current position to the target so that the trajectory does not intersect the estimated obstacle circles.

The plan is usually updated in a discrete manner. In our design, the planning algorithm is invoked every t_p seconds, and the planning algorithm determines

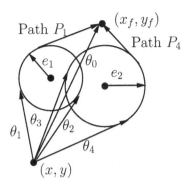

Figure 9.17: Path Planning While Avoiding Circular Obstacles

the control input θ that gives the direction for the robot motion. The direction stays unchanged until the next time the planning algorithm is invoked. We assume that the planning algorithm is captured by the function **plan** that takes as inputs (1) the current position (x, y), (2) the target position (x_f, y_f), (3) the first obstacle circle given by the center (x_o^1, y_o^1) and radius e_1, and (4) the second obstacle circle given by the center (x_o^2, y_o^2) and radius e_2 and returns the direction θ in which the robot should head.

The first step of the planning algorithm is to decide if the straight line from the current position (x, y) to the target (x_f, y_f) intersects any of the two estimated obstacle circles. If not, then the chosen direction is along this straight line. If it does, as is the case in figure 9.17, then the planner considers rays that are tangents from the current position (x, y) to the two obstacle circles. These are shown as directions θ_1 and θ_2 tangential to the first obstacle and directions θ_3 and θ_4 tangential to the second obstacle in figure 9.17. A direction that is tangential to one but intersects the other is discarded. Among the remaining choices, the direction that minimizes the distance to the target is chosen. In figure 9.17, the tangential directions θ_2 and θ_3 are not viable as they intersect with the other obstacle. The direction θ_1 corresponds to the path P_1 to the target, and the direction θ_4 specifies the path P_4 to the target. Since the length of the path P_1 is shorter than the path P_4, the planner returns the direction θ_1.

Note that the robot does not actually follow the path P_1; rather, it starts moving in the direction θ_1. When the planner is invoked again, if the estimates have improved by then, then it can revise its choice. In particular, in our example, as the robot moves along the direction θ_1, it would acquire an improved estimate of the first obstacle as a circle with a shorter radius; as a result, the robot would decrease the value of θ_1 in the clockwise direction and thereby head closer to the actual obstacle.

The function **plan** involves a sequence of floating-point calculations necessary to determine tangents and intersections. Such code is typically written in C or

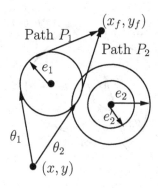

Figure 9.18: Impact of Improved Estimates via Coordination on Path Planning

MATLAB and the update description in the model-based design framework calls this function.

Coordination

To understand the impact of coordination, let us revisit the planning example of figure 9.17 as shown in figure 9.18. Suppose the robot R' is closer to the second obstacle and, thus, has a better estimate e'_2 of the radius of this obstacle. If the robot R' communicates this information to the robot R, then the robot R can simply update its value of the estimate e_2 to e'_2. Using this revised estimate, the planner now concludes that the direction θ_2 tangential to the first obstacle is a viable option since it does not intersect the circle with the center (x_o^2, y_o^2) and radius e'_2. The path P_2 corresponding to this choice is shorter than the path P_1 (see figure 9.18); as a result, the planner chooses the direction θ_2, leading to a more optimal solution.

The coordination strategy in this example is simple: every t_c seconds, the robot R sends its estimates e_1 and e_2 of the obstacles' radii to the robot R', and whenever it receives estimates (e'_1, e'_2) of the obstacles' radii from the other robot, it updates the estimate value e_1 to the minimum of its own current estimate e_1 and the received value e'_1, and the estimate value e_2 to the minimum of its own current estimate e_2 and the received value e'_2 (since a smaller radius is an improved estimate).

Hybrid Model

We proceed to describe the model of the robot as a hybrid process. We will describe the model for the robot R, and the model for the robot R' is symmetric and can be obtained by instantiation.

The hybrid process for the robot R is shown in figure 9.19. It uses the following variables:

- It has one input channel *in* of type (**real** × **real**) that is used to receive the estimates of the radii of the obstacles from the other robot.

- It has one output channel *out* of type (**real** × **real**) that is used to send the estimates of the radii of the obstacles to the other robot.

- The continuously updated state variables x and y, of type **cont**, model the position of the robot. These variables are initialized to x_0 and y_0, respectively.

- The discretely updated state variables e_1 and e_2, both of type **real**, capture the current estimates of the radii of the two obstacles. The initial estimates are obtained by executing the obstacle estimation algorithm and depend on the distance between the initial robot position and the centers of the two obstacles.

- The discretely updated state variable θ ranging over the interval $[-2\pi, 2\pi]$ models the direction in which the robot is currently moving. The initial direction is obtained by executing the function **plan**.

- The continuously updated state variable z_p, initialized to 0, is a clock variable that is used to enforce the timing constraint for invoking the planning algorithm every t_p seconds.

- The continuously updated state variable z_e, initialized to 0, is a clock variable that is used to enforce the timing constraint for updating the estimates using the vision data every t_e seconds.

- The continuously updated state variable z_c, initialized to 0, is a clock variable that is used to enforce the timing constraint for communicating the estimates every t_c seconds.

The process has two modes: Move and Stop. Initially, the mode is Move. During a timed action, the three clock variables z_p, z_e, and z_c increase at the rate 1, and the position variables x and y are updated at rates $v \cos \theta$ and $v \sin \theta$, respectively.

The switches from this mode are the following:

- When the robot reaches its target, captured by the condition $(x = x_f \wedge y = y_f)$, it switches to the mode Stop.

- When an input (e_1', e_2') is received on the input channel *in*, the obstacle estimates e_1 and e_2 are updated to the minimum of the current and the received values.

- When the clock z_p reaches the value t_p, the direction θ is updated by invoking the planning function **plan** based on the current estimates, and the clock z_p is reset to 0.

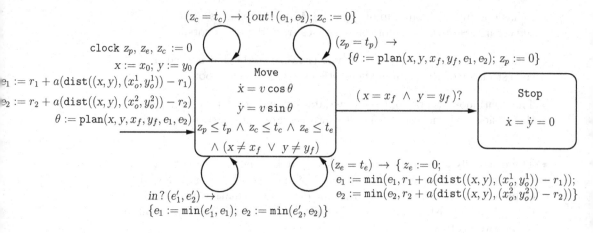

Figure 9.19: The Hybrid State Machine for the Robot

- When the clock z_c reaches the value t_c, the current values of the estimates e_1 and e_2 are transmitted on the output channel *out*, and the clock z_c is reset to 0.

- When the clock z_e reaches the value t_e, the current values of e_1 and e_2 are updated by executing the vision-based obstacle estimation algorithm, and the clock z_e is reset to 0. As discussed earlier, the effect of this algorithm is captured by computing the distance between the current robot position and the centers of the two obstacles and updating the estimate values if the revised estimates are better.

The continuous-time invariant of the mode Move ensures that when the conditions for updates corresponding to one of the discrete switches is satisfied, the elapse of time is interrupted to execute the corresponding discrete action. When the process is in the mode Stop, the robot simply waits there.

The desired system is the parallel composition of the two robots. The system description involves a large number of parameters. The system needs to be simulated many times to choose values for the parameters from possible choices. In particular, we would like to find out the value of t_c that determines how often the robots should communicate so that the communication actually improves the distance traveled.

Illustrative Execution

Figure 9.20 shows sample executions of the model obtained by simulating the model in STATEFLOW/SIMULINK. The initial position of the robot R is $(4.5, 2)$, the initial position of the robot R' is $(10, 2)$, and the target is at the position $(6, 10)$. The obstacle O_1 is centered at $(3.7, 7.5)$ and has a radius 0.9, and the

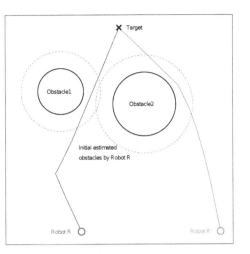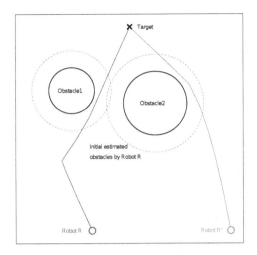

Figure 9.20: Illustrative Execution for Obstacle Avoidance

obstacle O_2 is centered at $(7, 7)$ and has a radius 1.25. The speed v is set to 0.5 units/sec, and the coefficient a used in the obstacle estimation is 0.12. The value of t_p, the time period for executing the planning algorithm, is 2 seconds, and the value of t_e, the time period for updating the obstacle estimates, is 2 seconds. The value of t_c, the time period for communicating the obstacle estimates, is different for the two executions: the left execution is obtained by setting t_c to 4 seconds, and the right execution is obtained by setting t_c to a high value.

Based on the initial estimates of the robot R, the two obstacles seem to overlap. As it gets closer to the obstacles, the estimates improve, suggesting a route to the target that passes between the two obstacles. For the robot R', the planned route does not exhibit a such qualitative change, but note that its estimate of the second obstacle constantly improves as it moves, leading to a curved trajectory. Observe that the distance traveled by the robot R is smaller in the left scenario as a result of communication. This is because it receives a better estimate of the second obstacle from the robot R' and switches its route a bit earlier thanks to this collaboration. In particular, the distance traveled by the robot R is 8.6480 with communication and is 8.8136 in the absence of communication (the distance traveled by the robot R' is 9.1550 in both the scenarios).

Exercise 9.8: For the problem of obstacle avoidance with coordination, consider the following optimization that reduces the needed computation. If a robot determines that the straight-line path from its current position to the target does not intersect any of the obstacles based on its current estimates of the obstacles, such a path cannot further be improved, and the robot can simply decide to move in this direction with no further planning. Modify the model of figure 9.19 so as to include this optimization. ■

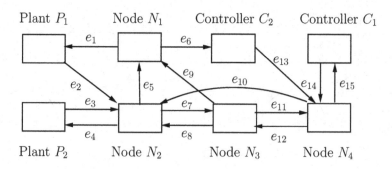

Figure 9.21: Example Multi-hop Control Network

9.2.3   Multi-hop Control Networks *

The classical architecture of a feedback control loop is shown in figure 6.1. In contrast, wireless networked control systems are spatially distributed systems where the communication among sensors, actuators, and computational units is supported by a shared wireless communication network. The deployment of such networked control systems in industrial automation results in flexible architectures and typically reduces the costs for installation, debugging, diagnostics, and maintenance when compared with the classical wired control loops. Design of controllers in the networked architecture faces new challenges. First, communication between a plant and the corresponding controller involves multiple hops and, thus, significant time delays. Second, multiple control loops may share the same network link, leading to mutual dependencies. Thus, the design of control laws for adjusting the plant inputs, the routing policies for transmission of messages through the network, and the scheduling policies for sharing network links must evolve in a synergistic manner. We now describe how to model such multi-hop control networks formally using the modeling concepts discussed in this textbook.

Example Network

Figure 9.21 shows a sample network. It consists of two plants P_1 and P_2 and their corresponding controllers C_1 and C_2. Messages among the plants and controllers are routed over a network consisting of four nodes N_1, N_2, N_3, and N_4. The links $e_1, e_2, \ldots e_{15}$ between different components are directed. For example, the output of the plant P_1 can be sent to the node N_2 over the link e_2, and the network can forward such messages to the controller C_1 using the links e_7, e_{11}, and e_{15}.

To be deployed in the context of control applications, the network must provide real-time guarantees regarding delivery of messages. In our set up, let us assume that one message can be delivered over a given link in a duration of time Δ. In other words, time is divided into slots with each slot of duration Δ time units.

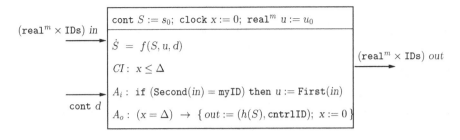

Figure 9.22: Plant Model in Multi-hop Control Network

At the beginning of each time slot, each node can send one message on each of its outgoing links. The message on each link can be received by the target component at the end of the time slot. Such a network is called a *time-triggered* network. The emerging WirelessHART standard for wireless networks provides such an abstraction and is being increasingly deployed within industrial process control.

Plant Model

The hybrid process modeling a plant in a multi-hop control network is shown in figure 9.22. The plant maintains state variables S that are updated continuously. The state is initialized to the value s_0 and evolves according to the differential equation $\dot{S} = f(S, u, d)$ during a timed action, where the variable u denotes the controlled input and the variable d denotes the uncontrolled input (or the disturbance). The disturbance d is a continuously updated external input signal.

The controlled input u, unlike in the models of continuous-time components in chapter 6, is updated only discretely, when the process receives a new value on the input channel in. Let us assume that the input u is an m-dimensional vector. The process model needs to store the value of this variable in its internal state. A message communicated over the network is a pair (v, id), consisting of a value v and the identifier id of the destination of the message. Let IDs denote the set of identifiers of all the plants and controllers that are connected by the network. The description of the plant process is then parameterized by its own identifier, denoted myID. The processing of an input over the channel in is then specified by the input task A_i: whenever it receives a message (v, id) on the input channel, it checks if the identifier id equals its own identifier; if so, it updates the value of the control input stored in the state variable u to the value v. If the channel in receives the sequence of values $v_1, v_2, \ldots$ destined for this plant at times $t_1, t_2, \ldots$ respectively, then the evolution of the variable u is a piecewise-constant signal whose value during the interval $[t_i, t_{i+1})$ is v_i, for each $i \geq 1$.

The function h maps the plant state to its output. Let us assume that the plant output is also an m-dimensional vector. Then every message exchanged on the network is of type $(\mathtt{real}^m \times \mathtt{IDs})$. The output of the plant is transmitted on the channel *out*. Since each network link can carry only one message every Δ time units, the plant should send a message every Δ time units on the channel *out*. To capture this timing constraint, we use a clock variable x. The clock is initialized to 0. The clock-invariant associated with the process is the condition $(x \leq \Delta)$, and the guard associated with the output task A_o responsible for sending messages on the channel *out* is $(x = \Delta)$. These two together ensure that the message is transmitted exactly every Δ time units. The value of the message to be transmitted is computed by applying the output map h to the plant state. The destination of the message is represented by the parameter `cntrlID`, which is the identifier of the controller responsible for this specific plant.

For the network shown in figure 9.21, we need two instances of the plant process. One is instantiated with $\mathtt{myID} = P_1$, $\mathtt{cntrlID} = C_1$, $in = e_1$, and $out = e_2$, and the other is instantiated with $\mathtt{myID} = P_2$, $\mathtt{cntrlID} = C_2$, $in = e_4$, and $out = e_3$. In each case, the model is completed by filling in the details of the dynamics f and the output map h.

Controller Model

The timed process modeling a controller in a multi-hop control network is shown in figure 9.23. The controller maintains an *estimate* of the state of the plant using the variables S'. The estimate is initialized to the initial plant state s_0.

The input to the controller is the channel *in* on which it receives the observed outputs of the plant communicated over the network. Recall that the messages are tagged with the destination identifier. Whenever the controller receives a message addressed to itself, it updates the state estimate S' based on the current estimate and the newly received plant observation. Based on this estimate, it computes an updated value of the control input $g(S')$ to be communicated back to the plant. This value is enqueued in the queue u that contains messages to be transmitted over the output channel.

A clock variable x is used to ensure that only one message is sent on the output channel every Δ time units. To achieve the desired behavior, we choose the clock invariant to state that "if an output message is waiting, then the clock should not exceed Δ." The output task is enabled when the queue is non-empty and the clock reaches Δ. Whenever a value is sent over the output channel, it is tagged with the identifier of the corresponding plant, denoted by the parameter `plantID`. The output task also resets the clock to 0. When the input task generates a new value to be transmitted, and thus to be enqueued in the output queue, it checks if the output queue is empty and, if so, resets the clock to 0 so that this new value will be transmitted after a delay of Δ time units.

If the controller receives an input only once every Δ time units, then the queue u contains one message most of the time. Consider the state in which the queue contains one message, the clock x equals Δ, and the process that sends messages

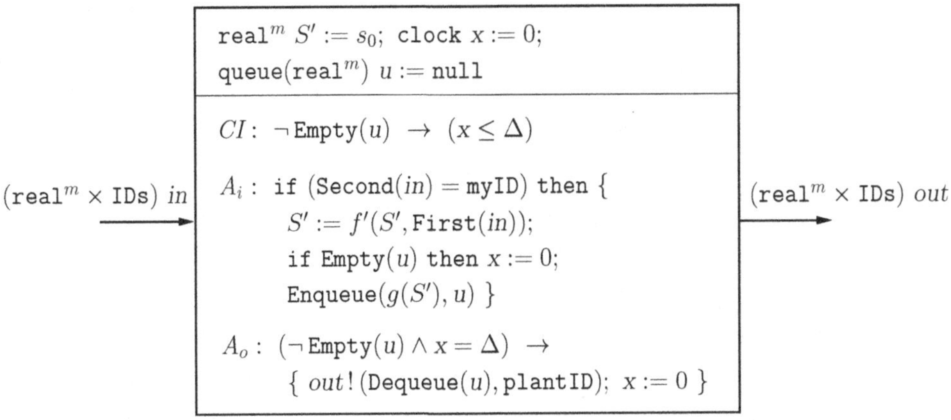

$$
\begin{array}{|l|}
\hline
\texttt{real}^m\ S' := s_0;\ \texttt{clock}\ x := 0; \\
\texttt{queue}(\texttt{real}^m)\ u := \texttt{null} \\
\hline
CI:\ \neg\,\texttt{Empty}(u)\ \rightarrow\ (x \le \Delta) \\
\\
A_i:\ \texttt{if}\ (\texttt{Second}(in) = \texttt{myID})\ \texttt{then}\ \{ \\
\qquad\quad S' := f'(S, \texttt{First}(in)); \\
\qquad\quad \texttt{if}\ \texttt{Empty}(u)\ \texttt{then}\ x := 0; \\
\qquad\quad \texttt{Enqueue}(g(S'), u)\ \} \\
\\
A_o:\ (\neg\,\texttt{Empty}(u) \wedge x = \Delta)\ \rightarrow \\
\qquad\quad \{\ out\,!\,(\texttt{Dequeue}(u), \texttt{plantID});\ x := 0\ \} \\
\hline
\end{array}
$$

$(\texttt{real}^m \times \texttt{IDs})\ in$ →   (to the left of box)

$(\texttt{real}^m \times \texttt{IDs})\ out$ →   (to the right of box)

Figure 9.23: Controller Model in Multi-hop Control Network

on the channel *in* is ready to transmit a message. In this case, both the input task A_i and the output task A_o are enabled and can execute in either order. No matter in which order they get executed, in the resulting state, the clock x is 0, and the queue contains one message that reflects the update of the controller's output in response to the value just received. If the controller is supplied inputs at a rate higher than once per Δ time units, then the number of messages waiting in the queue u will keep growing, and this scenario should be avoided.

For the network shown in figure 9.21, we need two instances of the controller process. One is instantiated with $\texttt{myID} = C_1$, $\texttt{plantID} = P_1$, $in = e_{15}$, and $out = e_{14}$, and the other is instantiated with $\texttt{myID} = C_2$, $\texttt{plantID} = P_2$, $in = e_6$, and $out = e_{13}$. In each case, the model is completed by filling in the details of the state estimator function f' and the control map g.

Network Routing

Given the set of plants, controllers, network nodes, and directed links connecting them, we need to determine how to route the messages from each plant to the corresponding controller and back. This problem can be formalized as computing paths between multiple source-destination pairs in a directed graph and is a classical network routing problem. Ideally, we would like all the routes to be mutually *disjoint*. In such a case, transmission of messages along different routes can proceed independently. Additionally, shorter routes mean shorter end-to-end delays in transmission of messages, and hence shorter routes are preferred.

In the example network shown in figure 9.21, we need to determine routes from the plant P_1 to the controller N_1, from the controller C_1 to the plant P_1, from the plant P_2 to the controller N_2, and from the controller C_2 to the plant P_2. A good choice of such routes is:

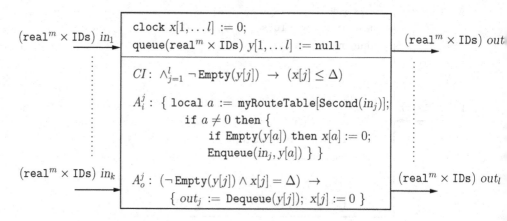

Figure 9.24: Network Node Model in Multi-hop Control Network

- the four-hop path e_2, e_7, e_{11}, e_{15} from the plant P_1 to the controller C_1,

- the three-hop path e_3, e_5, e_6 from the plant P_2 to the controller C_2,

- the four-hop path e_{14}, e_{12}, e_9, e_1 from the controller C_1 to the plant P_1, and

- the three-hop path e_{13}, e_{10}, e_4 from the controller C_2 to the plant P_2.

Note that these four routes are indeed disjoint, and the length of each route equals the length of the shortest path for the corresponding source-destination pair.

The problem of finding a path of shortest length between a single source-destination pair in a directed graph can be solved efficiently in time linear in the size of the graph using classical graph-search algorithms. However, finding disjoint paths among multiple source-destination pairs in a directed graph cannot be solved efficiently, and the problem is known to be NP-complete. In the context of multi-hop control networks, the size of the graph is typically not large (in current industrial process control, a typical graph consists of tens of nodes), and thus it is possible to explore different alternatives in an exhaustive manner to find the desired routes. When multiple sets of disjoint routes are possible, the routing strategy should prefer shorter paths. However, since a solution consists of routes between multiple source-destination pairs, two solutions may be incomparable: the path between a plant-controller pair may be shorter in one solution than in the other, but the path between another plant-controller pair may be shorter in the second solution than in the first. In such a case, the choice among the different solutions can be based on analyzing the overall performance of the entire system in conjunction with the design of control laws.

Modeling Network Node

The timed process modeling a generic network node is shown in figure 9.24. The description is parameterized by (1) the number k of incoming links, (2) the number l of outgoing links, and (3) a routing table `myRouteTable` that maps the set IDs of message destinations to one of the outgoing links. For a destination id in the set IDs, if `myRouteTable`[id] is a number j between 1 and l, then the message should be transmitted on the jth outgoing link, and if `myRouteTable`[id] is 0, then that means the node is not expecting messages sent to the destination id, and the message should be simply ignored.

The process has an input channel in_j, for $j = 1, \ldots k$, corresponding to each incoming link, and an output channel out_j for $j = 1, \ldots l$, corresponding to each outgoing link. The state of the process has a queue for each output channel: the messages to be transmitted on the output channel out_j are stored in the queue $y[j]$.

The processing of the messages received on the input channel in_j is captured by the input task A_i^j, for $j = 1, \ldots k$. Whenever the process receives a message, it uses the destination of the message, available as the second field of the incoming message, and the routing table to choose the output channel on which the message should be propagated. If the routing table entry is 0, then the node is not expecting such a message, and the message is simply dropped. Otherwise, the message is enqueued in the corresponding queue.

The timing constraint that the process should send only one message every Δ time units on each output channel is enforced in a manner analogous to the controller model. For each output channel out_j, the process has a clock variable x_j. The clock invariant ensures that, for each $j = 1, \ldots l$, if an output message is waiting to be transmitted on the jth output channel, then the corresponding clock should not exceed Δ. The output task A_o^j corresponding to the j-th output channel is enabled when the corresponding queue $y[j]$ is non-empty and the corresponding clock $x[j]$ reaches Δ. The clock corresponding to an output channel is reset to 0 every time a message is transmitted on this channel by the output task and also when an input task enqueues a message in the corresponding queue when it is empty.

For the network of figure 9.21, we need four instances of the network process. For the process corresponding to the network node N_3, the number k of incoming links is 2, and the number l of outgoing links is 3. The input channels in_1 and in_2 are renamed to the link names e_7 and e_{12}, respectively. The output channels out_1, out_2, and out_3 are renamed to the link names e_8, e_9, and e_{11}, respectively. According to the routes that we chose, at this node, messages sent to the controller C_1 should be forwarded on the link e_{11}, and messages sent to the plant P_1 should be forwarded on the link e_9. The node N_3 does not appear on the routes to the controller C_2 and to the plant P_2. Thus, the routing table for the node N_3 should be specified as `myRouteTable`[P_1] = 2, `myRouteTable`[P_2] = 0, `myRouteTable`[C_1] = 3, and `myRouteTable`[C_2] = 0.

System Model

The desired system corresponding to the multi-hop control network is the parallel composition of the instances of all the plants, controllers, and network nodes. If each link appears in at most one route, then the traffic flows smoothly through the network. In our example network, the plant P_1 sends an output value every Δ time units on the link e_2. A value v sent at time t is transmitted by the node N_2 at time $(t + \Delta)$ on the link e_7, then by the node N_3 at time $(t + 2\Delta)$ on the link e_{11}, and then by the node N_4 at time $(t + 3\Delta)$ on the link e_{15}. The controller C_1 updates its internal estimate in response to this value, and the corresponding control value v' is transmitted on the link e_{14} at time $(t + 4\Delta)$. This value is propagated by the node N_4 at time $(t + 5\Delta)$ on the link e_{12}, then by the node N_3 at time $(t + 6\Delta)$ on the link e_9, and then by the node N_1 at time $(t + 7\Delta)$ on the link e_1. For the subsequent interval of length Δ, the plant P_1 uses this value as its control input.

When there are no shared links among different routes, each control loop can be analyzed independently. Consider the closed-loop system consisting of a plant and its controller. The state of this system then consists of the plant state variables S and their estimates S' maintained by the controller. Let us assume that the route from the plant to the controller consists of k_1 hops and the route from the controller to the plant consists of k_2 hops. Then a plant observation transmitted at time t is received by the controller at time $[t + (k_1 - 1)\Delta]$ time; if the controller computes a new control value at time t, then it is received by the plant at time $(t + k_2\Delta)$ time.

Let $t_1 = \Delta$, $t_2 = 2\Delta$, ... be the sequence of times at which messages are processed. We know that in each interval $[t_i, t_{i+1})$, the control input for the plant stays constant. Let $\overline{d}$ be the input signal for the external disturbance. The state response of the system is then defined by the following rules:

- The state signal for the controller estimate $\overline{S'}(t)$ is a piecewise-constant signal. For the first k_1 slots, the controller does not receive any update, and thus, for $i < k_1$, the state s_i' during the interval $[t_i, t_{i+1})$ equals the initial state s_0. After this, at each time t_i, for $i \geq k_1$, the controller receives the plant output $(k_1 - 1)$ slots earlier, that is, the value $h(s_{i-k_1+1})$. Thus, for $i \geq k_1$, the state s_i' during the interval $[t_i, t_{i+1})$ equals $f'(s_{i-1}', h(s_{i-k_1+1}))$.

- The state signal for the plant state $\overline{S}(t)$ is a piecewise-continuous signal. For the first $(k_1 + k_2)$ slots, the controller does not receive any update for the controlled input. Thus, for $i < (k_1 + k_2)$, the state signal $\overline{S}(t)$ during the interval $[t_i, t_{i+1})$ corresponds to the solution of the initial value problem with initial state s_i and dynamics $f(S, u_0, d)$, in response to the disturbance signal $\overline{d}([t_i, t_{i+1}))$. After this, at each time t_i, for $i \geq (k_1+k_2)$, the plant receives the controller's message that reflects its computation k_2 slots earlier. Thus, for $i \geq (k_1+k_2)$, the state signal $\overline{S}(t)$ during the interval $[t_i, t_{i+1})$ corresponds to the solution of the initial value problem with initial state s_i and dynamics $f(S, g(s_{i-k_2}'), d)$. That is, the disturbance is given by the signal $\overline{d}([t_i, t_{i+1}))$, and the controlled input stays constant equal to the value obtained by applying the control map g to its estimated state k_2 slots earlier.

When all the functions f, h, f', and g are linear, the analysis techniques discussed in chapter 6 for linear systems can be adapted to compute the closed-form solution for the state response and to check properties such as stability.

Exercise 9.9: Recall the design of a cruise controller from section 6.3.3. Figure 6.18 shows the response of the car to a PI controller. In exercise 6.21, you considered the response to the same controller of the model of the car on a graded road shown in figure 6.8 for the input signal $\overline{\theta}(t) = [\sin(t/5)]/3$ (measured in radians). Now let us assume that the sensors measuring the speed communicate with the cruise controller over a (time-triggered) multi-hop network. In terms of the model discussed in section 9.2.3 assume that the output of the car is its velocity, the controller is the same PI controller used in section 6.3.3, the number of hops from the sensors to the controller is three, and the number of hops from the controller back to the plant is two. Using a simulation tool such as MATLAB, plot the velocity of the car using the parameters: the initial velocity v_0 is 0, the mass m is $100\,kg$, the coefficient of friction k is 50, the gravitational acceleration g is $9.8\,m/s^2$, the reference velocity r is $10\,m/s$, the proportional gain K_P is 600, the integral gain K_I is 40, and the time-step Δ of the network is 0.1 seconds. ∎

Exercise 9.10*: Suppose the given multi-hop control network with two plants and two controllers is such that the route from the plant P_1 to the controller C_1 shares a link with the route from the controller C_1 back to the plant P_1. In fact, the network is such that this sharing cannot be avoided (the routes involving the plant P_2 and its controller C_2 are disjoint). If we use exactly the same models for the plants, the controllers, and the network nodes as described in section 9.2.3, describe the behavior of the entire system, and how the closed-loop behavior of the two control loops be affected. Discuss a possible modification to the models so that this undesirable behavior is avoided (hint: what happens if the plant P_1 sends its outputs every two slots while all other components stay unchanged?). How will this modification affect the closed-loop behavior of the two control loops? ∎

9.3 Linear Hybrid Automata *

In chapter 7, we studied timed automata as a subclass of timed processes: a timed automaton restricts how clock variables are tested and updated, and these restrictions allow the development of symbolic reachability analysis using the data structure of difference-bounds matrices. With a similar motivation, we now consider the restriction of hybrid processes to a subclass known as *linear hybrid automata*. A linear hybrid automaton can be viewed as a generalization of a timed automaton. While a clock variable in a timed automaton can only be

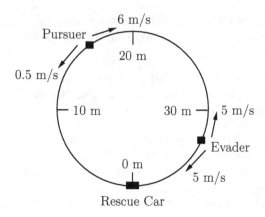

Figure 9.25: Pursuit Game

compared with a constant and reset to 0, continuously updated variables of a linear hybrid automaton are tested and updated using *affine* constraints. While a clock variable increases at the rate 1 during a timed action, a continuously updated variable in a linear hybrid automaton increases at a constant rate and, more generally, at a rate chosen from an interval with constant bounds. This structure then allows symbolic reachability analysis based on the representation of the sets of states by *polyhedra*.

Before we develop this model, it is worth emphasizing that the adjective "linear" in this context has a different meaning from its use in the classical model of "linear systems" that we studied in chapter 6. In a linear system, the rate of change of a state variable is a linear function of the system state, whereas in a linear hybrid automaton, the rate of change of a state variable is a constant or is bounded by a constant, which results in a state signal that is a linear function of time.

9.3.1   Example Pursuit Game

We illustrate the model with a two-player game of pursuit-evasion shown in figure 9.25. There is a pursuer in a golf cart chasing an evader on a circular track 40 meters long. The cart can travel upto 6 m/s in the clockwise direction but only upto 0.5 m/s in the counter-clockwise direction since it must use its reverse gear to travel counter-clockwise. The evader is on a bicycle and travels at 5 m/s in either direction. However, the evader makes a decision whether to change its direction only at fixed instances in time, separated by exactly 2 seconds. The goal of the evader is to avoid the pursuer. The evader has the added advantage that there is a rescue car at a fixed position on the track.

The evader uses a simple strategy: determine if the evader will win the race to the car if both players proceed clockwise at their respective full speeds, and if so, head clockwise and otherwise choose to move counterclockwise. The game, with this specific strategy for the evader, is shown as an extended-state machine in figure 9.26.

The continuously updated variable p models the position of the pursuer on the track measured in meters in a clockwise direction relative to the stationary car at position 0. Similarly, the continuously updated variable e models the position of the evader. The clock variable x measures the delay with respect to the most recent time instance when the evader chose the direction.

There are three modes: in the mode ClkW, the evader is moving in the clockwise direction on the track; in the mode CntrClkW the evader is moving in the counter-clockwise direction on the track; and in the mode Rescued, the evader has reached the car thus bringing the game to an end.

In the mode ClkW, the motion of the evader is described by the differential equation $\dot{e} = 5$. Regarding the evolution of the variable p that specifies the pursuer's position, we only know the bounds on the pursuer's speed in the two directions, and this is captured by the *differential inequality* $-0.5 \leq \dot{p} \leq 6$. This means that during a timed action of duration δ, the value of the variable p changes by an amount equal to δc for a constant c belonging to the interval $[-0.5, 6]$. The constraint $(0 \leq p \leq 40) \wedge (0 \leq e \leq 40) \wedge (x \leq 2)$ labeling the mode ClkW is the continuous-time invariant and ensures that all the three continuously updated variables stay within their respective ranges during a timed action.

The specification of the mode CntrClkW is similar. The only difference is that the evader moves in the counter-clockwise direction, and its evolution is captured by the differential equation $\dot{e} = -5$. In the mode Rescued, the game has ended, so all the position variables stay unchanged.

The decision logic is captured by the mode-switches. Consider the mode ClkW. Whenever the value of the variable e equals 0 or 40, the evader has reached the car, and the switch to the mode Rescued is enabled. When the clock variable x equals 2, the evader compares the times needed by the two players to reach the car moving clockwise at their respective maximum speeds. If the time $(40 - e)/5$ the evader needs is less than the time $(40 - p)/6$ the pursuer needs, then the guard condition $(6e - 5p > 40)$ of the self-loop from the mode ClkW is enabled, and the mode continues to be ClkW; otherwise the guard condition $(6e - 5p \leq 40)$ of the mode-switch from the mode ClkW to the mode CntrClkW is enabled. Note that if the pursuer is between the evader and the car along the clockwise direction, that is, the condition $(e < p)$ holds, the evader is guaranteed to choose to switch to mode CntrClkW. In either case, the clock x is reset to 0. Note that the track is circular, and thus the positions 0 and 40 coincide. To capture this, if the value of the variable p is increasing and reaches 40, then it must be reset to 0; symmetrically, if the value of the variable p is decreasing and reaches 0, then it must be updated to 40. This explains the left self-loop on the mode ClkW.

The mode-switches originating from the mode CntrClkW are symmetric.

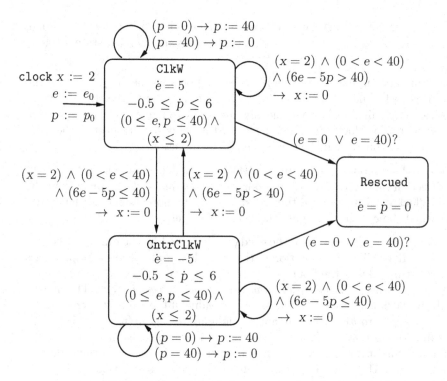

Figure 9.26: Linear Hybrid Automaton for the Pursuit Game

The initial position of the pursuer is p_0 and that of the evader is e_0. The clock is initialized to 2 so that the evader gets to make a decision about which direction to move at initial time.

During an execution, if the mode of the system ever becomes **Rescued**, then the evader wins by reaching the car. If the property $(e = p)$ becomes true at some time, then the evader loses. An execution may keep switching between the modes **ClkW** and **CntrClkW** forever with the property $(e \neq p)$ being true in all states belonging to the execution, and in such a case, the evader wins.

To illustrate the behavior of the model, consider the scenario with the initial positions $e_0 = 20$ and $p_0 = 1$. One possible execution from this initial state resulting in a win for the pursuer is shown below. A state is shown by listing the mode, followed by the values of the variables e, p, and x in that order:

$(\text{ClkW}, 20, 1, 2) \xrightarrow{\varepsilon} (\text{ClkW}, 20, 1, 0) \xrightarrow{2} (\text{ClkW}, 30, 0, 2) \xrightarrow{\varepsilon}$
$(\text{ClkW}, 30, 40, 2) \xrightarrow{\varepsilon} (\text{CntrClkW}, 30, 40, 0) \xrightarrow{2} (\text{CntrClkW}, 20, 40, 2) \xrightarrow{\varepsilon}$
$(\text{CntrClkW}, 20, 40, 0) \xrightarrow{2} (\text{CntrClkW}, 10, 39, 2) \xrightarrow{\varepsilon} (\text{CntrClkW}, 10, 39, 0) \xrightarrow{0.17}$
$(\text{CntrClkW}, 9.17, 40, 0.17) \xrightarrow{\varepsilon} (\text{CntrClkW}, 9.17, 0, 0.17) \xrightarrow{0.83} (\text{CntrClkW}, 5, 5, 1).$

In this scenario, the evader first moves clockwise for 2 seconds, during which time the pursuer moves counter-clockwise, resulting in a position where $e = 30$ and $p = 40$. Then the evader reverses direction moving counter-clockwise, during which time the pursuer stays stationary, resulting in a position where $e = 20$ and $p = 40$. The evader keeps moving counter-clockwise, during which time the pursuer moves counter-clockwise, resulting in a position where $e = 10$ and $p = 39$. The evader now still keeps moving counter-clockwise, but now the pursuer moves clockwise at full speed, and the two meet at 5 meters.

9.3.2 Formal Model

The formal definition of a linear hybrid automaton is a variation of the corresponding definitions of timed and hybrid processes. It consists of an asynchronous process some of whose state variables are of type cont and are updated continuously during a timed action. As in a timed process, we assume that all the input and output variables are updated only discretely. The dynamics of the continuously updated state variables is specified using a continuous-time invariant and a rate constraint.

The continuous-time invariant is a Boolean expression over the state variables, and a timed action is allowed only if all the states visited during the timed action satisfy the invariant. For the system of figure 9.26, the continuous-time invariant is:

$$(mode = \texttt{Rescued}) \;\vee\; [\,(0 \le e \le 40) \;\wedge\; (0 \le p \le 40) \;\wedge\; (x \le 2)\,].$$

The *linearity* restriction means that all the tests and updates involving the variables of type cont are affine expressions. More precisely, given variables $x_1, x_2, \ldots x_n$, an *affine test* is of the form $a_1 x_1 + a_2 x_2 + \cdots + a_n x_n \sim a_0$, where $a_0, a_1, \ldots a_n$ are (integer or real) constants, and $\sim$ is a comparison operation and can be either $<$, $\le$, $=$, $>$, or $\ge$. An *affine assignment* is of the form $x_i := a_0 + a_1 x_1 + a_2 x_2 + \cdots + a_n x_n$, where $a_0, a_1, \ldots a_n$ are (integer or real) constants. In a linear hybrid automaton, whenever an expression involving continuously updated variables appears in a guard, in the update code of a task, or in the continuous-time invariant, it must be an affine test, and every assignment to a continuously updated variable must be an affine assignment.

The rate constraint is a Boolean expression over the derivatives of the continuously updated variables and the discrete state variables. The expressions involving the derivatives must be affine. For the pursuit game of figure 9.26, the rate constraint is

$$(mode = \texttt{ClkW}) \;\wedge\; (\dot{e} = 5) \;\wedge\; (-0.5 \le \dot{p} \le 6) \;\wedge\; (\dot{x} = 1)$$
$$\vee\; (mode = \texttt{CntrClkW}) \;\wedge\; (\dot{e} = -5) \;\wedge\; (-0.5 \le \dot{p} \le 6) \;\wedge\; (\dot{x} = 1)$$
$$\vee\; (mode = \texttt{Rescued}) \;\wedge\; (\dot{e} = 0) \;\wedge\; (\dot{p} = 0) \;\wedge\; (\dot{x} = 1).$$

To execute a timed action in a state s, we choose a rate vector r, that is, a constant r_x, for every continuously updated variable x so that the rate constraint is satisfied when evaluated in the state s using the values r_x for the derivatives $\dot{x}$. Given a time value t, let $s + tr$ denote the state s' such that the value of a discrete variable in the state s' coincides with its value in the state s, and the value of a continuously updated variable x in the state s' equals $s(x) + t\,r_x$. Then a timed action of duration δ can be executed if the state $s + tr$ satisfies the continuous-time invariant for every time value t in the interval $[0, \delta]$, and the state resulting from the timed action is the state $s + \delta\,r$.

The formal definition is summarized below:

LINEAR HYBRID AUTOMATON

A *linear hybrid automaton HP* consists of (1) an asynchronous process P, where some of its state variables can be of type cont, and appear only in affine tests and affine assignments in the guards and updates of the tasks of P; (2) a *continuous-time invariant CI*, which is a Boolean expression over the state variables S, where the variables of type cont appear only in affine tests; and (3) a *rate constraint RC*, which is a Boolean expression over the discrete state variables and the derivatives of the continuously updated state variables that appear only in affine tests. Inputs, outputs, states, initial states, internal actions, input actions, and output actions of the linear hybrid automaton *HP* are the same as that of the asynchronous process P. Given a state s and a real-valued time $\delta > 0$, $s \xrightarrow{\delta} s + \delta r$ is a *timed action* of *HP*, for a rate vector r consisting of a constant r_x for every continuously updated state variable x, if (1) the expression RC is satisfied when for every continuously updated variable x, the derivative $\dot{x}$ is assigned the value r_x, and every discrete variable x is assigned the value $s(x)$; and (2) the state $s + tr$ satisfies the expression CI for all values $0 \le t \le \delta$.

Note that during a timed action, each continuously updated variable x evolves at a constant rate, and thus, during the duration of the timed action, the signal $\overline{x}$ is a linear function of time. In contrast, in a linear system, a typical differential equation is of the form $\dot{x} = a\,x$, and the corresponding signal $\overline{x}(t) = x_0 e^{at}$ is an exponential function of time.

As already seen in the example of the pursuit game, a rate constraint can be used to specify bounds on the rate of change of a variable. The definition of linear hybrid automata also allows constraints involving rates of two variables. For example, if the variables (x, y) denote the position of a robot in a plane, then the rate constraint

$$(1 \le \dot{x} \le 2) \wedge (\dot{x} = \dot{y})$$

specifies that the robot is moving along the diagonal (due to the constraint $\dot{x} = \dot{y}$) at a constant speed for which we know a lower and an upper bound (due to the constraint $1 \le \dot{x} \le 2$).

Notions such as executions and reachable states for a linear hybrid automaton are defined in the same way as for a hybrid process.

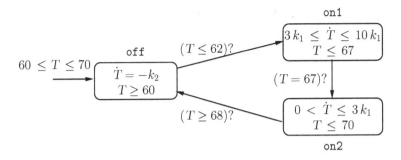

Figure 9.27: Linear Hybrid Automaton Model of a Thermostat

Note that since the rate constraint in the description of a linear hybrid automaton allows differential inequalities, syntactically it is not a hybrid process. However, the model of hybrid processes is strictly more expressive than linear hybrid automata as it is straightforward to capture the same behavior as that of a given linear hybrid automaton by introducing auxiliary variables corresponding to the rates that are updated discretely and nondeterministically. For example, to capture the differential constraint $(-0.5 \leq \dot{p} \leq 6)$ in the description of the automaton of figure 9.26, we introduce a discrete variable r_p of type real, set its value using the nondeterministic assignment $r_p := \text{choose}\{v \mid -0.5 \leq v \leq 6\}$ during a mode-switch, and specify the dynamics of the state variable p by the linear differential equation $\dot{p} = r_p$.

Exercise 9.11: Consider the linear hybrid automaton shown in figure 9.27 that models the behavior of a thermostat. Describe the possible executions of this model and explain how this model differs from the hybrid process of figure 9.1. ∎

Exercise 9.12*: Design an alternative strategy for the evader that can still be described as a linear hybrid automaton but is "better" than the strategy shown in figure 9.26. It is still required that the evader makes its decisions only every two seconds; once it decides to move clockwise or counter-clockwise, it moves in that direction at full speed. Thus, the only permissible change in the linear hybrid automaton modeling the revised strategy compared to the automaton of figure 9.26 is in the test $(6e - 5p > 40)$ that is used to determine whether to move clockwise or counter-clockwise. Your alternative strategy should be such that the evader wins the game starting from the initial position $e_0 = 20$ and $p_0 = 1$. Is your strategy optimal for the evader (that is, is it the case that for every initial position, if your strategy results in a loss for the evader, then in every alternative strategy also results in a loss for the evader from this initial position)? ∎

9.3.3 Symbolic Reachability Analysis

Now we develop an algorithm for the invariant verification problem for linear hybrid automata based on symbolic reachability analysis. The high-level algorithm is the same as the one discussed in section 3.4. To implement the symbolic search algorithm, we need a representation for regions, that is, for sets of states, that is suitable in the context of linear hybrid automata. For this purpose, let us assume that all the discrete variables have enumerated types as is the case for the pursuit game of figure 9.26. Then the linear hybrid automaton has only finitely many discrete states. The reachability algorithm then analyzes the discrete variables by enumerating their values and the continuously updated variables in a symbolic manner using affine constraints.

Affine Formulas

Let V be a set of variables that is partitioned into two sets: the set V_d of discrete variables of enumerated types and the set V_c of continuously updated variables of type **real**. A state over V then consists of a discrete state that assigns values to all the variables of enumerated types and an $|V_c|$-dimensional real-valued vector. A region A over (V_c, V_d) is then represented by a formula in the *disjunctive normal form* built using affine constraints over V_c and equality constraints over V_d.

Formally, the type **AffForm** is parameterized by the variable sets (V_c, V_d) and consists of affine formulas defined by the following rules:

- An atomic affine formula over (V_c, V_d) is an equality of the form $(x = d)$, where x is a discrete variable in V_d and d is a constant belonging to the type of the variable x, or an affine constraint of the form $(a_1 x_1 + a_2 x_2 + \cdots + a_n x_n \sim a_0)$, where $x_1, x_2, \ldots x_n$ are real-valued variables in V_c, $a_0, a_1, \ldots a_n$ are real numbers, and $\sim$ is a comparison operator from the set $\{<, \leq, =, >, \geq\}$.

- A conjunctive affine formula φ over (V_c, V_d) is a conjunction $\varphi_1 \wedge \varphi_2 \wedge \cdots \wedge \varphi_k$, where the conjuncts $\varphi_1, \varphi_2, \ldots \varphi_k$ are atomic affine formulas over (V_c, V_d).

- An affine formula A over (V_c, V_d) is a disjunction $\varphi_1 \vee \varphi_2 \vee \cdots \vee \varphi_l$, where the disjuncts $\varphi_1, \varphi_2, \ldots \varphi_l$ are conjunctive affine formulas over (V_c, V_d).

Note that the syntax of affine formulas does not allow negation explicitly, but it can be expressed. For example, the constraint $\neg(x = d)$, where x is a variable in V_d and d is a constant, is equivalent to the affine formula $(x = d_1) \vee (x = d_2) \vee \cdots \vee (x = d_a)$, if the constants $d_1, d_2, \ldots d_a$ are all the values belonging to the type of the variable x other than the value d. The constraint $\neg(a_1 x_1 + \cdots + a_n x_n \leq a_0)$, where $x_1, x_2, \ldots x_n$ are real-valued variables in V_c and $a_0, a_1, \ldots a_n$ are real numbers, is equivalent to the affine formula $(a_1 x_1 + \cdots + a_n x_n > a_0)$.

Symbolic Representation of a Linear Hybrid Automaton

Having fixed the representation of affine formulas, now we can encode the various components of a linear hybrid automaton using this symbolic representation. For a given linear hybrid automaton *HP*, let S_d denote the set of its discrete variables and let S_c be the set of its continuously updated real-valued variables. We are restricting attention to the case where each variable in the set S_d has an enumerated type. For the pursuit game of figure 9.26, the set S_d contains the variable *mode*, and the set S_c contains the variables *e, p,* and *x*.

The set of initial states of the automaton *HP* is represented by a formula *Init* of type AffForm over (S_c, S_d). For the pursuit game, assuming the initial position e_0 of the evader is 20 and the initial position p_0 of the pursuer is 10, the formula *Init* equals

$$(\textit{mode} = \texttt{ClkW}) \wedge (x = 2) \wedge (e = 20) \wedge (p = 10).$$

The transition relation of the asynchronous process corresponding to the automaton *HP* is represented by a formula *Trans* of type AffForm over $(S_c \cup S_c', S_d \cup S_d')$. Here, for every variable *x*, its primed version x' denotes the value of the variable *x* after executing the transition as explained in section 3.4. This transition relation captures all the input, output, and internal actions of the underlying asynchronous process and, thus, the set of all discrete transitions of the automaton *HP*. For the pursuit game, the formula *Trans* has a disjunct for every mode-switch of the automaton of figure 9.26. For instance, the disjunct corresponding to the mode-switch from the mode ClkW to the mode CntrClkW is

$$(\textit{mode} = \texttt{ClkW}) \wedge (x = 2) \wedge (e > 0) \wedge (e < 40) \wedge (6e - 5p \le 40) \wedge$$
$$(\textit{mode}' = \texttt{CntrClkW}) \wedge (x' = 0) \wedge (e' - e = 0) \wedge (p' - p = 0).$$

The continuous-time invariant of the automaton *HP* is represented by a formula *CI* of type AffForm over (S_c, S_d). The continuous-time invariant for the pursuit game is captured by:

$$(\textit{mode} = \texttt{Rescued}) \vee [(e \ge 0) \wedge (e \le 40) \wedge (p \ge 0) \wedge (p \le 40) \wedge (x \le 2)].$$

The rate constraint of the automaton *HP* is represented by a formula *RC* of type AffForm over $(\dot{S}_c, S_d)$. Here, for every real-valued variable *x*, the real-valued variable $\dot{x}$ represents its rate of change with time. The rate constraint for the pursuit game is given by:

$$[(\textit{mode} = \texttt{ClkW}) \wedge (\dot{e} = 5) \wedge (\dot{p} \ge -0.5) \wedge (\dot{p} \le 6) \wedge (\dot{x} = 1)]$$
$$\vee \quad [(\textit{mode} = \texttt{CntrClkW}) \wedge (\dot{e} = -5) \wedge (\dot{p} \ge -0.5) \wedge (\dot{p} \le 6) \wedge (\dot{x} = 1)]$$
$$\vee \quad [(\textit{mode} = \texttt{Rescued}) \wedge (\dot{e} = 0) \wedge (\dot{p} = 0) \wedge (\dot{x} = 1)].$$

In summary, the symbolic representation of a linear hybrid automaton *HP* with discrete state variables S_d and continuously updated state variables S_c consists of (1) an initialization formula *Init* of type `AffForm` over (S_c, S_d), (2) a transition formula *Trans* of type `AffForm` over $(S_c \cup S'_c, S_d \cup S'_d)$, (3) a continuous-time invariant *CI* of type `AffForm` over (S_c, S_d), and (4) a rate constraint *RC* of type `AffForm` over $(\dot{S}_c, S_d)$. Such a description can be compiled from the source language used to describe linear hybrid automata automatically.

Operations on Affine Formulas

The symbolic search techniques discussed in section 3.4 require the following operations on the data type for regions: union `Disj`, intersection `Conj`, set difference `Diff`, emptiness test `IsEmpty`, existential quantification `Exists`, and renaming of variables `Rename`. All these operations can be effectively implemented for the data type `AffForm` of affine formulas as discussed below.

For two affine formulas A and B, the union `Disj`(A, B) is simply the formula $A \vee B$, as it is guaranteed to be of type `AffForm`.

Consider two affine formulas A and B. The formula corresponding to `Conj`(A, B) cannot simply be the conjunction $A \wedge B$ since it need not be an affine formula in the required disjunctive normal form. However, the distributivity properties of the logical disjunction and conjunction operations can be used to implement the desired operation. If the formula A equals $\varphi_1 \vee \cdots \vee \varphi_a$, where each φ_i is a conjunctive affine formula, and the formula B equals $\psi_1 \vee \cdots \vee \psi_b$, where each ψ_j is a conjunctive affine formula, then `Conj`(A, B) is the affine formula

$$\bigvee_{1 \leq i \leq a, 1 \leq j \leq b} (\varphi_i \wedge \psi_j).$$

Note that each disjunct $(\varphi_i \wedge \psi_j)$ is a conjunctive affine formula, and the size of the resulting formula grows quadratically as it has $a \cdot b$ number of disjuncts.

The set-difference operation `Diff`(A, B) for two affine formulas can be implemented with some rewriting and is left as an exercise.

The renaming operation can be implemented by simple textual substitution. For an affine formula A, to rename a variable x in A to another name y, which does not occur in A, the result `Rename`(A, x, y) is obtained by replacing every occurrence of the variable x in the formula A by y.

To implement the operation `IsEmpty`, given a formula A of type `AffForm` over (V_c, V_d), we need to check if the variables can be assigned values so that the formula A is satisfied. Suppose the formula A is the disjunction $\varphi_1 \vee \cdots \vee \varphi_a$, where each φ_i is a conjunctive affine formula. Then the formula A is satisfiable exactly when one of the formulas φ_i is satisfiable, and satisfiability of each of these subformulas can be checked independently. Thus, it suffices to focus on testing satisfiability of a conjunctive affine formula. If it contains two conjuncts of the form $(x = d)$ and $(x = d')$, for a discrete variable x and two *distinct* constants d and d', then the formula cannot be satisfied. Otherwise the constraints involving discrete variables do not influence satisfiability. Thus, the core computational problem for checking satisfiability of affine formulas reduces to checking satisfiability of a conjunction of atomic affine constraints, that is, checking satisfiability of a formula of the form $\varphi_1 \wedge \cdots \wedge \varphi_k$, where each φ_i is an affine constraint of the form $(a_1 x_1 + a_2 x_2 + \cdots + a_n x_n \sim a_0)$. Checking satisfiability of a conjunction of such affine constraints is a classical problem in linear programming with a well-understood theoretical foundation and a variety of efficient implementations.

Quantifier Elimination

Finally, let us focus on the operation of existential quantification: given an affine formula A and a variable x, we want to compute the result $B = \mathtt{Exists}(A, x)$ so that the formula B is of type $\mathtt{AffForm}$, does not involve the variable x, and a state s satisfies the formula B exactly when there exists a value c for the variable x such that the state $s[x \mapsto c]$ satisfies the formula A. Let us assume that the variable x to be quantified is a real-valued variable, and the formula A is a conjunction $\varphi_1 \wedge \cdots \wedge \varphi_k$, where each φ_i is an affine constraint of the form $(a_1 x_1 + a_2 x_2 + \cdots + a_n x_n \sim a_0)$. This case captures the computational essence of the problem, and handling the general case is left as an exercise.

Consider a conjunct φ_i of the form $(a_1 x_1 + a_2 x_2 + \cdots + a_n x_n \sim a_0)$, where the variable x equals x_1. If the coefficient a_1 equals 0 (that is, the variable to be eliminated does not appear in the conjunct φ_i), then φ_i directly appears as a conjunct in the result B. If the coefficient a_1 is non-zero, then consider the expression e_i given by $(a_0 - a_2 x_2 - \cdots - a_n x_n)/a_1$. If the comparison operation $\sim$ is $\leq$ and the coefficient a_1 is positive, then we can rewrite the constraint φ_i as $(x \leq e_i)$, and in such a case, the expression e_i is an upper bound on the value of x. If the comparison operation $\sim$ is $\leq$ and the coefficient a_1 is negative, then we can rewrite the constraint φ_i as $(x \geq e_i)$, and in such a case, the expression e_i is a lower bound on the value of x. The other cases are similar but can lead to strict lower bound constraints of the form $(x > e_i)$ and strict upper bound constraints of the form $(x < e_i)$. Now we can eliminate the variable x from the constraints if we simply assert that every lower bound on x must be less than every upper bound on x. For instance, the constraints $(x \geq e_i)$ and $(x \leq e_j)$ lead to the implied constraint $(e_i \leq e_j)$, and the constraints $(x \geq e_i)$ and $(x < e_j)$ lead to the implied constraint $(e_i < e_j)$. If the conjunction of all such implied constraints is satisfiable, then the maximum of the lower bounds on x does not exceed the minimum of the upper bounds on x, and in such a case, it is possible to find a value of x that satisfies all the original constraints in A. Each implied constraint of the form $(e_i \leq e_j)$ or $(e_i < e_j)$ can be easily rewritten so that it is an atomic affine formula and contributes a conjunct to the desired result B.

To illustrate the quantifier elimination procedure, consider the conjunctive affine formula A given by:

$$(2x + 3y - 5z < 7) \wedge (6y + 8z \geq -2) \wedge (-x + y - 7z \leq 10) \wedge (3x + z \leq 0).$$

The first conjunct gives the strict upper bound constraint $x < (7 - 3y + 5z)/2$, the second conjunct does not constrain x, the third conjunct gives the lower bound constraint $x \geq (-10 + y - 7z)$, and the fourth conjunct gives the upper bound constraint $x \leq -z/3$. We eliminate x by requiring every lower bound not to exceed every upper bound and retaining the second conjunct. This leads to:

$$(6y + 8z \geq -2) \wedge [(-10 + y - 7z) \leq -z/3] \wedge [(-10 + y - 7z) < (7 - 3y + 5z)/2].$$

We then rewrite the last two conjuncts so that the formula is in the desired affine form:

$$(6y + 8z \geq -2) \wedge (3y - 20z \leq 30) \wedge (5y - 19z < 27).$$

Note that the number of atomic constraints in the result B can be quadratic in the number of atomic constraints in the input formula A. If we apply the existential quantification repeatedly, the number of constraints grows exponentially.

Image Computation: Discrete Transitions

The core of the symbolic search is *image computation*: given a region A over the state variables, we want to compute the region that contains all the states that can be reached from the states in A using one transition. If we focus on the discrete transitions, then the algorithm for image computation is identical to the one discussed in section 3.4. Given an affine formula A, we first conjoin it with *Trans*, a region over unprimed and primed state variables containing all the discrete transitions. The intersection $\text{Conj}(A, \textit{Trans})$ is a region over $S \cup S'$ and contains all the discrete transitions that originate in the states in A. Then we project the result onto the set S' of primed state variables by existentially quantifying the variables in S. Renaming each primed variable x' to x gives us the desired region. Thus, the discrete post-image of the region A of a linear hybrid automaton HP with the transition formula *Trans* is defined by

$$\text{DiscPost}(A, \textit{Trans}) = \text{Rename}(\text{Exists}(\text{Conj}(A, \textit{Trans}), S), S', S).$$

For the pursuit-game example, the discrete post-image of the initial region can be obtained using the above formula. The result of this computation is the affine formula A_1:

$$\text{DiscPost}(\textit{Init}, \textit{Trans}) = A_1 = (\textit{mode} = \text{ClkW}) \wedge (x = 0) \wedge (e = 20) \wedge (p = 10).$$

Image Computation: Timed Transitions

Our next goal is to compute a symbolic representation of the set of all states resulting from a timed action starting from a state in a given region of the hybrid automaton HP. Let A be a region of the hybrid automaton, let CI be its continuous-time invariant, and let RC be its rate constraint. The timed post-image B of the region A consists of all states s', such that there exists a state s belonging to the region A and a time duration δ and a rate vector r such that (1) the state s together with the rates r satisfy the formula RC; (2) for every $0 \le t \le \delta$, the state $s + rt$ satisfies the invariant CI; and (3) the final state s' equals $s + r\delta$. For simplicity, let us assume that the continuous-time invariant CI defines a *convex* set: if two states satisfy the formula CI, then so does every state that lies on the segment joining the two states. This is a typical case and holds for the pursuit-game example. In such a case, the condition (2) can be replaced by the simpler condition that the states s and s' at the beginning and the end of the timed action both satisfy CI.

The formula A uses the discrete variables S_d and continuously updated variables S_c. To get the desired result, we use auxiliary variables S'_c that denote the values of the continuously updated variables at the end of the timed action and the variable *inc* that denotes the duration of the timed action (assume that *inc* is not a variable of the hybrid automaton). We will first construct a formula B' over the variables S_d, S_c, S'_c, and *inc*. This formula captures all the timed transitions starting in the region A.

Since the region B' should capture all timed actions starting in the region A, A is a conjunct in B'. To ensure that the continuous-time invariant holds at the beginning of the timed action, the formula CI is also a conjunct of B'. To ensure that the continuous-time invariant holds at the end of the timed action, the formula $\mathtt{Rename}(CI, S_c, S'_c)$ obtained by renaming every continuously updated variable x to x' is also a conjunct of B'.

We now need a constraint that says the rates for variables are chosen according to the rate constraint, and the increment $(x' - x)$ in the value of a continuously updated variable x equals the value of the duration variable *inc* multiplied by the rate of change of x. A direct encoding of this constraint uses the multiplication operation and thus results in a non-linear constraint. We avoid this by exploiting the special structure of the rate formula RC: an atomic affine constraint of the form $(a_1\dot{x}_1 + \cdots + a_n\dot{x}_n \sim a_0)$ is replaced by the constraint $[a_1(x'_1 - x_1) + \cdots + a_n(x'_n - x_n) \sim a_0 \, inc]$. Observe that this modified constraint is equivalent to the original constraint since the constant rate of change $\dot{x}_i$ of a variable equals the change in its value divided by the duration, which is captured by the expression $(x'_i - x_i)/inc$. More precisely, given the affine formula RC over $(\dot{S}_c, S_d)$, let RC' be the affine formula over $(S'_c \cup \{inc\}, S_d)$ obtained by replacing an atomic affine constraint of the form

$$(a_1\dot{x}_1 + \cdots + a_n\dot{x}_n) \sim a_0$$

by the atomic affine constraint

$$(a_1x'_1 - a_1x_1 + \cdots + a_nx'_n - a_nx_n - a_0 \, inc) \sim 0.$$

The formula B' is the conjunction of the affine formulas A, CI, $\mathtt{Rename}(CI, S_c, S'_c)$, and RC'.

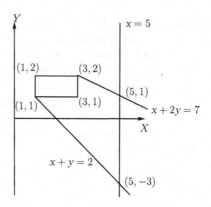

Figure 9.28: Example of Timed Post-image Computation

To obtain the desired timed post-image B from the formula B', we can use quantifier elimination to project out the starting values of the continuously updated variables along with the time duration and rename every primed variable x' back to x. Thus, the timed post-image $\texttt{TimedPost}(A, CI, RC)$ of the region A is defined by:

$$\texttt{Rename}(\texttt{Exists}(\texttt{Conj}(A, CI, \texttt{Rename}(CI, S_c, S'_c), RC'), S_c \cup \{\, inc \,\}), S'_c, S_c).$$

To illustrate the computation of the timed post-image, suppose there are two continuously updated variables x and y and no discrete variables. Suppose the region A is given by $(1 \le x \le 3) \wedge (1 \le y \le 2)$ (see the rectangle in figure 9.28). Suppose the continuous-time invariant is $(x \le 5)$, and the rate constraint is $(\dot{x} = 1) \wedge (-1 \le \dot{y} \le -0.5)$. This means that during a timed action, a state evolves along a line within the cone bounded by lines with slopes -1 and -0.5 as long as the value of x does not exceed 5. To compute the timed post-image, we first introduce the primed variables x' and y' and the duration variable inc. The transformed rate constraint RC' is

$$[(x' - x) = inc] \wedge [-inc \le (y' - y) \le -0.5\,inc].$$

The formula B' is the conjunction

$$(1 \le x \le 3) \wedge (1 \le y \le 2) \wedge (x \le 5) \wedge (x' \le 5) \wedge$$
$$(x' - x = inc) \wedge [-inc \le (y' - y) \le -0.5\,inc].$$

The desired region B is obtained by eliminating the variables x, y, and inc from the above formula and renaming x' to x and y' to y in the result. The final result is equivalent to the affine formula

$$(x \ge 1) \wedge (y \le 2) \wedge (x \le 5) \wedge (x + y \ge 2) \wedge (x + 2y \le 7).$$

In figure 9.28, the timed post-image of the rectangle A is the pentagon with the vertices $(1, 1)$, $(1, 2)$, $(3, 2)$, $(5, 1)$, and $(5, -3)$.

For the pursuit-game example, no timed action of a positive duration is possible from the initial state (since the clock x is 2), and the timed post-image of the region *Init* is *Init* itself. The timed post-image of the region A_1, which was obtained by applying the discrete post-image operation to *Init*, is the affine formula

$$(\, mode = \texttt{ClkW}\,) \wedge (\, 0 \le x \le 2\,) \wedge (\, e - 5x = 20\,) \wedge (\, 10 - 0.5x \le p \le 10 + 6x\,).$$

Iterative Image Computation

The post-image of a region A of a linear hybrid automaton is simply the union (or disjunction) of its discrete post-image and timed post-image:

$$\texttt{Post}(A, \textit{Trans}, \textit{CI}, \textit{RC}) = \texttt{Disj}(\texttt{DiscretePost}(A, \textit{Trans}), \texttt{TimedPost}(A, \textit{CI}, \textit{RC})).$$

To check whether a property φ over the state variables S is an invariant (or, equivalently, whether the negated property $\neg\varphi$ is reachable), we can now apply the symbolic breadth-first search algorithm of figure 3.18 based on the representation of regions as affine formulas.

The complexity of the representation, that is, the size of the affine formula representing the current set of reachable states, grows significantly with the number of iterations. Tools for symbolic reachability analysis of linear hybrid automata such as HyTech and SpaceEx incorporate a number of optimizations to improve the computational efficiency. When a linear hybrid automaton has n continuously updated variables, each atomic affine constraint is a hyperplane in the n-dimensional space of real-valued vectors. A conjunctive affine formula then is a polyhedron in the n-dimensional space. The optimizations for manipulating affine formulas are based on employing alternative representations for polyhedra and algorithms for simplifying such representations. For example, given the polyhedron as a conjunction of atomic affine constraints, one of the constraints can be omitted if it does not represent a bounding facet of the polyhedron.

Even when the linear hybrid automaton has only finitely many discrete states, the symbolic reachability algorithm may not terminate. As an example, consider the linear hybrid automaton of figure 9.29 with a single mode and two continuously updated variables. Both variables are initialized to 0. The variable x evolves at the rate 1, whereas the variable y changes at the rate 2. Initially, both variables are 0. The set of reachable states discovered in the first iteration of the symbolic search is the line segment joining $(0, 0)$ and $(1/2, 1)$. When y reaches 1, the discrete mode-switch updates the state to $(0, 1/2)$. In the next iteration, when y reaches 1, the value of x is $1/4$, and the discrete mode-switch updates the state to $(0, 1/4)$. This pattern repeats forever (see figure 9.29). The set of reachable states of this system contains infinitely many disconnected line segments: the kth line segment connects the states $(0, 1 - 1/2^k)$ and $(1/2^{k+1}, 1)$.

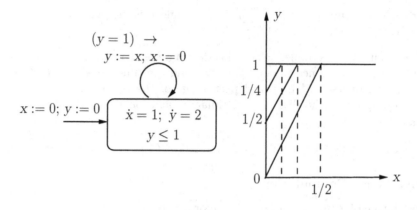

Figure 9.29: Example of Non-terminating Symbolic Image Computation

It is clear that the symbolic breadth-first search algorithm keeps iterating forever discovering shorter and shorter such segments. In contrast, as studied in section 7.3, for timed automata, all continuously updated variables increase at the same rate, and if the timed automaton has finitely many discrete states, the symbolic reachability algorithm based on the symbolic representation of clock-zones (or clock-regions) is guaranteed to terminate.

Analysis of the Pursuit Game

The symbolic reachability analysis can be used to determine whether the evader's strategy of figure 9.26 is a winning strategy for a given initial position. For this purpose, the symbolic breadth-first search algorithm of figure 3.18 is executed to check whether the property $(e - p = 0)$ is reachable: if the property is discovered to be reachable, then the pursuer wins the game, or else the evader wins the game.

For the initial position $e_0 = 20$ and $p_0 = 1$, after four iterations of the algorithm, the property $(e - p = 0)$ is found to be reachable. Thus, the pursuer wins in this case.

For the initial position $e_0 = 20$ and $p_0 = 10$, after five iterations, the algorithm terminates having computed all the reachable states, without finding any state where the property $(e - p = 0)$ holds. The set of reachable states is given by the formula:

$$
\begin{array}{lllll}
& mode = \mathtt{ClkW} & \wedge\, e = 20 & \wedge\, p = 10 & \wedge\, x = 2 \\
\vee\, & mode = \mathtt{ClkW} & \wedge\, e = 20 + 5x & \wedge\, 10 - 0.5x \le p \le 10 + 6x & \wedge\, 0 \le x \le 2 \\
\vee\, & mode = \mathtt{ClkW} & \wedge\, e = 30 + 5x & \wedge\, 9 - 0.5x \le p \le 22 + 6x & \wedge\, 0 \le x \le 2 \\
\vee\, & mode = \mathtt{Rescued} & \wedge\, e = 0 & \wedge\, 8 \le p \le 34 & \wedge\, x \ge 2\,.
\end{array}
$$

Now suppose, having fixed the initial position e_0 of the evader to be 20, we want to compute the set of initial positions p_0 of the pursuer for which the evader wins. This can be achieved using the same symbolic reachability algorithm but treating the initial position p_0 as another symbolic variable. Formally, we modify the linear hybrid automaton of figure 9.26 by adding a fourth continuously updated variable called p_0. The value of this variable does not change: in every mode, modify the rate constraint by adding the conjunct $(\dot{p}_0 = 0)$. The initial condition is now the formula

$$(\mathit{mode} = \texttt{ClkW}) \wedge (x = 2) \wedge (e = 20) \wedge (p - p_0 = 0).$$

Then we apply the iterative image computation algorithm until all the reachable states are computed. If the affine formula *Reach* over the variables *mode*, x, e, p, and p_0 represents all the reachable states, then the formula

$$\texttt{Exists}(\texttt{Conj}(\texttt{Reach}, (e - p = 0)), \{\mathit{mode}, p, x, e\})$$

is an affine formula containing only the variable p_0 that gives the constraint on the initial position of the pursuer for which the pursuer wins. Based on the computation using the tool HYTECH, this formula turns out to be $(0 \le p_0 \le 2) \vee (16 \le p_0 \le 40)$. This tells us that when the evader's initial position is 20, the evader wins exactly when the pursuer's initial position belongs to the interval $(2, 16)$.

Exercise 9.13: Given two formulas A and B of type AffForm, show how to implement the operation $\texttt{Diff}(A, B)$, that is, how to compute a formula that is (1) equivalent to the logical formula $A \wedge \neg B$, and (2) of type AffForm. ■

Exercise 9.14: Describe the procedure for implementing the existential quantification for the data type of affine formulas in its full generality: given an affine formula A over (V_c, V_d) and a variable $x \in V_c \cup V_d$, show how to obtain the affine formula for $\texttt{Exists}(A, x)$. ■

Exercise 9.15: Consider the region A given by the conjunctive affine formula

$$(-x_1 + 6x_4 \le 17) \wedge (3x_1 + 12x_3 < 1) \wedge (2x_1 - 3x_2 + 5x_4 \le 7) \wedge$$
$$(7x_2 - x_3 - 8x_4 > 0) \wedge (5x_1 + 2x_2 - x_3 > -5).$$

Calculate the region $\texttt{Exists}(A, x_1)$ as an affine formula. ■

Exercise 9.16: Suppose a linear hybrid automaton has two continuously updated variables x and y and no discrete variables. Consider the triangular region A with vertices $(0, 0)$, $(1, 2)$, and $(2, 1)$. Suppose the continuous-time invariant is $(x - y \le 4)$ and the rate constraint is $(\dot{x} = 1) \wedge (0.5 \le \dot{y} \le 1)$. Calculate the timed post-image of the region A and describe it as an affine formula. ■

Bibliographic Notes

The study of formal models for hybrid systems by combining discrete transition systems and differential/algebraic equations started in the early 1990s [MMP91]. Our model of hybrid processes is based on hybrid automata [ACH+95]. As examples of well-developed formal approaches to specification and verification of hybrid systems, see [Tab09] and [Pla10]. Modeling of embedded control systems using hybrid automata is now also supported by commercial modeling software; for example, Mathworks (see `mathworks.com`) supports modeling using the combination of STATEFLOW and SIMULINK.

Mathematical analysis of stability and control design for hybrid dynamical systems is an active topic of research in control theory (see [Bra95] and [LA14]).

The case study of the automated guided vehicle is based on [LS11], the modeling of obstacle avoidance appears in [AEK+99], the analysis of multi-hop time-triggered control network is based on [ADJ+11], and the pursuit game of figure 9.25 is from [AHW97].

The model of linear hybrid automata was introduced in [ACH+95], and the corresponding symbolic analysis algorithm was first implemented in the model checker HyTech [AHH96, HHW97] (see [FLGD+11] for recent and efficient techniques for analysis of linear hybrid automata).

Bibliography

[ACH+95]   R. Alur, C. Courcoubetis, N. Halbwachs, T.A. Henzinger, P.-H. Ho, X. Nicollin, A. Olivero, J. Sifakis, and S. Yovine. The algorithmic analysis of hybrid systems. *Theoretical Computer Science*, 138:3–34, 1995.

[AD94]   R. Alur and D.L. Dill. A theory of timed automata. *Theoretical Computer Science*, 126:183–235, 1994.

[ADJ+11]   R. Alur, A. D'Innocenzo, K.H. Johansson, G.J. Pappas, and G. Weiss. Compositional modeling and analysis of multi-hop control networks. *IEEE Trans. Automat. Contr.*, 56(10):2345–2357, 2011.

[AEK+99]   R. Alur, J. Esposito, M. Kim, V. Kumar, and I. Lee. Formal modeling and analysis of hybrid systems: A case study in multirobot coordination. In *FM'99 – World Congress on Formal Methods in the Development of Computer Systems*, LNCS 1708, pages 212–232. Springer, 1999.

[AH95]   K.J. Åström and T. Hägglund. *PID Controllers: Theory, Design, and Tuning*. Instrument Society of America, 1995.

[AH99a]   R. Alur and T.A. Henzinger. *Computer-Aided Verification*. 1999. Unpublished manuscript, available at `www.cis.upenn.edu/~alur/CAVBook.pdf`.

[AH99b]   R. Alur and T.A. Henzinger. Reactive modules. *Formal Methods in System Design*, 15(1):7–48, 1999.

[AHH96]   R. Alur, T.A. Henzinger, and P.-H. Ho. Automatic symbolic verification of embedded systems. *IEEE Transactions on Software Engineering*, 22(3):181–201, 1996.

[AHW97]   R. Alur, T.A. Henzinger, and H. Wong-Toi. Symbolic analysis of hybrid systems. In *Proceedings of the 37th IEEE Conference on Decision and Control*, 1997.

[AM06]   P.J. Antsaklis and A.N. Michel. *Linear Systems*. Birkhäuser, 2006.

[BBC+10]   A. Bessey, K. Block, B. Chelf, A. Chou, B. Fulton, S. Hallem, C.-H.
            Gros, A. Kamsky, S. McPeak, and D.R. Engler. A few billion lines
            of code later: Using static analysis to find bugs in the real world.
            Commun. ACM, 53(2):66–75, 2010.

[BCD+92]   J.R. Burch, E.M. Clarke, D.L. Dill, L.J. Hwang, and K.L. McMil-
            lan. Symbolic model checking: 10^{20} states and beyond. *Information
            and Computation*, 98(2):142–170, 1992.

[BCE+03]   A. Benveniste, P. Caspi, S.A. Edwards, N. Halbwachs, P. Le Guer-
            nic, and R. de Simone. The synchronous languages 12 years later.
            Proceedings of the IEEE, 91(1):64–83, 2003.

[BDL+11]   G. Behrmann, A. David, K.G. Larsen, P. Pettersson, and W. Yi.
            Developing UPPAAL over 15 years. *Software – Practice and Experi-
            ence*, 41(2):133–142, 2011.

[BG88]   G. Berry and G. Gonthier. The synchronous programming language
            ESTEREL: Design, semantics, implementation. Technical Report
            842, INRIA, 1988.

[BGK+96]   J. Bengtsson, W.D. Griffioen, K.J. Kristoffersen, K.G. Larsen,
            F. Larsson, P. Pettersson, and W. Yi. Verification of an audio
            protocol with bus collision using UPPAAL. In *Computer Aided Ver-
            ification, 8th International Conference (CAV)*, LNCS 1102, pages
            244–256, 1996.

[BHSV+96]   R. Brayton, G. Hachtel, A. Sangiovanni-Vincentelli, F. Somenzi,
            A. Aziz, S. Cheng, S. Edwards, S. Khatri, Y. Kukimoto, A. Pardo,
            S. Qadeer, R. Ranjan, S. Sarwary, T. Shiple, G. Swamy, and
            T. Villa. VIS: A system for verification and synthesis. In *Computer
            Aided Verification: 8th International Conference (CAV)*, LNCS
            1102, pages 428–432. Springer-Verlag, 1996.

[BK08]   C. Baier and J.-P. Katoen. *Principles of Model Checking*. MIT
            Press, 2008.

[BKSY12]   D. Bustan, D. Korchemny, E. Seligman, and J. Yang. SystemVer-
            ilog Assertions: Past, present, and future SVA standardization ex-
            perience. *IEEE Design & Test of Computers*, 29(2):23–31, 2012.

[BLR11]   T. Ball, V. Levin, and S.K. Rajamani. A decade of software model
            checking with SLAM. *Commun. ACM*, 54(7):68–76, 2011.

[BM07]   A.R. Bradley and Z. Manna. *The Calculus of Computation – Deci-
            sion Procedures with Applications to Verification*. Springer, 2007.

[Bra95]   M. S. Branicky. *Studies in Hybrid Systems: Modeling, Analysis,
            and Control*. PhD thesis, Massachusetts Institute of Technology,
            1995.

[Bry86]   R.E. Bryant. Graph-based algorithms for Boolean-function manip-
           ulation. *IEEE Transactions on Computers*, C-35(8), 1986.

[BSW69]   K.A. Bartlett, R.A. Scantlebury, and P.T. Wilkinson. A note on
           reliable full-duplex transmission over half-duplex links. *Commun.
           ACM*, 12(5):260–261, 1969.

[Büc62]   J.R. Büchi. On a decision method in restricted second-order arith-
           metic. In *Proceedings of the International Congress on Logic,
           Methodology, and Philosophy of Science 1960*, pages 1–12. Stan-
           ford University Press, 1962.

[But97]   G.C. Buttazo. *Hard Real-time Computing Systems: Predictable
           Scheduling Algorithms and Applications*. Kluwer Academic Pub-
           lishers, 1997.

[CCGR00]   A. Cimatti, E. M. Clarke, F. Giunchiglia, and M. Roveri. NUSMV:
           A new symbolic model checker. *Software Tools for Technology
           Transfer*, 2(4):410–425, 2000.

[CE81]   E.M. Clarke and E.A. Emerson. Design and synthesis of synchro-
           nization skeletons using branching time temporal logic. In *Proc.
           Workshop on Logic of Programs*, LNCS 131, pages 52–71. Springer-
           Verlag, 1981.

[CES09]   E.M. Clarke, E.A. Emerson, and J. Sifakis. Model checking: Algo-
           rithmic verification and debugging. *Commun. ACM*, 52(11):74–84,
           2009.

[CGP00]   E.M. Clarke, O. Grumberg, and D.A. Peled. *Model Checking*. MIT
           Press, 2000.

[CM88]   K.M. Chandy and J. Misra. *Parallel Program Design: A Founda-
           tion*. Addison-Wesley, 1988.

[CPHP87]   P. Caspi, D. Pilaud, N. Halbwachs, and J. Plaice. Lustre: A declar-
           ative language for programming synchronous systems. In *Proceed-
           ings of the 14th Annual ACM Symposium on Principles of Pro-
           gramming Languages (POPL)*, pages 178–188, 1987.

[CVWY92]   C. Courcoubetis, M.Y. Vardi, P. Wolper, and M. Yannakakis. Mem-
           ory efficient algorithms for the verification of temporal properties.
           Formal Methods in System Design, 1:275–288, 1992.

[Dij65]   E.W. Dijkstra. Solution of a problem in concurrent programming
           control. *Commun. ACM*, 8(9):569, 1965.

[Dil89]   D.L. Dill. Timing assumptions and verification of finite-state con-
           current systems. In J. Sifakis, editor, *Automatic Verification Meth-
           ods for Finite State Systems*, LNCS 407, pages 197–212. Springer–
           Verlag, 1989.

[Dil96]   D.L. Dill. The Mur*phi* verification system. In *Computer Aided Verification, 8th International Conference (CAV)*, LNCS 1102, pages 390–393, 1996.

[EF06]   C. Eisner and D. Fisman. *A Practical Introduction to PSL.* Springer, 2006.

[Eme90]   E.A. Emerson. Temporal and modal logic. In J. van Leeuwen, editor, *Handbook of Theoretical Computer Science*, volume B, pages 995–1072. Elsevier Science Publishers, 1990.

[FLGD+11] G. Frehse, C. Le Guernic, A. Donzé, S. Cotton, R. Ray, O. Lebeltel, R. Ripado, A. Girard, T. Dang, and O. Maler. SpaceEx: Scalable verification of hybrid systems. In *Proc. 23rd International Conference on Computer Aided Verification (CAV)*, LNCS 6806, pages 379–395. Springer, 2011.

[FLP85]   M.J. Fischer, N.A. Lynch, and M. Paterson. Impossibility of distributed consensus with one faulty process. *Journal of the ACM*, 32(2):374–382, 1985.

[FMPY06] E. Fersman, L. Mokrushin, P. Pettersson, and W. Yi. Schedulability analysis of fixed-priority systems using timed automata. *Theoretical Computer Science*, 354(2):301–317, 2006.

[FPE02]   G.F. Franklin, J.D. Powell, and A. Emami-Naeini. *Feedback Control of Dynamic Systems*. Prentice Hall, 2002. Fourth Edition.

[Fra86]   N. Francez. *Fairness*. Springer-Verlag, 1986.

[Hal93]   N. Halbwachs. *Synchronous Programming of Reactive Systems.* Kluwer Academic Publishers, 1993.

[Har87]   D. Harel. Statecharts: A visual formalism for complex systems. *Science of Computer Programming*, 8:231–274, 1987.

[Her91]   M. Herlihy. Wait-free synchronization. *ACM Trans. Program. Lang. Syst.*, 13(1):124–149, 1991.

[HHW97]   T.A. Henzinger, P.-H. Ho, and H. Wong-Toi. HYTECH: A model checker for hybrid systems. *Software Tools for Technology Transfer*, 1(1-2):110–122, 1997.

[HNSY94]   T.A. Henzinger, X. Nicollin, J. Sifakis, and S. Yovine. Symbolic model-checking for real-time systems. *Information and Computation*, 111(2):193–244, 1994.

[Hoa69]   C.A.R. Hoare. An axiomatic basis for computer programming. *Commun. ACM*, 12(10):576–580, 1969.

[Hoa85]   C.A.R. Hoare. *Communicating Sequential Processes*. Prentice-Hall, 1985.

[Hol97]   G.J. Holzmann. The model checker SPIN. *IEEE Transactions on Software Engineering*, 23(5):279–295, 1997.

[Hol04]   G.J. Holzmann. *The SPIN Model Checker: Primer and Reference Manual*. Addison-Wesley, 2004.

[Hol13]   G.J. Holzmann. Landing a spacecraft on Mars. *IEEE Software*, 30(2):83–86, 2013.

[HP85]   D. Harel and A. Pnueli. On the development of reactive systems. In *Logics and Models of Concurrent Systems*, volume F-13 of *NATO Advanced Summer Institutes*, pages 477–498. Springer-Verlag, 1985.

[HR04]   M. Huth and M.D. Ryan. *Logic in Computer Science: Modelling and Reasoning about Systems*. Cambridge University Press, 2004. Second Edition.

[HS06]   T.A. Henzinger and J. Sifakis. The embedded systems design challenge. In *FM 2006: 14th International Symposium on Formal Methods*, LNCS 4085, pages 1–15, 2006.

[HW95]   P.-H. Ho and H. Wong-Toi. Automated analysis of an audio control protocol. In *Proceedings of the Seventh Conference on Computer-Aided Verification*, LNCS 939, pages 381–394. Springer-Verlag, 1995.

[IBG+11]   F. Ivancic, G. Balakrishnan, A. Gupta, S. Sankaranarayanan, N. Maeda, H. Tokuoka, T. Imoto, and Y. Miyazaki. DC2: A framework for scalable, scope-bounded software verification. In *Proc. 26th IEEE/ACM Intl. Conf. on Automated Software Engineering*, pages 133–142, 2011.

[JPAM14]   Z. Jiang, M. Pajic, R. Alur, and R. Mangharam. Closed-loop verification of medical devices with model abstraction and refinement. *Software Tools for Technology Transfre (STTT)*, 16(2):191–213, 2014.

[Kah74]   G. Kahn. The semantics of simple language for parallel programming. In *IFIP Congress*, pages 471–475, 1974.

[KLSV10]   D.K. Kaynar, N.A. Lynch, R. Segala, and F.W. Vaandrager. *The Theory of Timed I/O Automata*. Synthesis Lectures on Distributed Computing Theory. Morgan & Claypool Publishers, 2010. Second Edition.

[Kop00]   H. Kopetz. *Real-Time Systems: Design Principles for Distributed Embedded Applications.* Kluwer Academic Publishers, 2000.

[KSLB03]   G. Karsai, J. Sztipanovits, A. Ledeczi, and T. Bapty. Model-integrated development of embedded software. *Proceedings of the IEEE,* 91(1):145–164, 2003.

[LA14]   H. Lin and P.J. Antsaklis. *Hybrid Dynamical Systems: An Introduction to Control and Verification.* Number 1 in Foundations and Trends in Systems and Control. 2014.

[Lam87]   L. Lamport. A fast mutual exclusion algorithm. *ACM Transactions on Computer Systems,* 5(1):1–11, 1987.

[Lam94]   L. Lamport. The temporal logic of actions. *ACM Transactions on Programming Languages and Systems,* 16(3):872–923, 1994.

[Lam02]   L. Lamport. *Specifying Systems: The TLA+ Language and Tools for Hardware and Software Engineers.* Addison-Wesley, 2002.

[Lee00]   E. A. Lee. What's ahead for embedded software. *IEEE Computer,* pages 18–26, 2000.

[Liu00]   J.S. Liu. *Real-Time Systems.* Prentice Hall, 2000.

[LL73]   C. Liu and J. Layland. Scheduling algorithms for multiprogramming in a hard real-time environment. *Journal of the ACM,* 20(1), 1973.

[LP95]   E.A. Lee and T.M. Parks. Dataflow process networks. *Proceedings of the IEEE,* 83(5):773–801, 1995.

[LPY97]   K. Larsen, P. Pettersson, and W. Yi. UPPAAL in a nutshell. *Springer International Journal of Software Tools for Technology Transfer,* 1(1-2):134–152, 1997.

[LS11]   E.A. Lee and S.A. Seshia. *Introduction to Embedded Systems, A Cyber-Physical Systems Approach.* 2011. Available at http://LeeSeshia.org.

[LSC+12]   I. Lee, O. Sokolsky, S. Chen, J. Hatcliff, E. Jee, B. Kim, A.L. King, M. Mullen-Fortino, S. Park, A. Roederer, and K.K. Venkatasubramanian. Challenges and research directions in medical cyber-physical systems. *Proceedings of the IEEE,* 100(1):75–90, 2012.

[LT87]   N.A. Lynch and M. Tuttle. Hierarchical correctness proofs for distributed algorithms. In *Proceedings of the Seventh ACM Symposium on Principles of Distributed Computing,* pages 137–151, 1987.

[LV02]   E.A. Lee and P. Varaiya. *Structure and Interpretation of Signals and Systems.* Addison Wesley, 2002.

[Lyn96]   N.A. Lynch. *Distributed Algorithms*. Morgan Kaufmann, 1996.

[Mar03]   P. Marwedel. *Embedded System Design*. Kluwer, 2003.

[McM93]   K.L. McMillan. *Symbolic Model Checking: An Approach to the State Explosion Problem*. Kluwer Academic Publishers, 1993.

[Mil89]   R. Milner. *Communication and Concurrency*. Prentice-Hall, 1989.

[MMP91]   O. Maler, Z. Manna, and A. Pnueli. From timed to hybrid systems. In *Real-Time: Theory in Practice, REX Workshop*, LNCS 600, pages 447–484. Springer, 1991.

[MP81]   Z. Manna and A. Pnueli. Verification of concurrent programs: Temporal proof principles. In *Logics of Programs*, LNCS 131, pages 200–252. Springer, 1981.

[MP91]   Z. Manna and A. Pnueli. *The Temporal Logic of Reactive and Concurrent Systems: Specification*. Springer-Verlag, 1991.

[Pet81]   G.L. Peterson. Myths about the mutual exclusion problem. *Information Processing Letters*, 12(3), 1981.

[Pet82]   G.L. Peterson. An O(n log n) unidirectional algorithm for the circular extrema problem. *ACM Trans. Program. Lang. Syst.*, 4(4):758–762, 1982.

[PJ04]   S. Prajna and A. Jadbabaie. Safety verification of hybrid systems using barrier certificates. In *Hybrid Systems: Computation and Control, 7th International Workshop*, LNCS 2993, pages 477–492, 2004.

[Pla10]   A. Platzer. *Logical Analysis of Hybrid Systems - Proving Theorems for Complex Dynamics*. Springer, 2010.

[Pnu77]   A. Pnueli. The temporal logic of programs. In *Proceedings of the 18th IEEE Symposium on Foundations of Computer Science*, pages 46–77, 1977.

[Pto14]   C. Ptolemaeus, editor. *System Design, Modeling, and Simulation Using Ptolemy II*. Ptolemy.org, 2014. available at `ptolemy.org/books/Systems`.

[QS82]   J.P. Queille and J. Sifakis. Specification and verification of concurrent programs in CESAR. In *Proceedings of the Fifth International Symposium on Programming*, LNCS 137, pages 195–220. Springer, 1982.

[RS94]   K. Ramamritham and J.A. Stankovic. Scheduling algorithms and operating systems support for real-time systems. *Proceedings of the IEEE*, 1(82):55–67, 1994.

[SAÅ+04]   L. Sha, T.F. Abdelzaher, K.-E. Årzén, A. Cervin, T.P. Baker,
            A. Burns, G.C. Buttazzo, M. Caccamo, J.P. Lehoczky, and A.K.
            Mok. Real time scheduling theory: A historical perspective. *Real-
            Time Systems*, 28(2-3):101–155, 2004.

[SH92]   B. Shahian and M. Hassul. *Computer-Aided Control System Design
            Using MATLAB*. Prentice Hall, 1992.

[Sif13]   J. Sifakis. Rigorous system design. *Foundations and Trends in
            Electronic Design Automation*, 6(4):293–362, 2013.

[Sip13]   M. Sipser. *Introduction to the Theory of Computation*. Cengage
            Learning, 2013. Third Edition.

[SLMR05]   J.A. Stankovic, I. Lee, A.K. Mok, and R. Rajkumar. Opportunities
            and obligations for physical computing systems. *IEEE Computer*,
            38(11):23–31, 2005.

[SV07]   A. Sangiovanni-Vincentelli. Quo Vadis SLD: Reasoning about
            trends and challenges of system-level design. *Proceedings of the
            IEEE*, 95(3):467–506, 2007.

[Tab09]   P. Tabuada. *Verification and Control of Hybrid Systems*. Springer,
            2009.

[Tho90]   W. Thomas. Automata on infinite objects. In J. van Leeuwen,
            editor, *Handbook of Theoretical Computer Science*, volume B, pages
            133–191. Elsevier Science Publishers, 1990.

[TT09]   A. Taly and A. Tiwari. Deductive verification of continuous dy-
            namical systems. In *IARCS Annual Conference on Foundations of
            Software Technology and Theoretical Computer Science*, LIPIcs 4,
            pages 383–394, 2009.

[VW86]   M.Y. Vardi and P. Wolper. An automata-theoretic approach to
            automatic program verification. In *Proceedings of the First IEEE
            Symposium on Logic in Computer Science*, pages 332–344, 1986.

[Wan04]   F. Wang. Efficient verification of timed automata with BDD-like
            data structures. *Software Tools for Technology Transfer*, 6(1):77–
            97, 2004.

[WEE+08]   R. Wilhelm, J. Engblom, A. Ermedahl, N. Holsti, S. Thesing,
            D.B. Whalley, G. Bernat, C. Ferdinand, R. Heckmann, T. Mitra,
            F. Mueller, I. Puaut, P.P. Puschner, J. Staschulat, and P. Sten-
            ström. The worst-case execution-time problem: Overview of meth-
            ods and survey of tools. *ACM Trans. Embedded Comput. Syst.*,
            7(3), 2008.

Index

Printed in the United States
by Baker & Taylor Publisher Services